AF581671

Drought Risk Management in Reflect Changing of Meteorological Conditions

Drought Risk Management in Reflect Changing of Meteorological Conditions

Editors

Andrzej Wałęga

Agnieszka Ziernicka-Wojtaszek

MDPI • Basel • Beijing • Wuhan • Barcelona • Belgrade • Manchester • Tokyo • Cluj • Tianjin

Editors
Andrzej Wałęga
University of Agriculture in Kraków
Poland

Agnieszka Ziernicka-Wojtaszek
University of Agriculture in Kraków
Poland

Editorial Office
MDPI
St. Alban-Anlage 66
4052 Basel, Switzerland

This is a reprint of articles from the Special Issue published online in the open access journal *Atmosphere* (ISSN 2073-4433) (available at: https://www.mdpi.com/journal/atmosphere/special_issues/Drought_Management_Meteorological_Conditions).

For citation purposes, cite each article independently as indicated on the article page online and as indicated below:

LastName, A.A.; LastName, B.B.; LastName, C.C. Article Title. *Journal Name* **Year**, *Volume Number*, Page Range.

ISBN 978-3-0365-2838-0 (Hbk)
ISBN 978-3-0365-2839-7 (PDF)

Contents

About the Editors

Andrzej Wałęga is a Doctor at the Department of Sanitary Engineering and Water Management at the University of Agriculture in Krakow, Poland, where he teaches hydrology, modelling of hydrological processes and water management in urban catchments. He holds a Ph.D. M.S. and an M.Eng. from the University of Agriculture in Krakow, focusing on Environmental Engineering. Since 2005, he has been working as a researcher and lecturer. Currently, he is the President Association of Polish Hydrologists. He was the expert in a project studying hydraulic modeling against floods—1st stage, from preparation to implementation—of the EU directive No 2007/60EC, and was the leader in the study of hydraulic modeling against floods—2nd stage—supporting the Georgian institution at the Polish Center of International Aid. He is a guest editor for the journals, *Sustainability, Atmosphere, Water* (MDPI), concerning Intelligent Automation and Soft Computing, and a member of the Editorial Board for the journal, *Sustainability*. He has co-authored over 190 journals and conference papers, books and book chapters, and technical guides related to hydrology and water management.

Agnieszka Ziernicka-Wojtaszek is a Doctor of Agricultural Sciences in the field of protecting and shaping the environment, specialization meteorology and climatology. She is head of the Department of Climatology, Ecology, and Air Protection within the Faculty of Agricultural Engineering and Land Surveying at the University of Agriculture in Krakow. Her scientific interests include agro-climatic regionalization of Poland, air temperature and precipitation variability in Poland, the evaluation of the qualities and resources of the atmospheric environment for the purpose of shaping the environment, the water needs of cultivated plants, the impact of the weather and climate on agro-technical activity and on the yielding of crops, extreme and harmful to agriculture meteorological phenomena, present-day climate change and the agro ecological results of global warming, city and suburban climate, mountain climate, geography of tourism, and agro tourism. She has co-authored over 95 journal and conference papers, books and book chapters related to meteorology and climatology.

MDPI

Editorial

Preface to Drought Risk Management to Reflect Changing Meteorological Conditions

Andrzej Wałęga [1,*] and Agnieszka Ziernicka-Wojtaszek [2]

1 Department of Sanitary Engineering and Water Management, University of Agriculture in Krakow, 30-059 Krakow, Poland

2 Department of Ecology, Climatology and Air Protection, University of Agriculture in Krakow, 30-059 Krakow, Poland; agnieszka.ziernicka-wojtaszek@urk.edu.pl

* Correspondence: andrzej.walega@urk.edu.pl

Citation: Wałęga, A.; Ziernicka-Wojtaszek, A. Preface to Drought Risk Management to Reflect Changing Meteorological Conditions. *Atmosphere* **2021**, *12*, 1660. https://doi.org/10.3390/atmos12121660

Received: 26 November 2021
Accepted: 8 December 2021
Published: 10 December 2021

Publisher's Note: MDPI stays neutral with regard to jurisdictional claims in published maps and institutional affiliations.

Drought is one of the main extreme meteorological and hydrological phenomena which influence both the functioning of ecosystems and many important sectors of human economic activity. Throughout the world, various direct changes in meteorological and climatic conditions, such as air temperature, humidity, and evapotranspiration can be observed. They have a significant influence upon the shaping of the phenomenon of drought. Land cover and land use can also be indirect factors influencing evapotranspiration as well as, by the same token, the water balance in the water catchment area. They can also influence the course of the process of drought. The observed climate change, manifested mainly by increases in temperature, in turn influences evapotranspiration, which may cause an intensification in terms of both the degree and frequency of drought. Drought is related to changes in the hydrological regime, and to the decrease in water resources. Its results can be observed in various sectors, related—among others—to a demand for water for people, agriculture (a vital issue in areas with large irrigation systems), and industry. It can also prove problematic for water ecosystems. To reflect the aforementioned information, reasonable drought risk management is indispensable. It can ease water-demand-related problems in various sectors of human activity.

This book contains 11 chapters presenting some of the main lines of research concerning the phenomenon of drought, with particular emphasis placed upon international experiences in a few countries. The chapters constitute a collection of articles published in a Special Issue of the MDPI journal *Atmosphere*, entitled *Drought Risk Management to Reflect Changing Meteorological Conditions*, in 2021. The readers of this book wish to express their appreciation for the excellent job and effort carried out by the MDPI editorial team, as well as for the quality and scope of research presented by the 54 authors who contributed to the success of this Special Issue.

This book presents original research on various problems linked with drought. The first paper [1] presented the positive role of green roofs in the local climate. In the next paper [2], data-driven techniques (artificial neural network (ANN), wavelet-based ANN (WANN), radial function-based support vector machine (SVM-RF), linear function-based SVM (SVM-LF), and multi-linear regression (MLR) models) for pan evaporation were analyzed in an aspect role to input data on the accuracy estimation of this parameter. In the next paper [3], the trend of dry and wet spells was analyzed at 150 rain gauges in the period 1970–2018 in the Mediterranean region of North Africa. The influence of drought on the ecosystem, mainly expressed by the carbon dioxide dynamic, was analyzed at two Norway spruce forest sites representing two contrasting climatic conditions—a cold and humid climate and a moderately warm and dry climate [4]. Additionally, large-scale results of drought were described in this book. In [5], it was shown that extreme droughts over North America are associated with a long warm and dry period of weather and the development of moderate ridges over Central USA and over Eastern Europe and Western Russia are driven by the occurrence of prolonged blocking episodes, as well as surface processes. Complex analysis of the meteorological and hydrological droughts expressed by the Standardized Precipitation Index (SPI) and the Standardized Runoff Index (SRI) in various time scales were presented in [6]. Drought can indirectly influence habitat conditions in rivers by decreasing the refuge habitats needed by freshwater fish, diminishing the fish abundance, and influencing the spatial and temporal variation in the fish assemblage structure [7]. In [8], a technique for forecasting drought based on an Autoregressive Integrated Moving Average that can be used as effective tool in drought analyses was presented. The drought phenomenon is changing over time. As shown in [9], in the Polish Carpathians in 1991–2020, the dry month frequency in the last decade was much higher than in previous decades, especially in the cold half-year. In [10], new techniques for measuring and monitoring clay soil caused by drought-rewetting cycles were shown. Finally, [11] presented drought in one of the warmest years in Poland, 2019, where there was a visible, strong variability of drought expressed as climatic water balance (CWB) and the strong influence of drought on the agricultural sector was shown.

We hope this collection of articles will be useful for both scientists and practitioners helping in drought risk management, in the context of the change in meteorological conditions brought about by ongoing climate change.

Author Contributions: All four guest editors (A.W. and A.Z.-W.) contributed to this editorial. All authors have read and agreed to the published version of the manuscript.

Funding: This editorial received no external funding.

Acknowledgments: The editors would like to thank the authors from countries all over the world for their valuable contributions, the reviewers for their constructive comments and suggestions that helped to improve the manuscripts, and Calvin Li from the editorial office for his excellent support in processing and publishing this issue.

Conflicts of Interest: The authors declare no conflict of interest.

References

1. Köhler, M.; Kaiser, D. Green Roof Enhancement on Buildings of the University of Applied Sciences in Neubrandenburg (Germany) in Times of Climate Change. *Atmosphere* **2021**, *12*, 382. [CrossRef]
2. Kumar, M.; Kumari, A.; Kumar, D.; Al-Ansari, N.; Ali, R.; Kumar, R.; Kumar, A.; Elbeltagi, A.; Kuriqi, A. The Superiority of Data-Driven Techniques for Estimation of Daily Pan Evaporation. *Atmosphere* **2021**, *12*, 701. [CrossRef]
3. Achite, M.; Krakauer, N.Y.; Wałęga, A.; Caloiero, T. Spatial and Temporal Analysis of Dry and Wet Spells in the Wadi Cheliff Basin, Algeria. *Atmosphere* **2021**, *12*, 798. [CrossRef]
4. Mensah, C.; Šigut, L.; Fischer, M.; Foltýnová, L.; Jocher, G.; Acosta, M.; Kowalska, N.; Kokrda, L.; Pavelka, M.; Marshall, J.D.; et al. Assessing the Contrasting Effects of the Exceptional 2015 Drought on the Carbon Dynamics in Two Norway Spruce Forest Ecosystems. *Atmosphere* **2021**, *12*, 988. [CrossRef]
5. Lupo, A.R.; Kononova, N.K.; Semenova, I.G.; Lebedeva, M.G. A Comparison of the Characteristics of Drought during the Late 20th and Early 21st Centuries over Eastern Europe, Western Russia and Central North America. *Atmosphere* **2021**, *12*, 1033. [CrossRef]
6. Kubiak-Wójcicka, K.; Pilarska, A.; Kamiński, D. The Analysis of Long-Term Trends in the Meteorological and Hydrological Drought Occurrences Using Non-Parametric Methods—Case Study of the Catchment of the Upper Noteć River (Central Poland). *Atmosphere* **2021**, *12*, 1098. [CrossRef]
7. Bănăduc, D.; Sas, A.; Cianfaglione, K.; Barinova, S.; Curtean-Bănăduc, A. The Role of Aquatic Refuge Habitats for Fish, and Threats in the Context of Climate Change and Human Impact, during Seasonal Hydrological Drought in the Saxon Villages Area (Transylvania, Romania). *Atmosphere* **2021**, *12*, 1209. [CrossRef]
8. Kumar, P.; Shah, S.F.; Uqaili, M.A.; Kumar, L.; Zafar, R.F. Forecasting of Drought: A Case Study of Water-Stressed Region of Pakistan. *Atmosphere* **2021**, *12*, 1248. [CrossRef]
9. Bokwa, A.; Klimek, M.; Krzaklewski, P.; Kukułka, W. Drought Trends in the Polish Carpathian Mts. in the Years 1991–2020. *Atmosphere* **2021**, *12*, 1259. [CrossRef]
10. Burnol, A.; Foumelis, M.; Gourdier, S.; Deparis, J.; Raucoules, D. Monitoring of Expansive Clays over Drought-Rewetting Cycles Using Satellite Remote Sensing. *Atmosphere* **2021**, *12*, 1262. [CrossRef]
11. Ziernicka-Wojtaszek, A. Summer Drought in 2019 on Polish Territory—A Case Study. *Atmosphere* **2021**, *12*, 1475. [CrossRef]

MDPI

Article

Green Roof Enhancement on Buildings of the University of Applied Sciences in Neubrandenburg (Germany) in Times of Climate Change

Manfred Köhler [1,*] and Daniel Kaiser [2]

[1] University of Applied Sciences Neubrandenburg, Brodaer Str. 2-4, 17033 Neubrandenburg, Germany
[2] Lehr- und Versuchsanstalt für Gartenbau und Arboristik e.V., 14979 Großbeeren, Germany; kaiser@lvga-bb.de
* Correspondence: Koehler@hs-nb.de

Abstract: The reduction in evaporative surfaces in cities is one driver for longer and hotter summers. Greening building surfaces can help to mitigate the loss of vegetated cover. Typical extensive green roof structures, such as sedum-based solutions, survive in dry periods, but how can green roofs be made to be more effective for the longer hot and dry periods to come? The research findings are based on continuous vegetation analytics of typical extensive green roofs over the past 20 years. -Survival of longer dry periods by fully adapted plants species with a focus on the fittest and best adapted species. -Additional technical and treatment solutions to support greater water storage in the media in dry periods and to support greater plant biomass/high biodiversity on the roofs by optimizing growing media with fertilizer to achieve higher evapotranspiration (short: ET) values. The main findings of this research: -The climate benefits of green roofs are associated with the quantity of phytomass. Selecting the right growing media is critical. -Typical extensive green roof substrates have poor nutrition levels. Fertilizer can significantly boost the ecological effects on CO_2 fixation. -If the goal of the green roof is a highly biodiverse green roof, micro-structures are the right solution.

Keywords: extensive green roofs; climate change; summer drought; urban vegetation; phytomass; fertilizer; biodiversity; blue green infrastructure

Citation: Köhler, M.; Kaiser, D. Green Roof Enhancement on Buildings of the University of Applied Sciences in Neubrandenburg (Germany) in Times of Climate Change. *Atmosphere* **2021**, *12*, 382. https://doi.org/10.3390/atmos12030382

Academic Editor: Andrzej Walega

Received: 10 February 2021
Accepted: 5 March 2021
Published: 14 March 2021

Publisher's Note: MDPI stays neutral with regard to jurisdictional claims in published maps and institutional affiliations.

1. Introduction

Drought is related to massive deforestation in many parts of the world [1]. The daily new ground sealing around the world has not stopped, but new tree plantation initiatives work against this trend to stop further drying out of our planet. Successful initiatives of plantings are active in Australia, under the term of "land-care" [2]; Mongolia [3]; and the current Green Wall movement [4] in North Africa. There is still a long way to go until forest cover values, such as those that existed in central Europe 2000 years ago, of above 80% are reached, in contrast to today with about 30% forest cover [5]. The increasing world population is one reason for deforestation [6]. Drought in cities is connected with the low amount of evaporative green areas in cities [7,8]. The consequences are longer hot and dry summer periods that cause a number of environmental and health problems for residents. According to Kravcik et al. [9], evaporative surfaces are key instruments to combat global warming in cities. Wherever it is possible, urban forestry should be the first choice to enhance ecosystem services by planting trees, like the "trillion tree initiative" [10]. However, the second best choice is to green building surfaces, which are normally un-vegetated.

Green roofs and vegetated green facades are tools to support decentralized local water cycles and are widely used to combat the urban heat island effect. The current situation about green roofs shows that only a small amount of roofs are greened. Most of them are shallow growing media with about a 10-cm depth. Today, there is a gap between the potential greening of roofs and the number of projects that have currently been realized. Currently, the cities with the highest rates of roof top greenery are Singapore [11], Chicago [12], and the German cities Stuttgart and Berlin [13]. The coverage rates of green roofs are between 3% and 8%, while vertical green systems, also known as living walls or green facades, are well below 1% green coverage of all buildings [14]. However, the potential for roof greening is up to 50% of all buildings. Today, in the search for more solutions in cities to deal with urban drought and provide more cooling strategies to mitigate the increase in extreme high temperature values, green roofs can potentially function as spaces to cool down urban surfaces.

Green roofs are instruments with many additional benefits, which are supported by the structures of growing media, an extra drainage layer, and vegetation cover [15]. The extra urban green space that becomes available for recreation and the greater biodiversity are further positive reasons to invest in such technologies. Cristiano et al. [16] conducted a literature review of the water–energy–food–ecosystem nexus, pointing out the multiple benefits of green roofs in contributing to the sustainable development goals of the United Nations.

The current situation in Germany is 15% coverage with intensive roof gardens and about 85% coverage with extensive green roofs [13]. Extensive green roofs normally have low rates of evapotranspiration in summer due to the low water storage capacity in the typical 10-cm-thick layer of growing media. This report highlights some details of the construction (growing media and retention layer) as well some treatments to optimize the functionality of extensive green roofs in the future [17].

The thesis in this publication is the need to shift from extensive green roofs that can survive long dry periods but that only have low evapotranspiration rates to semi-intensive green roofs with better ecological performance that form part of the blue-green infrastructure in cities [18]. Green roofs are underestimated in their potential to contribute against climate change. Cock and Larson summarized 179 peer-reviewed surveys to provide evidence of the multi-disciplinary ecological functionality. They concluded that greater efficiency is possible with some updates in some roof construction details [19].

In a meta-study, Shafique et al. [20] analyzed the efficiency of CO_2- sequestration of green roofs. They found two effects: a direct effect, such as uptake via photosynthesis, and an indirect effect, like extra insulation value for the building. Both have lower heating/cooling demands as a positive consequence. Most of these surveys were based on model calculation, and more local experimental works over longer periods are recommended. The thesis in this report is that more evaporation is connected with more phytomass in general, and this can have positive effects on greater CO_2 fixation; what kind of effect will have this on the other aim of greater plant biodiversity?

The tests in this publication involved varying the types of growing media and media depth:

-Is more evaporative phytomass produced on green roofs by selecting the right growing media and adapted plant species?

-Does irrigation or fertilization have positive effects? How do these treatments influence the goal of greater plant biodiversity on roofs? There are two theses: on the one hand, more fertilizer support more phytomass, but what about the rare species: do they prefer poor media? Is a lot of biomass counterproductive to biodiversity?

-How much better does thicker growing media for green roofs perform concerning CO_2 fixation as an indicator [21]? Today, solutions for more opportunities for CO_2 fixation are searched for.

2. Experiments

Two buildings of the University of Applied Sciences in Neubrandenburg were constructed as research and demonstration roofs. Building 2 of this campus complex was opened in 1999 with a research green roof site of about 2000 m^2 while building 3 opened in 2001 with around 1000 m^2 of green roof space. The details of the roofs can be seen on

Google maps with the GPS Coordinates: Degree:Minutes:Seconds); see the sites: building 2: 53:33:23N 13:14:44E and building 3: 53: 33:15N 13:14:43E.

Preliminary reports presented the results of the continuous 20-year climate measurements on these roofs. The evidence of the positive influence of the microclimate of the vegetation layer is seen in a significant reduction in the surface temperature [22]. A further paper explains how different retention layers can capture larger volumes of rain in a comparison between conventional growing media and various retention layers. This also contributes to greater and more sustained evapotranspiration to mitigate summer heat and reduces run-off from the building. In the optimized cases, the run-off from a building with a green roof can approach zero [23].

This paper presents the results for the increased efficiency of green roofs as a result of the amount of phytomass produced as an indicator of greater storage of CO_2, a measurable variable in times of climate change. The plant performance on a typical 10-cm-thick layer of growing media was compared to a thicker layer of 30 cm (see Figure 1a,b).

(a) (b)

Figure 1. (**a**) Example of one of the 39 planter boxes with the 30-cm growing media. (**b**) One of the microhabitat structures on the roof of building 2, the west part of the green roof. The semi-shade situation supports some endangered plants, such as the grass *Briza media*, enabling it to survive on the extensive 10-cm roof media for 20 years.

Research design

In contrast to many other green roof research studies, the focus here was the long-term performance of typical FLL-Standard [24] green roofs under real roof conditions. The second aspect was a complete green roof not a small test installation. Easy accessibility and the inclusion into teaching programs has allowed frequent observation of changes. The use of market leader products, such as in the German FLL guidelines, which have existed since 1990, has helped to improve these technical standards. Additionally, some further growing media and treatments, such as irrigation and fertilization, has helped to extend the existing knowledge. At the beginning, in 1999, the basic aims were to achieve 60% vegetation cover on a very shallow layer. In the last years, new upcoming questions were the enhancement of biodiversity and CO_2 fixation under hotter and dryer summer periods in central Europe. The basic research design can be described on both buildings as follows:

-Building 2 had three commercial growing media, identified as Zi, Blä, and Op, with a 10-cm depth plus a drainage layer. The additional 26 planter boxes with a 30-cm-thick layer of the extensive growing media from two main deliverers of green roof materials, ZinCo and Optigruen with the abbreviation Zi and Op, were the test installations. These materials followed the necessary requirements after [16], such as a granulometric distribution and the minimum water holding capacity. The primary vegetation was similar, using grass seeds and sedum cuttings.

-Building 3 had the 10-cm growing media Op-2 (product name "Optima Tiefgarage schwer") and Ulo (expanded slate, grain size 2–11 mm, brand name "Thüringer Bläh-schiefer"). The vegetation layer on top of the growing media was prefabricated turf mats on the west, north, and south sides while the east side used both growing media with a selection of sedum cuttings on top. An additional 13 planter boxes with a 30-cm media depth of Op-2 were used, representing semi-intensive green roofs with some grass seeds and some herbal plantings in the beginning.

Research methods

In the years up to 2010, the team investigated the floral and vegetative components of the test plots using methods derived from the Braun-Blanquet methods [25]. On each sub-plot of the roof, a complete species list was produced by noting the percentage of the area covered by each vascular higher plant genus. Additional information included the total coverage value of the indicator group of sedum and grasses.

The green roof turf mats on building 3 were a pre-grown professional turf layer with a mixture of grasses and sedum species. Such industrially produced mats have been used in many comparable projects throughout Germany. The plant species development is a role model for many typical extensive green roofs. On building 3, one question examined how these well-suited vegetation layers performed compared to green roofs started with sedum cuttings only, as an alternative approach to provide a cheaper vegetation layer on flat roofs. In the following text, the west, north, and south sides with turf mats are compared to the east side with the sedum cuttings (see Figure 2a,b).

(a)

(b)

Figure 2. (**a**) Building 2, May 2001: The turf mats were placed on the north plots, as was done on the west and south plots. The growing media can be seen in the background: light red Op, dark in the front: Ulo. (**b**) Sedum cutting area on the east roof in August 2002: Right: Ulo, sedum developed slowly, left: Sedum and several spontaneously sown grasses covered the Op successfully. Maintenance work on the weather station caused some open areas in the vegetation cover.

In contrast to this, a vegetation layer completely based on sedum cuttings and grass seeds was constructed on building 2 in 1999.

The test plots on building 3 were divided in half, with one half using the professional Op growing medium (Optima-extensive media) and the other half Ulo (Ulopor—expanded slate 2–11 mm sizes) growing medium. Op, similar to the Zi (Zinco growing media on building 2), is a professional growing medium that satisfies all the growing media requirements of FLL 2018 – Standard [24] in regard to the water retention capacity, grain size distribution, and basic fertilizer. On building 2, the Blä substrate was used as a test. This is simply crushed expanded clay as a recycled product with no additional nutrition, merely being a supplement for extensive green roofs. The same applies to Ul (Ulopor). This expanded slate has good performance in terms of its water retention capacity, but its dark color means that it heats up considerably in summer, much more than the typical gravel layer of 16/32-mm stones. On the east side the vegetation reached the maximum coverage of 80% on Op and 75% on Ulo in 2002, the second year. This was ultimately a comparison between the different layers to learn more about the long-term performance of the materials for improving the growing media mixtures and to use the results to contribute to the updates of the FLL guideline [24].

Harvest on 10-cm media

In 2017, from the growing media test plot Zi, Blä, Op, Ulo, and Op-2, all above-ground and root mass was harvested, with three replicates for each test plot. All phytomass material was divided into the following groups: above-ground vascular plants shoots, roots, mosses, and lichens. The material was dried to a constant weight. The total carbon was 50% of the dry mass. The CO_2 concentration was calculated by multiplying with 3.65 to provide a guide value [26–29].

Harvest on 30-cm media

On building 2, the 26 planter boxes on the roof terraces were used as phytomass test plots. Using scissors, the above-ground phytomass was harvested in 2011, 2012, 2013, 2014, 2015, and 2017. The similar 13 boxes on building 3 were treated in the same way in 2013, 2014, 2015, and 2017. The oven-dried material allowed an initial estimate of the annual growth rate on these materials.

Fertilizer and Irrigation

Since 2011, on building 3, subplots were established, as shown in Table 1, to investigate the effects of additional irrigation and subplots of the same size were established to investigate the effects of fertilizer use. These subplots were compared with ongoing monitoring of the plant lists in the normal roof positions.

Table 1. The sizes of the 10-cm depth research subplots on building 3 from 2001; on each part of the roofs, the following sized areas were selected for treatments with additional irrigation and fertilizer, and the "normal" comparison plots for the four test plots (in m^2).

Exposure	Media 1: Ulo	Media 2: Op-2
West	8	12
North shaded	6	6
North sun	6a	6
South	5.5	5.5
East	11	11

The different sizes are due to the roof construction.

Additional irrigation was applied once a year in May using 10 L/m^2 tap water. The fertilization subplots were fertilized with 10 g/m^2 Kristalon 16-11-16 NPK plus MG on the same day. Figure 3a shows the effect of the fertilizer on the eastern part of the roof and Figure 3b gives an impression of the western part with the sedum mats. *Sedum sexangulare* was stimulated to flower intensively while *Allium schoenoprasum* and *Petrorhagia saxifraga* showed significantly better coverage and performance.

Statistics

For the statistics, SPSS_Vers.27 procedures, such as descriptive statistics, dependent *t*-tests with two variables each, tests of significance, cluster analysis, and analysis of variances (ANOVA), were completed using the datasets.

(a) (b)

Figure 3. (**a**) View to the eastern roof section, left: growing media Op, right: Ul-media. The line demarcating fertilizer application in June 2020 can be clearly seen, about 4 weeks after the application. (**b**) Western roof area, the fertilized area on the Op media is in the foreground while the Ulo section is in the background. The more intense blooming can be seen on 20 June 2020, four weeks after the fertilizer application.

3. Results

The effects of using different growing media, substrate depths, and fertilizers were significant and are demonstrated in the following section by the statistical improvements. The additional application of 10 L/m^2 water had no impact on the vegetation and will not be described in detail.

3.1. Cluster Analysis 2001–2020: Test Based on Plant Species Observation

The annual application of fertilizer affected the visual outcome as demonstrated in Figure 2.

Test 1: Analysis of the first year on building 3 showed the development of the plant species richness in the turf mats in Table 2. The results, supported by the cluster/correlation analyses, are shown in Table 2, with the species mix on all turf mats with no additional treatments being highly significantly similar with 0.896***, as demonstrated by the examples of the turf mats on the Ulo-West and Op-West test plots. Both mats in the year 2020 were also highly significant, with 0.963***.

Test 2: How did the growing media influence the number of species and the coverage value? In general, this was demonstrated here by the north test plots in 2001. The Op media in all cases showed higher species richness, as shown by way of the example in Table 3, with 25 species on the Ulo medium and 29 species on the Op medium.

At all times in 2001 and 2020, the Op media showed higher coverage values. In general, the number of observed species decreased over the past 20 years, but in contrast, the fertilizer test plots showed higher coverage values in this time.

Table 2. Cluster analysis: correlation based on the number of species in 2001 and 2020 on the west roof, 10-cm media. High significance between species in 2001 on both media (Ulo and Opti). Also high significance in 2020 on both media. Time is more important than the different media.

Test Plots	Ulo_West 2001	Opti_West_ 2001	Ulo_West 2020	Op_West 2020
Ulo_West_2001	1.000	0.896 ***	0.151	0.223
Ulo_West_2020	0.151	0.131	1.000	0.963 ***

Table 3. Changes in the number of species from 2001 to 2020, 10-cm media.

Name	Number of Species	Coverage Value in %
Ulo_Norm_N_2001	25	88
Op_Norm_N_2001	29	94
Ulo_Norm_Shade 2020	27	94
Ulo_Fert_N_Shade_2020	14	98
Ulo_Norm_N_Sun_2020	15	81
Ulo_Fert_N_Sun_2020	13	96

In Table 4, the similarity is demonstrated by the Pearson index. Again, the similarity of the mats on the north side in 2001 was highly significant. The species similarity remained significant until 2020 in the shade on both test plots of "normal" and "fertilizer". This is demonstrated by the high grass coverage. In contrast to this, the south section differed, with higher numbers of sedum species. Shade is a significant factor determining the plant species mix.

Test 3: Species development of the cuttings, eastern exposure test plots.

Table 4. Correlation and significance (two-tailed) of the north test plots, 10-cm media, based on the number of similar plant species, in a comparison between 2001 and 2020. High significance ** on the 5‰ level between both media Ulo and Op 2001, it remain similar until 2020 on the shade plots. The sunny parts are significant different from these.

Name	Criteria	Ulo_Norm_ N_ 2001	Op_Norm_ N_ 2001	Ulo_Norm_ N_Shade 2020	Ulo_Fert_ N__Shade_ 2020	Ulo_Norm_ N_Sun_ 2020	Ulo_Fert_ N_Sun 2001
Ulo_Norm_N_2001	Pearson correlation	1	0.948 **	0.954 **	0.957 **	0.259	0.210
	Sig.		0.000	0.000	0.000	0.471	0.617
	Species	25	24	15	8	10	8

The dark material surface of Ulo has significantly higher temperatures during summer than other comparable professional roof growing media. This limited the plant growth of the seedlings and cuttings. Additionally, spontaneous plants found it difficult to establish themselves in this first year. Table 5 show the low number of species in the first year, with 9 (Ulo) and 14 (Op) species and low coverage values of 9% and 14%. These low rates were ultimately determined by the growth of *Petrorhagia saxifraga* on both east test plots, but these plants only covered a small area. In addition, the sedum cuttings must first establish themselves on the test plots. The vegetation on both areas needed a few years to achieve the estimated coverage value of 60%.

Table 5. Changes in the number of species on the east area with sedum cuttings from 2001 to 2020, 10-cm media.

Name	Number of Species	Coverage Values in %
Ulo_Norm_East_2001	9	11
Ulo_Norm_East_2020	16	69
Ulo_Fert_East 2020	10	98
Op_Norm_East_2001	14	14
Op_Norm_East_2020	18	83
Op_Fert_East_2020	12	88

Like the other exposure, there were differences between the fertilized and normal areas in terms of the plant species diversity. The effect of the fertilizer on all test plots is to reduce the species richness while increasing the coverage values.

Table 6 shows the correlation on the seeded eastern plots. The differences in early 2001 between the plant mix on Ulo and Op indicate widely different development. This was apparent in 2001, with very low vegetation coverage and fewer species compared to the turf mat on the other sites. For many years, the Ulo test plots in particular had low performance values. Finally, all indicators in 2020 showed that in the long run, similar development was achieved for both media.

Table 6. Correlation and significance (2-tailed) of east test plots, 10-cm media, based on the number of similar plant species, in a comparison between 2001 and 2020. Significance here means: the fertilizer equalize differences between the different vegetation stands also on the east plots.

Name	Criteria	Ulo Norm 2001	Op Norm 2001	Ulo Norm 2020	Ulo Fert 2020	Op Norm 2020	Opt Fert 2020
Ulo Norm 2001	Pearson correlation	1	−0.359	[1]	[1]	[1]	[1]
	Sig.		0.342	[1]	[1]	[1]	[1]
	Species	9	9	2	1	2	2
Ulo Fert 2020	Pearson correlation	[1]	[1]	0.821 **	1	0.862 *	0.869 *
	Sig.			0.004	[1]	0.013	0.025
	Species	1	2	10	10	7	6

[1] There are not enough similar species for the correlation pairs, and significance expression cannot be calculated.

The hierarchical cluster in Figure 4 demonstrates the similarity of the eastern test plots. From the perspective of plant species development, these seeded plots are valuable observation areas. Considering the aim of achieving full coverage on the roof as soon as possible, it is important to note here that this process took nearly 10 years. These aspects must be evaluated in terms of the aims of green roofs and the risk of wind or water erosion on green roof areas that are not fully covered.

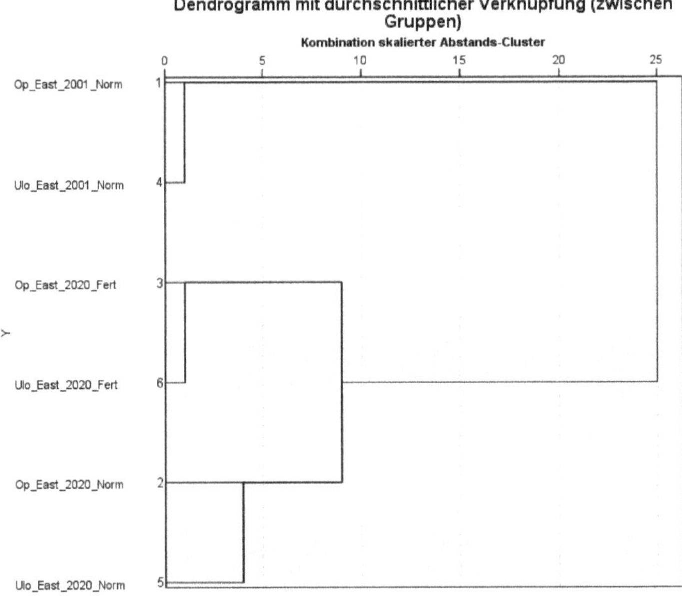

Figure 4. Dendrogram of the similarities between the eastern test plots from 2001 to 2020.

3.2. Phytomass in Relation to the Type of Media and the Media Depth

Test 4: focused on the phytomass on various growing media depths. The question is whether a greater growing media depth supports a higher biomass with better performance, e.g., in regard to CO_2 fixation.

3.2.1. The 10-cm Plots

In 2017, phytomass was harvested from all representative parts from each of the different roof orientations and exposures, with three replicates. Table 7 shows the dry matter values for the fractional biomass analysis from the vascular higher plants, mosses, and lichens. The total carbon is the sum of the shoots and roots together. All areas were fully developed over more than 10 years. One visible difference was the spontaneous growth of mosses and lichens, which represent a very special type of extensive green roof. These roofs are located in a region in northeastern Germany with clean air, and the composition of the vegetation on these roofs resembles that of typical poor sandy ground conditions, as is typical in the surrounding Müritz National Park. These dry-adapted lichens, mostly from the group of Cladonia, are endangered plants and are seldom found in ground-level habitats. The green roof is a retreat area for these plants. They are susceptible to foot traffic on dry roof conditions in summer. All values in Table 7 were calculated for an area of 1 m^2. In the calculation, the shoot and root phytomass of the vascular plants were calculated. The annual growth rate decreased after the establishment of full coverage with vegetation.

Table 7. Examples of the harvested dry phytomass in g/m^2 with SD from various media with 10-cm depth. Calculation for 1 m^2; analysis based on the mean of 3 harvesting plots each.

Exposure	Zi [1]	Blä [1]	Opti [1]	Ulo-N-Sun [1]	Ulo-N-Shade [1]	Op-N Sun
Vasc.plants	109 ± 6.5	183 ± 92	394 ± 145	243 ± 55	334 ± 88	411 ± 152
Mosses	1 ± 0.3	2 ± 0.5	487 ± 234	174 ± 135	846 ± 339	173 ± 94
Lichens	1448 ± 167	1080 ± 357	1 ± 0.3	444 ± 148	8 ± 7	721 ± 368
Roots	1319 ± 156	1456 ± 509	3301 ± 1740	2855 ± 993	3157 ± 305	2056 ± 415
Dry matter [3]	2876 ± 43	2719 ± 597	4182 ± 1656	3717 ± 832	4345 ± 314	3362 ± 177
Total C	1438 ± 22	1359 ± 298	2091 ± 828	1859 ± 416	2173 ± 159	1681 ± 88
CO_2	5249 ± 79	4961 ± 1090	7633 ± 3023	6784 ± 1518	7930 ± 579	6135 ± 322

Exposure	Op-North Shade [1]	Ulo-East [1]	Op-East
Vasc.plants	417 ± 84	132 ± 48	461 ± 86
Mosses	534 ± 306	17 ± 16	201 ± 81
Lichens	5 ± 4	110 ± 37	229 ± 211
Roots	2940 ± 333	656 ± 166	3550 ± 965
Dry matter [3]	3896 ± 302	915 ± 247	4442 ± 1001
Total C	1948 ± 151	458 ± 123	2221 ± 500
CO_2	7110 ± 550	1670 ± 450	8106 ± 1826

[1] Zi, Blä, Opti = on building 2, est. 1999, Op, Ulo: = on building 2, est. 2001. [3] Total dry organic matter in plants. Data from further plots are available.

Table 8 shows the correlation between vascular plant phytomass and its importance for CO_2 fixation. Broadly speaking, the more phytomass, the greater the effects, whereas mosses and lichens are counter-productive in that they suppress the growth of endangered higher plant species in sites with low-nutrient media while only making a minimal contribution to CO_2 fixation.

Table 8. Correlation and significance of harvested plants, n = 27 samples.

Pearson Correlation	CO_2	Vasc. Higher Plants	Mosses	Lichens	Roots
CO_2 fixation	1.0	0.661	0.233	−0.220	0.952
Sign. CO_2 1-tailed		0.000	−0.121	0.135	0.000 [1]

The dependence of CO_2 fixation on the phytomass follows a regression line as presented in Figure 5. More plants cover means higher CO_2-fixation in general.

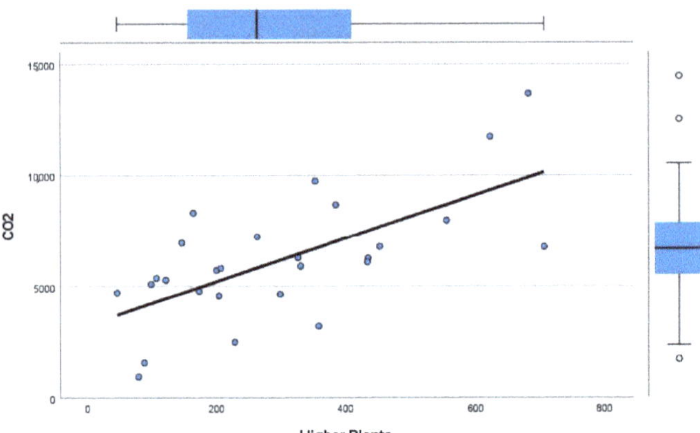

Figure 5. Regression—CO_2 fixation dependent on the phytomass of the higher plants from the 27 samples. All numbers are in grams.

The better growth rate on the 10-cm media is related to the different growing media used in these tests. Figure 6 shows the median values for CO_2 fixed by the plants, which varies considerably, with the lowest values of about 2000 g/m^2 on the Ulo media compared to average values of about 6000 g/m^2. In isolated cases, values that are twice as high are possible. If CO_2 fixation is to be the primary aim of the roof, careful choice of the media plus fertilizer is one solution. Calculating based on the approximately 8.5 million m^2 new green roofs realized in Germany [13], this will result in carbon storage of between 17,000 and 51,000 t CO_2/year after a few years.

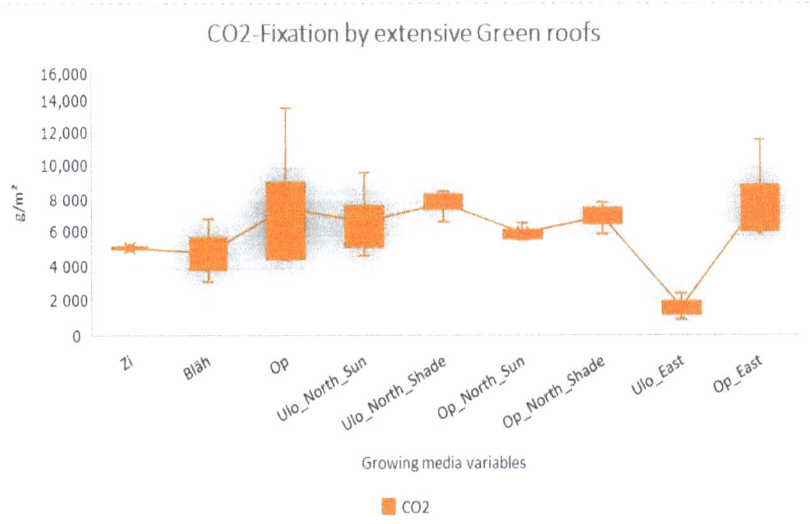

Figure 6. Comparison of the CO_2 fixation by the total phytomass (shoots and roots) of the vascular higher plants in g/m^2 on the nine different growing media, 10-cm layer, 3 replicates each.

3.2.2. The 30-cm Plots

These dry matter calculations were based on the 30-cm depth substrates, and the aboveground phytomass was harvested as described in Section 2. The 3 groups of 13 planter boxes were used in the following statistics. In the years between planting in 1999/2001 up to the first harvests in 2011 on building 2 and in 2013 on building 3, no harvesting of the plants influenced the typical growth of these plots. Working on the principle that extensive green roofs do not need maintenance, we allowed them to grow untouched for the first 10 years. The decline in the appearance was the initial reason for starting maintenance, which included mowing of the meadow. This means that the first years of harvesting, 2011 and 2013, respectively, have higher values than the annual productivity in the following years; see Table 9 with the standard deviation values.

Table 9. Examples of the harvested dry aboveground phytomass in g/m^2 and SD from the 30-cm-depth planter boxes, calculated as dry matter/m^2.

Year	Media	N	x-Mean	SD
2011	Op	13	385	72
	Zi	13	196	44
2012	Op	13	291	113
	Zi	13	34	14
2013	Op	13	168	35
	Zi	13	70	26
	Op-2	13	344	233
2014	Op	13	429	76
	Zi	13	206	50
	Op-2	13	131	58
2015	Op	13	399	85
	Zi	13	220	50
	Op-2	13	209	57
2017	Op	13	476	118
	Zi	13	223	80
	Op-2	13	95	45

The 13 boxes with the same growing media should have similar growth rates, but as can be seen in Figure 7, the values for the phytomass vary widely. These variations are caused by several factors but are ultimately due to impacts by the users of the roof. The extensive roof growing media is poor media with no additional fertilizer. The annual aboveground productivity on this 30-cm media is between 100 and 400 g/m^2. The boxes on the Op media on building 2 showed the best rates. The annual productivity of the dry mass ranges between 100 and 400 g/m^2. The 8.5 million m^2 of new green roofs in Germany [13] will result in fixation of between 1551 and 6205 t CO_2/year.

3.3. *Effect of Irrigation and Fertilization on Vegetation Cover and Biodiversity*

In 2011, on building 3, plots were marked to create fixed areas for annual irrigation and fertilization and to compare these areas with the normal situation without any extra treatments (see Table 1).

The test questions that were to be answered with this survey were:

In the vegetation analysis, the data were interpreted as follows. Specifically, these are the coverage values for the vascular plants, the total number of species, as well as the comparison of coverage values for all sedum species. The sedum coverage is an indicator of typical extensive green roofs. The coverage of all grasses together is an indicator for the shady parts of the north side and particularly the section of the roof shaded by the elevated part of building 3.

It was observed that the performance of sedum in general decreased after the first 10 years due to a lack of fertilization and irrigation. The number of plant species decreased on all test plots. In the following dependent paired *t*-test procedures, only the observations for 2011–2020 are interpreted.

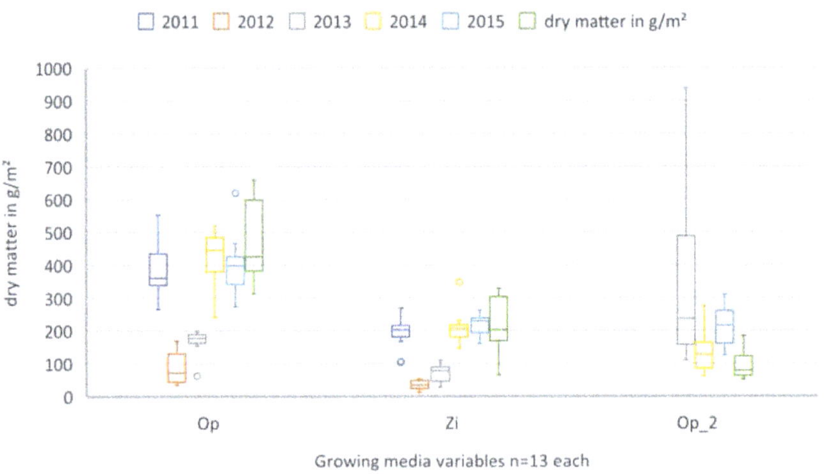

Figure 7. Aboveground annual harvested phytomass on the 30-cm boxes. Each of the three test plots represent 13 replicates at each date. The Op and Zi boxes are on building 2, and Op_2 is on building 3. All data refer to dry matter.

3.3.1. Eastern Test Plots

The eastern part of the roof with the initial plantings of sedum cuttings developed differently at all times compared to the other the test areas on the west, north, and south sides with the turf mats. In 2011, all test plots achieved the minimum value of 60%. Table 10 shows the changes in the normal areas; Ulo values were lower than the Op values. The effect of fertilizer is greater on the very poor Ulo media compared to the better performing Op media.

Table 10. Greening with sedum cuttings, n = 9 years, from 2011 to 2020. The coverage and number of species of the test plots with fertilizer (Fert.) and without (Norm.) for the eastern exposure. Mean value, standard deviation, and kurtosis. The negative kurtosis value indicates that the distribution is characterized by weaker marginal areas than the normal distribution.

Criteria	East Test Plots	x-Mean	SD	Kurtosis
Cover	Ulo_Norm [1]	71.6	11.1	−1.9
Cover	Ulo_Fert [1]	90.2	6.5	−1.1
Cover	Op_Norm [1]	90.1	7.9	−1.9
Cover	Op_Fert [1]	92.3	6.2	−0.2
Species	Ulo_Norm [1]	14	4.2	−0.2
Species	Ulo_Fert [1]	8	1.5	−1.1
Species	Op_Norm [1]	17	2.7	0.99
Species	Op_Fert [1]	10	2.4	−1.2
Cover Sedum	Ulo_Norm [1]	25.5	8.5	−1.2
Cover Sedum	Ulo_Fert [1]	42.6	5.4	−1.1
Cover Sedum	Op_Norm [1]	32.6	11.9	−1.5
Cover Sedum	Op_Fert [1]	40	25	−1.1

[1] Ulo = Ulopor media, Op = Optima media, Fert = Fertilizer, Norm = normal without fertilizer, East = Eastern section of the research roof.

The sedum coverage values increased on both media. The number of plant species decreased with the use of fertilizer.

The following Table 11 shows the statistics for the pairwise interpretation. In general, if the T value is not 0, the significance is high. On both media (Ulo, Op), the fertilizer significantly and positively impacted the vegetation cover. In addition, the decrease in the number of species was significant. The sedum coverage was only significantly enhanced on the Ulo test plot. On Op, the sedum coverage was not significantly more extensive because of the fertilizer but they bloomed better, as can be seen in Figure 2a.

Table 11. Greening with sedum cuttings, n=9 years, from 2011 to 2020 (df 8), coverage and number of species for the test plots with fertilizer (Fert.) and without (Norm.) of the eastern position. Pairwise T and the level 1‰ of significance differences between fertilized and non-fertilized plots in relation to cover values on Ulo. The minus symbols in relation to the Sedum cover; a reduction of these plants in relation to other plants.

Criteria	East Test Plots—Pairs	Mean	T	df	Significance (2-Sided)
Cover	Ulo_Fert [1] – Op_Fert [1]	−2.7	−0.89	8	0.401
Cover	Ulo_Fert [1] – Ulo_Norm [1]	18.7	3.55	8	0.008 ***
Cover	Op_Norm [1] – Op_Fert [1]	−2.1	−1.04	8	0.33
Species	Ulo_Fert [1] – Op_Fert [1]	−1.8	−2.1	8	0.07
Species	Ulo_Norm [1] – Ulo_Fert [1]	6	5.0	8	0.001 ***
Species	Op_Norm [1] –Op_Fert [1]	8	7.7	8	0.000 ***
Cover Sedum	Ulo_Fert [1] – Op_Fert [1]	2.6	0.32	8	0.76
Cover Sedum	Ulo_Norm [1] – Ulo_Fert [1]	−17	−4.4	8	0.002 ***
Cover Sedum	Op_Norm [1] – Op_Fert [1]	−7	−1.6	8	0.146

[1] Ulo = Ulopor media, Op = Optima media, Fert = Fertilizer, Norm = normal without fertilizer, East = Eastern section of the research roof.

3.3.2. Western test plots

The turf mat layout followed the same model as the eastern test plot. The basic mean values are shown in Table 12. The differences between the fertilized and non-fertilized sections on the Ulo test plots is less dramatic, with an 11% difference compared to the eastern test plot with the cuttings with a difference of 21%. This difference is significant, as can be seen in Table 13. The reduction in the number of species is also significant for both media, as Ulo dropped from 20 to 12 and Op dropped from 27 to 16, as both tables show. Again, the sedum coverage was significantly influenced by fertilizer usage but only on the Ulo media did it result in a significant change in the coverage value.

Table 12. Greening with vegetation mats, n = 9 years, from 2011 to 2020. The coverage and number of species for the test plots with fertilizer (Fert.) and without (Norm.) for the western position. Mean value, standard deviation, and kurtosis.

Criteria	West Test Plots	x-Mean	SD	Kurtosis
Cover	Ulo_Norm [1]	85.1	6.77	0.09
Cover	Ulo_Fert [1]	96.4	4.6	5.7
Cover	Op_Norm [1]	92.6	3.6	0.94
Cover	Op_Fert [1]	98.1	0.93	3.3
Species	Ulo_Norm [1]	20	4.6	−1.6
Species	Ulo_Fert [1]	12	1.7	−1.6
Species	Op_Norm [1]	27	3.6	−0.72
Species	Op_Fert [1]	16	2.4	0.72
Cover Sedum	Ulo_Norm [1]	49.9	2.6	2.6
Cover Sedum	Ulo_Fert [1]	62.4	17.0	−1.3
Cover Sedum	Op_Norm [1]	50.3	5.6	−2.1
Cover Sedum	Op_Fert [1]	58.8	10.9	1.9

[1] Ulo = Ulopor media, Op = Optima media, Fert = Fertilizer, Norm = normal without fertilizer, West = Western section of the research roof.

Table 13. Greening with turf mats, n = 9 years, from 2011 to 2020 (df 8), coverage and number of species for the test plots with fertilizer (Fert.) and without (Norm) of the western position. Pairwise *t*-test and the different level of significance. The significance indicate the effects of fertilizer on the cover values, negative sign, indicate reduction of the values over the years.

Criteria	West Test Plots Pairs	Mean	T	df	Significance (2-Sided)
Cover	Ulo_Norm [1] – Op_Norm [1]	−7.4	−2.79	8	0.024 *
Cover	Ulo_Fert [1] – Ulo_Norm [1]	11.3	7.21	8	0.000 ***
Cover	Ulo_Fert [1] – Op_Fert [1]	−1.67	−5.53	8	0.35
Cover	Op_Fert [1] – Op_Norm [1]	5.6	4.86	8	0.001 ***
Species	Ulo_Fert [1] – Op_Fert [1]	−4.2	−4.81	8	0.001 ***
Species	Ulo_Fert [1 -] – Ulo_Norm [1]	−8.78	−4.25	8	0.003 ***
Species	Op_Fert [1] – Op_Norm [1]	−11.4	−8.27	8	0.000 ***
Species	Ulo_Norm [1] – Op_Norm [1]	−6.89	−8.04	8	0.000 ***
Cover Sedum	Ulo_Norm [1] –Ulo_Fert [1]	−12.56	−2.31	8	0.050 *
Cover Sedum	Ulo_Fert [1] – Op_Fert [1]	3.67	0.734	8	0.484
Cover Sedum	Op_Norm [1] – Op_Fert [1]	−8.44	−1.81	8	0.11

[1] Ulo = Ulopor, Op = Optima, Fert = Fertilizer, Norm = normal without fertilizer, West = Western section of the roof.

3.3.3. Northern Test Plots

In the first years of this study, the northern test plots differed between the very shady areas near the building "North_Shade" and those areas further away from the building with full sun, described here as "North Sun". This full-sun area differed from the southern test plots due to the additional heat reflected from the elevated building parts, with remarkably fewer grasses inside and a high dominance of the sedum cover. Table 14 shows that all north areas showed coverage values of above 90%. Consequently, the influence of the fertilizer was not significant in any case, as can be seen in Table 15 in the comparison of both Ulo Norm-Fertilizer pairs. Again, the reduction in the number of the plant species was significant for both growing media.

Table 14. Greening with vegetation mats, n = 9 years, from 2011 to 2020. Coverage and number of species for the test plots with fertilizer (Fert.) and without (Norm.) of the northern position; mean value, standard deviation, and kurtosis. Shade: in the shade of an elevated section of the roof.

Criteria	North Test Plots	x-Mean	SD	Kurtosis
Cover	Ulo_Norm_Shade [1]	91.3	3.24	0.78
Cover	Ulo_Fert_Shade [1]	91.7	3.67	0.404
Cover	Op_Norm_Shade [1]	94.3	3.27	−1.68
Cover	Op_Fert_Shade [1]	98.2	1.19	0.77
Species	Ulo_Norm_Shade [1]	27	3.53	0.71
Species	Ulo_Fert_Shade [1]	16	3.74	−0.11
Species	Op_Norm_Shade [1]	24	2.74	2.2
Species	Op_Fert_Shade [1]	16	2.6	0.05
Cover	Ulo_Norm_Sun [1]	91.3	3.24	0.786
Cover	Ulo_Fert_Sun [1]	90.1	5.01	0.14
Cover	Op_Norm_Sun [1]	94.6	2.1	3.0
Cover	Op_Fert_Sun [1]	97.6	1.24	1.52
Species	Ulo_Norm_Sun [1]	15	1.1	0.02
Species	Ulo_Fert_Sun [1]	11	1.33	−1.97
Species	Op_Norm_Sun [1]	11	0.53	−2.57
Species	Op_Fert_Sun [1]	11	0..53	−2.57

[1] Ulo = Ulopor, Op = Optima, Fert = Fertilizer, Norm = normal without fertilizer, N = Northern section of the roof.

3.3.4. Southern Test Plots

The southern section of the roof is much smaller than the northern section. It gets full sun all day plus the reflection from the metal façade of the elevated section of the building engineering area on the roof. This section of the roof also has an integrated air conditioning outlet that blows hot dry air over the southern green roof. The combination of these factors simulates extreme summer drying conditions for the green roof cover. The great performance of all the sedums growing here is remarkable. These succulents not only survive, but they also perform exceptionally well in these hot dry conditions. On the other hand, nearly all grasses are completely gone on this exposure.

The mean values of the southern plots are summarized in Table 16. With and without fertilizer, the coverage values are all above 90% and in general are the best of the test plots analyzed here. The sedum coverage values are above 70%, and with fertilizer they reach values above 80%. Again, the number of species decreased with the use of fertilizer. Table 17 presents the level of significance for these paired tests with the fertilizer test plots.

Table 15. Greening with turf mats, n = 9 years, from 2011 to 2020 (df 8), coverage and number of species for the test plots with fertilizer (Fert.) and without (Normal) of the northern position: North Shade: in the shade of an elevated section of the building. Sun_N: North, outside this shade. Pairwise *t*-test and the level of significance. Example explanation line 1 of Table 15: The minus sign in category "means" symbols lower cover values of the first element of the pair. As example: Athough Ulo and Op were fertilized, over the time, Ulo has all the times lower cover values than Op on the highest level of 1‰ level.

Criteria	North Test Plot Pairs	Mean	T	df	Significance (2-Sided)
Cover	Ulo_Fert_Shade [1] –Op_Fert_Shade [1]	−6.6	−4.8	8	0.001 ***
Cover	Ulo_Fert_Shade [1] – Ulo_Norm_Shade [1]	0.33	1.1	8	0.347
Cover	Op_Fert_Shade [1] – Op_Norm_Shade [1]	3.89	3.0	8	0.017 **
Cover	Ulo_Fert_Sun [1] – Op_Fert_Sun [1]	−7.4	−4.40	8	0.002 ***
Cover	Ulo_Fert_Sun [1] – Ulo_Norm_Sun [1]	−1.2	−0.53	8	0.61
Cover	Op_Fert_Sun [1] – Op_Norm_Sun [1]	3.00	3.84	8	0.005 ***
Species	Ulo_Norm_Sun [1] – Ulo_Fert_Sun [1]	3.8	6.34	8	0.000 ***
Species	Op_Norm_Sun [1] – Op_Fert_Sun [1]	−0.11	−0.55	8	0.594
Species	Ulo_Norm_Shade [1] –Ulo_Fert_Shade [1]	10.78	7.07	8	0.000 ***
Species	Ulo_Norm_Shade [1] – Op_Norm_Shade [1]	3.11	3.18	8	0.013 **
Species	Op_Norm_Shade [1] –Op_Fert_Shade [1]	7.67	5.84	8	0.000 ***
Species	Ulo_Norm_Sun [1] – Ulo_Fert_Sun [1]	3.78	6.34	8	0.000 ***
Species	Ulo_Norm_Sun [1] – Op_Norm_Shade [1]	3.78	9.43	8	0.000 ***
Species	Op_Norm_Sun [1] – Op_Fert_Sun [1]	−0.11	0.56	8	0.594

[1] Ulo = Ulopor, Op = Optima, Fert = Fertilizer, Norm = normal without fertilizer, N = Northern section of the roof.

Table 16. Greening with vegetation mats, n = 9 years, from 2011 to 2020. Coverage and number of species, and the coverage of sedum only for the test plots with fertilizer (Fert.) and without (Norm.) for the southern position; mean value, standard deviation, and kurtosis.

Criteria	South Test Plots	x-Mean	SD	Kurtosis
Cover	Ulo_Norm [1]	92.9	2.4	−2.1
Cover	Ulo_Fert [1]	98.1	1.69	−1.73
Cover	Op_Norm [1]	93.7	2.45	−1.14
Cover	Op_Fert [1]	96.9	3.2	1.5
Species	Ulo_Norm [1]	13	1.7	−0.008
Species	Ulo_Fert [1]	9	2.1	−1.91
Species	Op_Norm [1]	12	1.1	−1.14
Species	Op_Fert [1]	9	2.12	−1.91
Cover Sedum	Ulo_Norm [1]	74	5.28	−1.08
Cover Sedum	Ulo_Fert [1]	80	5.5	−1.23
Cover Sedum	Op_Norm [1]	72	8.12	1.4
Cover Sedum	Op_Fert [1]	81	18.4	−0.54

[1] Ulo = Ulopor, Op = Optima, Fert = Fertilizer, Norm = normal without fertilizer, South = Southern section of the roof.

Table 17. Greening with turf mats, n=9 years, from 2011 to 2020 (df 8), coverage and number of species for the test plots with fertilizer (Fert.) and without (Norm.) of the southern position. Pairwise *t*-test and the level of significance. As explanation e.g. in line 1; also the south plots has a profit in cover value by the fertilizer, in this case on the 2‰-level. On the other hand, the fertilizer reduces the number of species. The Sedum cover is not significant influenced by the fertilizer.

Criteria	South Test Plot Pairs	Mean	T	df	Significance (2-Sided)
Cover	Ulo_Fert [1] – Ulo_Norm [1]	5.22	−4.5	8	0.002 ***
Cover	Ulo_Fert [1] – Op_Fert [1]	1.22	1.4	8	0.194
Cover	Op_Norm [1] – Op_Fert [1]	−3.22	−3.4	8	0.009 ***
Species	Ulo_Norm [1] – Ulo_Fert [1]	4.0	7.24	8	0.000 ***
Species	Ulo_Norm [1] – Op_Norm [1]	0.89	1.512	8	0.169
Species	Op_Norm [1] – Op_Fert [1]	1.22	2.82	8	0.023 *
Cover Sedum	Ulo_Norm [1] –Ulo_Fert [1]	5.67	−3.44	8	0.009 ***
Cover Sedum	Ulo_Fert [1] – Op_Fert [1]	−1.11	−0.191	8	0.853
Cover Sedum	Op_Norm [1] – Op_Fert [1]	−8.89	−1.53	8	0.164

[1] Ulo = Ulopor, Op = Optima, Fert = Fertilizer, Norm = normal without fertilizer, South = Southern section of the roof.

4. Discussion

Effective solutions are needed to stop the increase in average temperatures around the globe and to reach the targets in the Paris climate agreement in the next few years [30]. On the macro-scale, as explained in Fang et al. [31], for the landmass of China, drought, temperature, and global warming are significantly connected to the existing vegetation cover. Many studies have shown that cities around the world are in general drier than their surroundings, as can be seen from examples from China [32] and 70 cities in Europe [33]. In the search for solutions on a city-wide scale, it is apparent that any decentralized greenery improves the nearby living environment of its citizens [34]. Global warming is a multi-factorial issue with an energy–water nexus. Evapotranspiration is one main energetic factor, while CO_2 is the accepted lead indicator from a political perspective to measure success in tackling climate change. However, not all factors can be explained by successful CO_2 reduction. Green roofs in general and detailed technical solutions to improve the performance of green roofs as described in this publication are methods to adapt to and mitigate climate change. On a brighter note, energetic and chemical procedures are connected and need more holistic solutions.

The economic lockdowns during the COVID-19 pandemic were reported to lead to a 7% reduction in CO_2 emissions in late 2020 and the start of 2021 [35]. Additional lockdowns are economically difficult and not considered acceptable for longer periods. More extensive and decentralized methods have to be considered. More CO_2 fixation is required, and green roofs can contribute to this. Their impact is the step from no green roof to a green roof with relatively stable values of a total amount of about 6 kg/m^2 as demonstrated in this survey. This value results from the total phytomass of shoots and roots of plants after a number of years of establishment, in this case 17 years. If we take this value as a baseline, extensive green roofs in Germany will result in about 51,000 t/CO_2 fixation per year with about 8,500,000 m^2 of new green roofs every year. If this annual amount needs to be increased, higher vegetation values could be achieved with a thicker layer of growing media combined with annual fertilization. This survey showed annual growth rates on 30-cm media of about 100 to 600 g/m^2 dry matter each year. However, this implied some maintenance of some form, such as mowing of the annual growth. Several earlier surveys measured the CO_2 sequestration of green roofs. Our results correspond to the wide range of sequestration values observed, which usually only focus on the aboveground material, mostly sedum [36], with variation between 64 and 381 g/m^2 x years. Grasses in general perform better, as can be seen in a study from Japan [21] for *Cynodon dactylon* with sequestration of 2.5 kg CO_2/m^2*year (fertilized and optimally irrigated roof modules) and *Sedum aizoon* (non-irrigated test plot of 1.2 kg CO_2/m^{2}*year). The green roof rates are comparable to typical dry meadow habitats on the ground.

The CO_2 fixation is an energetic procedure to tackle global warming. Finally, the size, quality, and distribution of the green roofs are important factors to have countable effects on the city scale [37].

Climate change is resulting in longer dry periods and lower humidity. Green roofs offer a range of benefits as demonstrated by several research projects over the last 20 years [38,39]. These reviews reveal that not all the results are comparable because the methods vary widely and most of the research was conducted over short time frames.

This 20-year survey confirmed that not all results could be achieved by a single roof project. What is important is the right selection of plant species, such as that shown in Table 18, which will help to achieve full vegetation coverage with many positive ecological effects.

Table 18. Preferred plant mixtures on the various roof spots of this survey.

Test Plots	North-Shade	West_East-North-Sun	South—Extreme Dry Hot Conditions
Plant preferences	Grasses	Mix of various life forms	Sedum in many variation

Which plants are perfectly adapted to the expected climate changes? This survey was based on test plots with turf mats containing a generalized plant species mix. The test plots studied ranged from shade to extreme full sun, with some additional stress caused by high levels of solar reflection, air conditioner exhaust air outlets, and water stress.

An additional important factor is correctly choosing the right substrate. The test plot for this experiment shows that the type of growing medium is always critical for the biodiversity development over the years [40]. In most cases, highly biodiverse green roofs remain highly biodiverse over the years and support a wide range of species of birds and invertebrates [41]. As seen with this green roof research in Neubrandenburg, a significant quantity of mosses and lichens grow on very poor media. They can make up to a quarter of the total phytomass, resulting in up to 1.5 L/m^2 *day evapotranspiration [42]. On the one hand, the abundant presence of mosses and lichens characterizes areas with low pollution levels while also providing additional CO_2 fixation. On the other hand, the massive growth of mosses and lichens displaces the typical or endangered higher plants that would otherwise be expected here [43].

Semi-intensive roof construction supports a wider range of plant species, such as dry-adapted prairie plants, with higher phytomass production and ultimately higher CO_2 fixation [44]. The increasing number of urban agriculture sites may also be a solution for productive roof systems with massive phytomass production. Apart from the typical productive roof substrates, the competitiveness of hydroponic substrates, which have been used with increasing success in intensive green roofing, has improved in recent years [45]. The phytomass and species richness of typical extensive green roofs are comparable to dry meadows in a low mountain range [46]. Similar findings coming from North America, where more CO_2 could be fixed by a green roof and intensive roof gardens can be the choice [47].

This survey also shows that over a time frame of 20 years, the number of plant species significantly decreases. If this is to be avoided, some maintenance work is helpful. Additional conclusions that can be drawn from this work are:

-If a biodiverse roof is the target, variation in the growing media and media depth and additional micro niches are necessary. The research design on this roof focused on replicated research plots for statistically comparable sites.

-The study provided information about the benefits of turf mats compared to sedum cuttings.

Table 19 compares the benefits of both methods, revealing that turf mats fulfill all requirements from the beginning, but they are more expensive. The effort for the initial maintenance is quite similar. The differences are apparent in the better CO_2 fixation.

Table 19. Turf mat benefits compared to sedum cuttings as the primary vegetation cover, with results from this survey.

Test Plots	Turf Mats	Sedum Cuttings
Investment	High	Low
Energy effort	High	Low
Maintenance duties	Irrigation first year	Irrigation first year
Vegetation coverage 60%	Immediately	After 2–3 years
Plant biodiversity, Year 1	25–29	9–14
Plant biodiversity, Year 20	15–27	16–18
CO_2 fixation	High	Low

Fertilizer supports plant growth and plant performance in general [48] but not the local biodiversity. One single extra irrigation of 10 L/m^2 in early summer days had no significant effect on the plant performance of these dry-resistant herbs. In contrast to this, Du et al. [49] recommended emergency irrigation if shrubs are on green roofs. In times of increasing drought, the selection of the right plant species becomes an important issue. Lists with drought-tolerant plants have to be done on the regional levels, and the details in the microhabitat design support significant biodiversity [50]. Additionally, the focus on biological plant traits can help to develop an understanding regarding which species can be selected better in future times of climate change conditions [51]. In our survey, the importance of lichens of the genus *Cladonia* are highlighted as an important easy-to-care plant group and low-pollution condition, as they are in Neubrandenburg, a similar observation to a green roof working group in Canada [52]. Under the conditions of drought and climate change, prairie plants from North America become an alternative to be established in extensive green roofs under drought conditions for Central Europe [53]. Finally, the acceptance of successful growing weeds has to be accepted in natural concepts, and in many cases they can survive under extreme climatic conditions [54]. The need for local and regional biodiversity green roofs should be anchored in the related norms or guidelines in future [55].

Removing forest vegetation around the planet has significantly increased water shortages [56]. Heavier rainfall is also a consequence of the removal of vegetation in many places. This survey shows that minimal additional water has no influence on typical green roofs. Graduated tests of rain intensity and duration show the typical behavior, with extensive green roofs able to easily handle rain up to 30 L/m^2 [57], with the right extra drainage elements below providing support [23]. Green roofs are effective in all climate zones around the world [58]. Green roofs work like a dewfall sink in cities. In comparison to typical bitumen roofs, green roofs capture the morning dew, which helps adapted plants to survive and is also a decentralized contribution against drought in cities [59].

Green roof technology has reached the international politics against global warming. They have become elements of the renovation strategy for building stocks and the EU biodiversity strategy 2030 [60]. The results presented here can contribute to ensuring more effective green roofs as part of the future wave of renovation [61] to successfully handle the coming climate challenges. Our current knowledge of green roofs reveals opportunities to construct new quality green spaces, which can be equivalent to vegetated spaces on the ground in terms of functionality and usability [62].

As recommendations for future research, this survey has shown that there is a need for more detailed research with more variations in growing media and structures to ensure that green roof spaces are more effective instruments of green infrastructure or systems for ecosystem services [63]. Edible cities are one of the new ecosystem services to which green roof spaces can contribute to in the future [64].

In addition, it is essential that methods between research institutions are aligned so that results can be more easily compared [65]. More long-term ecosystem studies can ensure better understanding of the variation in the integrated biological solutions. As a result of global warming and the adaptation of plants to growing on artificial urban surfaces, these systems react dynamically, unlike technical solutions that remain static.

5. Conclusions

This study modified the typical extensive green roof with a 10-cm media depth and low maintenance in the directions of deeper growing media with 30 cm. Further treatments were fertilization and irrigation. The aim was to enhance the CO_2 fixation of the roof vegetation. However, what are the effects on vegetation cover and biodiversity? The summarized conclusions of these couple of years:

-Green roofs perform better if maintenance, such as extra fertilizer, are applied.

-Deeper media can capture more water, have higher vegetation cover, and finally more CO_2 fixation.

-On the other hand, the biodiversity decreases by higher fertilizer rates. If biodiversity is the main target of the green roof project, more habitat niches and media variation are recommended as results from this research to achieve this.

Green roofs can be established in higher quantities in all cities around the world. Although the best citizen/green roof values today exist in Stuttgart/Germany with 4 m^2/citizen [13], and much more is possible if more regulation in legal plans is fixed.

Green roofs are one element in the strategies against global warming and can bring back a contact to man-made nature to citizens.

Typical extensive roof vegetation is well adapted to meet future climate change challenges, but detailed planning can help to develop green roofs to meet more specific aims, such as:

-Climate-adaptive roofs with higher plant biomass.

-Biodiverse green roofs with many micro-niches for high plant species richness.

-Green roofs as elements of blue-green infrastructure with optimized water storage capacity.

-Green roofs for urban agriculture with humus media and/or hydroponic plant growth systems.

Green roofs are a small but visible factor bringing more evaporation surfaces into the anthropogenic urban desert and into the struggle against the drought.

Author Contributions: Conceptualization, M.K; methodology, M.K.; D.K.; Measurements M.K.; D.K.; validation, M.K.; D.K.; formal analysis, M.K.; D.K.; investigation, M.K.; D.K.; resources, M.K.; data curation, M.K.; D.K.; writing—original draft preparation, M.K.; D.K.; writing—review and editing, M.K.; D.K.; visualization, M.K.; D.K.; project administration, M.K.; All authors have read and agreed to the published version of the manuscript.

Funding: This research received no external funding for the research; it is part of the Green roof lab of the University of Applied Sciences. We acknowledge support for the Article Processing Charge from the Deutsche Forschungsgemeinschaft (DFG, German Research Foundation, 414051096) and the Open Access Publication Fund of the Hochschule Neubrandenburg (Neubrandenburg University of Applied Sciences).

Informed Consent Statement: Not applicable.

Data Availability Statement: The basic data sets of this research are archived in the University of Applied Sciences Neubrandenburg.

Acknowledgments: Special thanks to Marion Japp, Schwerin, for language improvement.

Conflicts of Interest: The authors declare no conflict of interest.

References

1. Rom, W.N.; Pinkerton, K.E. Introduction: Consequences of Global Warming to Planetary and Human Health. In *Climate Change and Global Public Health*; Respiratory Medicine. Humana; Pinkerton, K.E., Rom, W.N., Eds.; Springer: Cham, Switzerland, 2021. [CrossRef]
2. Youl, R. *Landcare in Victoria*; Spicers Paper: Melbourne, Australia, 2006; p. 194.
3. Wüstenfrauen. Available online: https://wooxl.wordpress.com/2010/10/23/chinas-wustenfrauen/ (accessed on 28 February 2021).
4. The Great Green Wall. Available online: https://www.greatgreenwall.org/about-great-green-wall/ (accessed on 28 February 2021).
5. Mutke, J.; Quandt, D. Der Deutsche Wald. Seine Geschichte, seine Ökologie. Wald. *Forsch. Lehre* **2018**, *8*, 650–652.
6. DeFries, R.; Rudel, T.; Uriarte, M.; Hansem, M. Deforestation driven by urban population growth and agricultural trade in the twenty-first century. *Nat. Geosci.* **2010**, *3*, 178–181. [CrossRef]
7. Clinton, N.; Yu, L.; Fu, H.; He, C.; Gong, P. Global-Scale Associations of Vegetation Phenology with Rainfall and Temperature at a High Spatio-Temporal Resolution. *Remote Sens.* **2014**, *6*, 7320–7338. [CrossRef]
8. Kabano, P.; Lindley, S.; Harris, A. Evidence of urban heat island impacts on the vegetation growing season length in a tropical city. *Landsc. Urban Plan.* **2021**, *206*. [CrossRef]
9. Kravcik, M.; Pokorny, J.; Kohutiar, M.; Toth, E. *Water for the Recovery of the Climate—A New Water Paradigm*; Krupa Print: Žilina, Slovakia, 2007.
10. Trillion Tree Plantings. Available online: https://trilliontrees.org.au/ (accessed on 1 March 2021).
11. Tan, P.Y. *Vertical Garden City*; Straits Times Press: Singapore, 2013; p. 192.
12. Loder, A. Greening the City: Exploring Health, Well-Being, Green Roofs, and the Perspective of Nature in the Workplace. Ph.D. Thesis, University of Toronto, Toronto, ON, Canada, 2011.
13. Mann, G.; Gohlke, R.; Wolff, F. *Marktreport Dach-, Fassaden- und Innenraumbegrünung Deutschland*; BUGG-Publication: Saarbrücken, Germany, 2020; p. 72.
14. Dover, J.W. *Green Infrastructure*; Routledge: London, UK; New York, NY, USA, 2015; p. 337. [CrossRef]
15. Oberndorfer, E.; Lundholm, J.; Brass, B.; Coffmann, R.; Doshi, H.; Dunnett, N.; Gaffin, S.; Köhler, M.; Liu, K.; Rowe, B. Green Roofs as Urban Ecosystems: Ecological Structures, Functions, and Services. *BioScience* **2007**, *57*, 823–833. [CrossRef]
16. Cristiano, E.; Deidda, R.; Viola, F. The role of green roofs in urban Water-Energy-Food-Ecosystem nexus: A review. *Sci. Total. Environ.* **2021**, 756. [CrossRef]
17. Andersson, E.; Langemeyer, J.; Borgström, S.; Mcphearson, T.; Haase, D.; Kronenberg, J.; Barton, D.N.; Davis, M.; Naumann, S.; Röschel, L.; et al. Enabling Green and Blue Infrastructure to Improve Contributions to Human Well-Being and Equity in Urban Systems. *BioScience* **2019**, *69*, 566–574. [CrossRef]
18. Cirkel, D.G.; Voortman, B.R.; Van Veen, T.; Bartholomeus, R.P. Evaporation from (Blue-) Green Roofs: Assessing the Benefits of a Storage and Capillary Irrigation System Based on Measurements and Modeling. *Water* **2018**, *10*, 1253. [CrossRef]
19. Cook, L.M.; Larsen, T.A. Towards a performance-based approach for multifunctional green roofs: An interdisciplinary review. *Build. Environ.* **2021**, 188. [CrossRef]
20. Shafique, M.; Xue, X.; Luo, X. An overview of carbon sequestration of green roofs in urban areas. *Urban For. Urban Green.* **2020**, *47*. [CrossRef]
21. Kuronuma, T.; Watanabe, H.; Ishihara, T.; Kou, D.; Toushima, K.; Ando, M.; Shindo, S. CO2 Payoff of Extensive Green Roofs with Different Vegetation Species. *Sustainability* **2018**, *10*, 2256. [CrossRef]
22. Köhler, M.; Kaiser, D. Evidence of the Climate Mitigation Effect of Green Roofs–A 20-Year Weather Study on an Extensive Green Roof (EGR) in Northeast Germany. *Buildings* **2019**, *9*, 157. [CrossRef]
23. Kaiser, D.; Köhler, M.; Schmidt, M.; Wolf, F. Increasing Evapotranspiration on Extensive Green Roofs by Changing Substrate Depths, Construction, and Additional Irrigation. *Buildings* **2019**, *9*, 173. [CrossRef]
24. Lösken, G.; Ansel, W.; Backhaus, T.; Bartel, Y.-C.; Bornholdt, H.; Bott, P.; Henze, M.; Hokema, J.; Köhler, M.; Krupka, B.W.; et al. *Dachbegrünungsrichtlinien–Richtlinien für Planung, Bau und Instandhaltung von Dachbegrünungen*; Fll publishing: Bonn, Germany, 2018; p. 150.
25. Kopecký, K.; Dostálek, J.; Frantík, T. The use of the deductive method of syntaxonomic classification in the system of vegetational units of the Braun-Blanquet approach. *Vegetatio* **1995**, *117*, 95–112. [CrossRef]
26. De Graaf, M.A.; Van Groeninen, K.J.; Six, J.; Hungate, B.; Van Kessel, C. Interactions between plant growth and soil nutrient cycling under elevated CO2: A meta-analysis. *Glob. Chang. Biol.* **2006**, *12*, 2077–2091. [CrossRef]
27. Poorter, H.; Navas, M.L. Plant growth and competition at elevated CO2: On winners, losers and functional groups. *New Phytol.* **2003**, *157*, 175–198. [CrossRef]
28. Körner, C. Plant CO_2 responses: An issue of definition, time and resource supply. *New Phytol.* **2006**, *172*, 393–411. [CrossRef] [PubMed]
29. Natureoffice. Wir Realisieren Klimalösungen. Available online: https://www.natureoffice.com/ecocent (accessed on 6 January 2021).
30. Hein, J.; Guarin, A.; Frommé, E.; Pauw, P. Deforestation and the Paris climate agreement: An assessment of REDD+ in the national climate action plans. *For. Policy Econ.* **2018**, *90*, 7–11. [CrossRef]
31. Fang, W.; Huang, S.; Huang, Q.; Huang, G.; Wang, H.; Leng, G.; Wang, L.; Li, P.; Ma, L. Bivariate probabilistic quantification of drought impacts on terrestrial vegetation dynamics in mainland China. *J. Hydrol.* **2019**, *577*, 10. [CrossRef]

32. Li, Q.; Zhang, H.; Liu, X.; Huang, J. Urban heat island effect on annual mean temperature during the last 50 years in China. *Theor. Appl. Climatol.* **2004**, *79*, 165–174. [CrossRef]
33. Ward, K.; Lauf, S.; Kleinschmit, B.; Endlicher, W. Heat waves and urban heat islands in Europe: A review of relevant drivers. *Sci. Total Environ.* **2019**, *569*, 527–539. [CrossRef]
34. UNFCCC. Long-Term Low Greenhouse Gas Emission Development Strategy of the EU and Its Member States. 2020. Available online: https://unfccc.int/documents/210328 (accessed on 10 January 2021).
35. IDW-News. Rekord Rückggang der Globalen CO_2-Immissionen. 2020. Available online: https://nachrichten.idw-online.de/2020/12/11/rekord-rueckgang-der-globalen-CO2-emissionen-dank-corona/ (accessed on 10 January 2021).
36. Getter, K.L.; Rowe, D.B.; Robertson, G.P.; Cregg, B.M.; Andersen, J.A. Carbon Sequestration Potential of Extensive Green Roofs. *Environ. Sci. Technol.* **2009**, *43*, 7564–7570. [CrossRef] [PubMed]
37. Heusinger, J.; Weber, S. Surface energy balance of an extensive green roof as quantified by full year eddy-covariance measurements. *Sci. Total. Environ.* **2017**, *577*, 220–230. [CrossRef] [PubMed]
38. Stangl, R.; Medl, A.; Scharf, B.; Pitha, U. Wirkungen der Grünen Stadt. Studie zur Abbildung des Aktuellen Wissenstands im Bereich Städtischer Begrünungsmaßnahmen. In Bundesministerium für Verkehr, Innovation und Technologie (Hrsg.), Berichte aus Energie- und Umweltforschung 12/2019. Available online: https://nachhaltigwirtschaften.at/resources/sdz_pdf/schriftenreihe-2019-12-wirkungen-gruene-stadt.pdf (accessed on 11 January 2021).
39. Manso, M.; Teotonio, I.; Silva, C.M.; Cruz, C.O. Green roof and green wall benefits and costs: A review of the quantitative evidence. *Renew. Sustain. Energy Rev.* **2021**, *135*, 110111. [CrossRef]
40. Ksiazek-Mikenas, K.; Herrmann, J.; Menke, S.B.; Köhler, M. If You Build It, Will They Come? Plant and Arthropod Diversity on Urban Green Roofs over Time. *Urban Nat.* **2018**, *1*, 52–72.
41. Filazzola, A.; Shrestha, N.; MacIvor, S. The contribution of constructed green infrastructure to urban biodiversity: A synthesis and meta-analysis. *J. Appl. Ecol.* **2019**, 1–13. [CrossRef]
42. Heijmans, M.M.P.D.; Arp, W.J.; Chapin, F.S., III. Controls on moss evaporation in a boreal black spruce forest. *Glob. Biogeochem. Cycles* **2004**, *18*. [CrossRef]
43. Drake, P.; Grimshaw-Surette, H.; Heim, A.; Lundholm, J. Mosses inhibit germination of vascular plants on an extensive green roof. *Ecol. Eng.* **2019**, *117*, 111–114. [CrossRef]
44. Baraldi, R.; Neri, L.; Costa, F.; Facini, O.; Rapparini, F.; Carriero, G. Ecophysiological and micromorphological characterization of green roof vegetation for urban mitigation. *Urban For. Urban Green.* **2019**, *37*, 24–32. [CrossRef]
45. Rufí-Salís, M.; Calvo, M.J.; Petit-Boix, A.; Villalba, G.; Gabarrell, X. Exploring nutrient recovery from hydroponics in urban agriculture: An environmental assessment. *Resour. Conserv. Recycl.* **2020**, *155*, 104683. [CrossRef]
46. Schröder, R.; Kiehl, K. Extensive roof greening with native sandy dry grassland species: Effects of different greening methods on vegetation development over four years. *Ecol. Eng.* **2020**, *145*, 105728. [CrossRef]
47. Whittinghill, L.J.; Rowe, B.D.; Schutzki, R.; Cregg, B.M. Quantifying carbon sequestration of various green roof and ornamental landscape systems. *Landsc. Urban Plan.* **2014**, *123*, 41–48. [CrossRef]
48. McAlister, R.; Rott, A.S. Up on the roof and down in the dirt: Differences in substrate properties (SOM, potassium, phosphorus and pH) and their relationships to each other between sedum and wildflower green roofs. *PLoS ONE* **2019**, *14*, e0225652. [CrossRef]
49. Du, P.; Arndt, S.K.; Farrell, C. Is plant survival on green roofs related to their drought response, water use or climate of origin? *Sci. Total. Environ.* **2019**, *667*, 25–32. [CrossRef]
50. Lundholm, J.T.; Marlin, A. Habitat origins and microhabitat preferences of urban plant species. *Urban Ecosyst.* **2006**, *9*, 139–159. [CrossRef]
51. Mechelen, C.; Dutoit, T.; Kattge, J.; Hermy, H. Plant trait analysis delivers an extensive list of potential green roof species for Mediterranean France. *Ecol. Eng.* **2014**, *67*, 48–59. [CrossRef]
52. Heim, A.; Lundholm, J.T. Cladonia lichens on extensive green roofs: Evapotranspiration, substrate temperature, and albedo. *F1000Research* **2013**, *2*, 274. [CrossRef]
53. Liu, J.; Shresth, P.; Skabelund, L.R.; Todd, T.; Decker, A.; Kirkham, M.B. Growth of prairie plants and sedums in different substrates on an experimental green roof in Mid-Continental USA. *Sci. Total. Environ.* **2019**, *697*, 134089. [CrossRef] [PubMed]
54. Vanstockem, J.; Somers, B.; Hermy, M. Weeds and gaps on extensive green roofs: Ecological insights and recommendations for design and maintenance. *Urban For. Urban Green.* **2019**, *46*, 126484. [CrossRef]
55. Catalano, C.; Laudicina, V.A.; Badalucco, L.; Guarino, R. Some European green roof norms and guidelines through the lens of biodiversity: Do ecoregions and plant traits also matter? *Ecol. Eng.* **2018**, *115*, 15–26. [CrossRef]
56. Wiltshire, A.; Gornall, J.; Booth, B.; Dennis, E.; Falloon, P.; Kay, G.; McNeall, D.; McSweeney, C.; Betts, R. The importance of population, climate change and CO2 plant physiological forcing in determining future global water stress. *Glob. Environ. Chang.* **2013**, *23*, 1083–1097. [CrossRef]
57. Li, X.X.; Cao, J.J.; Xu, P.X.; Fei, L.; Dong, Q.; Wang, Z.L. Green roofs: Effects of plant species used on runoff. *Land Degrad. Dev.* **2018**, 1–11. [CrossRef]
58. Barriuso, F.; Urbano, B. Green roofs and walls design intended to mitigate climate change in urban areas across all continents. *Sustainability* **2021**, *13*, 2245. [CrossRef]

59. Heusinger, J.; Weber, S. Comparative microclimate and dewfall measurements at an urban green roof versus bitumen roof. *Build. Environ.* **2015**, *92*, 713–723. [CrossRef]
60. Cuffe, C. Draft Report on Maximising the Energy Efficiency Potential of the EU Building Stock. 2020. Available online: https://www.euractiv.com/wp-content/uploads/sites/2/2020/04/Cuffe_report.pdf (accessed on 10 January 2021).
61. European Commission. A Renovation Wave for Europe -Greening Our Buildings, Creating Jobs, Improving Lives. 2020. Available online: https://ec.europa.eu/energy/sites/ener/files/eu_renovation_wave_strategy.pdf (accessed on 10 January 2021).
62. Alizade, B.; Hitchmough, J. A review of urban landscape adaptation to the challenge of climate change. *Int. J. Clim. Chang. Strateg. Manag.* **2019**, *11*, 178–194. [CrossRef]
63. Pradhan, S.; Helal, M.I.; Al-Ghamdi, S.G.; Mackey, H.R. Performance evaluation of various individual and mixed media for greywater treatment in vertical nature-based systems. *Chemosphere* **2020**, *245*, 125564. [CrossRef] [PubMed]
64. Artmann, M.; Sartison, K.; Vavra, J. The role of edible cities supporting sustainability transformation—A conceptual multi-dimensional framework tested on a case study in Germany. *J. Clean. Prod.* **2020**, *255*, 120220. [CrossRef]
65. Bousselot, J.; Russell, V.; Tolderlund, L.; Celik, S.; Retzlaff, B.; Morgan, S.; Buffam, I.; Coffman, R.; Williams, J.; Mitchell, M.E.; et al. Green Roof Research in North America: A Recent History and Future Strategies. *J. Living Arch.* **2020**, *7*, 27–64. [CrossRef]

Article

The Superiority of Data-Driven Techniques for Estimation of Daily Pan Evaporation

Manish Kumar [1], Anuradha Kumari [1,*], Deepak Kumar [1], Nadhir Al-Ansari [2,*], Rawshan Ali [3], Raushan Kumar [4], Ambrish Kumar [5], Ahmed Elbeltagi [6] and Alban Kuriqi [7,*]

1 Department of Soil and Water Conservation Engineering, College of Technology, G.B. Pant University of Agriculture & Technology, Pantnagar 263145, Uttarakhand, India; ct52623d@gbpuat.ac.in (M.K.); deepakkumar.swce@gbpuat-tech.ac.in (D.K.)
2 Department of Civil, Environmental and Natural Resources Engineering, Lulea University of Technology, 97187 Lulea, Sweden
3 Department of Petroleum, Koya Technical Institute, Erbil Polytechnic University, Erbil 44001, Kurdistan, Iraq; rawshan.ali@epu.edu.iq
4 Department of Farm Machinery and Power Engineering, College of Technology, G.B. Pant University of Agriculture & Technology, Pantnagar 263145, Uttarakhand, India; 52473_raushankumar@gbpuat-tech.ac.in
5 College of Agricultural Engineering, Dr. Rajendra Prasad Central Agriculture University, Pusa 848125, Bihar, India; dean.cae@rpcau.ac.in
6 Agricultural Engineering Department, Faculty of Agriculture, Mansoura University, Mansoura 35516, Egypt; ahmedelbeltagy81@mans.edu.eg
7 CERIS, Instituto Superior Técnico, Universidade de Lisboa, 1049-001 Lisboa, Portugal
* Correspondence: ct49481d@gbpuat.ac.in (A.K.); nadhir.alansari@ltu.se (N.A.-A.); alban.kuriqi@tecnico.ulisboa.pt (A.K.)

Citation: Kumar, M.; Kumari, A.; Kumar, D.; Al-Ansari, N.; Ali, R.; Kumar, R.; Kumar, A.; Elbeltagi, A.; Kuriqi, A. The Superiority of Data-Driven Techniques for Estimation of Daily Pan Evaporation. *Atmosphere* **2021**, *12*, 701. https://doi.org/10.3390/atmos12060701

Academic Editors: Anthony R. Lupo and Alexander V. Chernokulsky

Received: 13 April 2021
Accepted: 26 May 2021
Published: 30 May 2021

Abstract: In the present study, estimating pan evaporation (E_{pan}) was evaluated based on different input parameters: maximum and minimum temperatures, relative humidity, wind speed, and bright sunshine hours. The techniques used for estimating E_{pan} were the artificial neural network (ANN), wavelet-based ANN (WANN), radial function-based support vector machine (SVM-RF), linear function-based SVM (SVM-LF), and multi-linear regression (MLR) models. The proposed models were trained and tested in three different scenarios (Scenario 1, Scenario 2, and Scenario 3) utilizing different percentages of data points. Scenario 1 includes 60%: 40%, Scenario 2 includes 70%: 30%, and Scenario 3 includes 80%: 20% accounting for the training and testing dataset, respectively. The various statistical tools such as Pearson's correlation coefficient (PCC), root mean square error (RMSE), Nash–Sutcliffe efficiency (NSE), and Willmott Index (WI) were used to evaluate the performance of the models. The graphical representation, such as a line diagram, scatter plot, and the Taylor diagram, were also used to evaluate the proposed model's performance. The model results showed that the SVM-RF model's performance is superior to other proposed models in all three scenarios. The most accurate values of PCC, RMSE, NSE, and WI were found to be 0.607, 1.349, 0.183, and 0.749, respectively, for the SVM-RF model during Scenario 1 (60%: 40% training: testing) among all scenarios. This showed that with an increase in the sample set for training, the testing data would show a less accurate modeled result. Thus, the evolved models produce comparatively better outcomes and foster decision-making for water managers and planners.

Keywords: pan evaporation; ANN; WANN; SVM-RF; SVM-LF; Pusa station

1. Introduction

Estimating pan evaporation (PE) is essential for monitoring, surveying, and managing water resources. In many arid and semi-arid regions, water resources are scarce and seriously endangered by overexploitation. Therefore, the precise estimation of evaporation becomes imperative for the planning, managing, and scheduling irrigation practices. Evaporation happens if there is an occurrence of vapor pressure differential between two surfaces, i.e., water and air. The most general and essential meteorological parameters that influence the rate of evaporation are relative humidity, temperature, solar radiation, the deficit of vapor pressure, and wind speed. Thus, for the estimation of evaporation losses, these parameters should be considered for the precise planning and managing of different water supplies [1,2].

In the global hydrological cycle, the evaporation stage is defined as transforming water from a liquid to a vapor state [3]. In recent decades, evaporation losses have increased significantly, especially in semi-arid and arid regions [4,5]. Many factors, such as water budgeting, irrigation water management, hydrology, agronomy, and water supply management require a reliable evaporation rate estimation. The water budgeting factor has been modeled on estimates and the responses of cropping water to varying weather conditions. The daily evaporation of the pan (Epan) was considered a significant parameter. It was widely used as an index of lake and reservoir evaporation, evapotranspiration, and irrigation [6].

It is usually calculated in one of two ways, either (a) directly with pan evaporimeters or (b) indirectly with analytical and semi-empirical models dependent on climatic variables [7,8]. However, the calculation has proved sensitive to multiple sources of error, including strong wind circulation, pan visibility, and water depth measurement in the pan, for various reasons, including physical activity in and around the pan, water litter, and pan construction material and pan height. It can also be a repetitive, costly, and time-consuming process to estimate monthly pan evaporation (EP_m) using direct measurement. As a result, in the hydrological field, the introduction of robust and reliable intelligent models is necessary for precise estimation [9–14].

Several researchers have used meteorological variables to forecast E_{pan} values, as reported by [15–18]. Since evaporation is a non-linear, stochastic, and complex operation, a reliable formula to represent all the physical processes involved is difficult to obtain [19]. In recent years, most researchers have commonly acknowledged the use of artificial intelligence techniques, such as artificial neural networks (ANNs), adaptive neuro-fuzzy inference method (ANFIS), and genetic programming (G.P.) in hydrological parameter estimation [15,20–22]. In estimating E_{pan}, Sudheer et al. [23] used an ANN. They found that the ANN worked better than the other traditional approach. For modeling western Turkey's daily pan evaporation, Keskin et al. [24] used a fuzzy approach. To estimate regular E_{pan}, Keskin and Terzi [25] developed multi-layer perceptron (MLP) models. They found that the ANN model showed significantly better performance than the traditional system. Tan et al. [26] applied the ANN methodology to model hourly and daily open water evaporation rates. In regular E_{pan} modeling, Kisi and Çobaner [27] used three distinct ANN methods, namely, the MLP, radial base neural network (RBNN), and generalized regression neural network (GRNN). They found that the MLP and RBNN performed much better than GRNN. In a hot and dry climate, Piri et al. [28] have applied the ANN model to estimate daily E_{pan}. Evaporation estimation methods discussed by Moghaddamnia et al. [19] were implemented based on ANN and ANFIS. The ANN and ANFIS techniques' findings were considered superior to those of the analytical formulas. The fuzzy sets and ANFIS were used for regular modeling of E_{pan} by Keskin et al. [29] and found that the ANFIS method could be more efficiently used than fuzzy sets in modeling the evaporation process. Dogan et al. [30] used the approach of ANFIS for the calculation of evaporation of the pan from the Yuvacik Dam reservoir, Turkey. Tabari et al. [31] looked at the potential of ANN and multivariate non-linear regression techniques to model normal pan evaporation. Their findings concluded that the ANN performed better than non-linear regression. Using linear genetic programming techniques, Guven and Kişi [20] modeled regular pan evaporation by gene-expression programming (GEP), multi-layer perceptrons (MLP), radial basis neural networks (RBNN), generalized regression neural networks (GRNN), and Stephens–Stewart (SS) models. Two distinct evapotranspiration models have been used and found that the subtractive clustering (SC) model of ANFIS produces reasonable accuracy with less computational amounts than the ANFIS-GP ANN models [32].

A modern universal learning machine proposed by Vapnik (1995) [33] is the support vector machine (SVM), which is applied to both regression [30,34] and pattern recognition. An SVM uses a kernel mapping device to map the input space data to a high-dimensional feature space where the problem is linearly separable. An SVM's decision function relates to the number of support vectors (S.V.s) and their weights and the kernel chosen a priori, called the kernel [1,21]. Several kinds of kernels are Gaussian and polynomial kernels that may be used [10]. Moreover, artificial neural networks (ANN), wavelet-based artificial neural networks (WANN), support vector machine (SVM) were applied at different combinations of input variables by [23]. Their results showed that ANN, which contains three variables of air temperatures and solar radiation, produces root mean square error (RMSE) of 0.701, mean absolute error (MAE) of 0.525, correlation coefficient (R) of 0.990, and Nash–Sutcliffe efficiency (NSE) of 0.977 had better performances in comparison with WANN and SVR.

In principle, wavelet decomposition emerges as an efficient approximation instrument [18]; that is to say, a set of bases can approximate the random wavelet functions. To approximate E_{pan}, researchers used ANN, WANN, radial function-based support vector machine (SVM-RF), linear function-based support vector machine (SVM-LF), and multilinear regression (MLR) models of climatic variables.

There have been many studies on the estimation of E_{pan} based on weather variables using data-driven methods. However, the estimation of E_{pan} based on lag-time weather variables, which can be obtained easily, is not standard. After testing different acceptable combinations as input variables, the same inputs were used in artificial intelligence processes. In the proposed study, the main objective is to (1) model E_{pan} using ANN, WANN, SVM-RF, SVM-LF, and MLR models under different scenarios and (2) to select the best-developed model and scenario in E_{pan} estimation based on statistical metrics. The document's format is as follows. Section 2 contains the study's materials and methods: Section 3 gives the statistical indexes and methodological properties. The models' applicability to evaporation prediction and the results are discussed in Section 4. The conclusion is found in Section 5.

2. Materials and Methods

2.1. Study Area and Data Collection

Pusa is located in the Samastipur district of Bihar state, with latitude 25°46′ N and 86°10′ E. The location map of the study area is shown in Figure 1. Pusa lies 53 m above mean sea level in a hot sub-humid agro-ecological region in the middle of the Gangetic plain. The study area is located near the Burhi Ganadak river, a tributary of the Ganges river. The study area is famous for the Dr. Rajendra Prasad Central Agricultural University, a backbone of the study area's development. The average rainfall for Pusa is 1270 mm, of which 80% of the total rain falls during the monsoon season. The study area is fully covered by the area of the southwest monsoon, which starts in June and eases off in September. The maximum temperature varies from 32 to 38 °C during May and June. The minimum temperature varies from 6 to 9 °C during December and January. The main crops grown in the study area are wheat, maize, paddy, green gram, lentil, potato, and brinjal.

Meteorological data of the study area were gathered from the official "Dr. RPCAU" website (https://www.rpcau.ac.in, accessed on 13 April 2021), Pusa, Bihar. This included maximum and minimum temperatures (T_{max} and T_{min}, °C), relative humidity (RH-1, percent) at 7 a.m. and at 2 p.m. (RH-2, percent), wind speed (WS, km/h), bright sunshine hours (SSH, h) and daily pan evaporation (E_{Pan}, mm). For modeling pan evaporation, five years daily data set between the month 1 June to 30 September means that a total of 610 datasets have been used as input. The same is used for output [35].

Figure 2 displays the climate parameters determined in a box-and-whisker plot between June 2013 and September 2017 (i.e., five-year duration), indicating minimum, first quartile, median, third quartile, and maximum values.

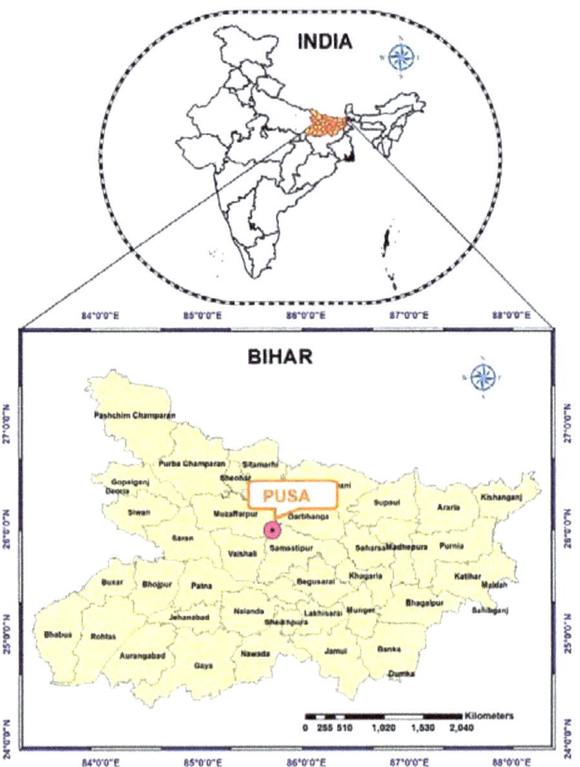

Figure 1. Location map of the study area.

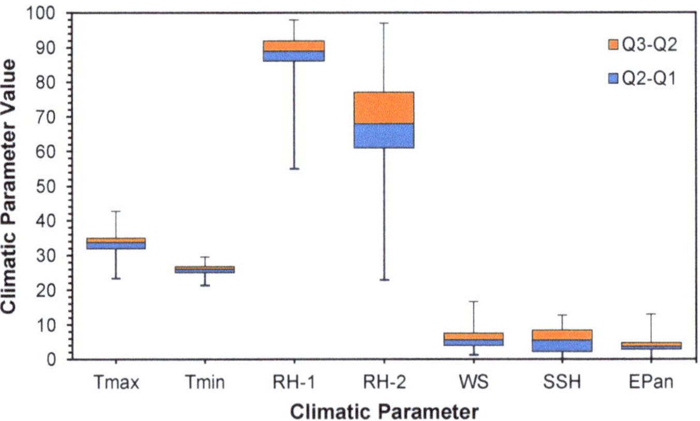

Figure 2. Box-and-whisker plot of climatic parameters in the study area.

The box-and-whisker plot shows that the relative humidity, measured at 7 a.m. and 2 p.m., respectively, demonstrates the highest variability among other meteorological parameters.

2.2. Statistical Analysis

Table 1 presents the statistical analysis of maximum and minimum temperatures (T_{max} and T_{min}, °C), relative humidity (RH-1, percent) at 7 a.m. and at 2 p.m. (RH-2, percent), wind speed (WS, km/h), bright sunshine hours (SSH, h) and daily pan evaporation (E_{Pan}, mm). The statistical analysis includes mean, median, minimum, maximum, standard deviation (Std. Dev.), kurtosis, and skewness values from 2013 to 2017. The given data is moderate to highly skewed; due to this problem, there has been a considerable negative effect on model performance. The standard deviation for the datasets shows that the values that are farther from zero mean that the variability in the data is higher. Hence, the variation of data from the mean value is higher. The statistical characteristics from the kurtosis values depict the platykurtic and leptokurtic nature of the climatic parameters, where kurtosis values are less than or greater than 3.

Table 1. Statistical constraints of climatic parameters from 2013 to 2017 in the study area.

Statistical Parameters	Mean	Median	Minimum	Maximum	Std. Dev.	Kurtosis	Skewness
T_{max} (°C)	33.58	33.80	23.40	42.70	2.43	1.30	−0.11
T_{min} (°C)	25.87	26.00	21.40	29.60	1.31	0.37	−0.51
RH-1 (%)	88.42	89.00	55.00	98.00	5.39	4.27	−1.33
RH-2 (%)	68.83	68.00	23.00	97.00	12.17	0.65	−0.22
WS (km/h)	6.03	5.70	1.20	16.70	2.63	0.82	0.85
SSH (h)	5.36	5.55	0.00	12.70	3.50	−1.20	−0.02
E_{Pan} (mm)	3.85	3.70	0.00	13.00	1.67	2.34	0.89

Table 2 depicts the inter-correlation between climatic variables at the given station. Thus, it can be observed that all climate parameters have a significant association with the E_{Pan} at a significance level of 5%.

Table 2. Intercorrelation values between climatic parameters in the study area.

Climatic Variable	T_{max}	T_{min}	RH-1	RH-2	WS	SSH	E_{Pan}
T_{max}	1.00						
T_{min}	0.32	1.00					
RH-1	−0.43	−0.29	1.00				
RH-2	−0.51	−0.15	0.48	1.00			
WS	−0.07	0.02	−0.19	0.00	1.00		
SSH	0.68	0.28	−0.42	−0.51	0.05	1.00	
E_{Pan}	0.58	0.11	−0.30	−0.34	0.19	0.51	1.00

2.3. *Data-Driven Techniques Used*

2.3.1. Artificial Neural Network

The ANN methodology is a tool used to replicate the problem-solving mechanism of the human brain. ANNs are incredibly robust at modeling and simulating linear and non-linear systems. The ANN's feed-forward back-propagation techniques were highly emphasized among ANNs because their lower level of difficulty in the present study were also used [36,37]. ANN consists of the input layer, output layer, and hidden layers between the input and output layers. Each node within a layer is connected to all the following layer nodes. Only those nodes within one layer are connected to the following layer nodes [29]. Each neuron receives processes and sends the signal to make functional relationships between future and past events. These layers are attached with the interconnected weight W_{ij} and W_{jk} between the layers of neurons. The typical structure using input variables is shown in Figure 3.

For this analysis, only one hidden layer network was used since it was considered dynamic enough to forecast meteorological variables. There are some transfer functions required to create an artificial neural network neuron. Transfer functions are needed to establish the input–output relationship for each neuron layer. In this analysis, Levenberg–Marquardt was used to train the model. A hyperbolic tangent sigmoid transfer function was used to measure a layer's output from its net input. The neural network learns by changing the connection weights between the neurons. By using a suitable learning algorithm, the connection weights are altered using the training data set. The number of hidden layers is typically determined by trial and error. A comprehensive ANN overview is available [25,38,39].

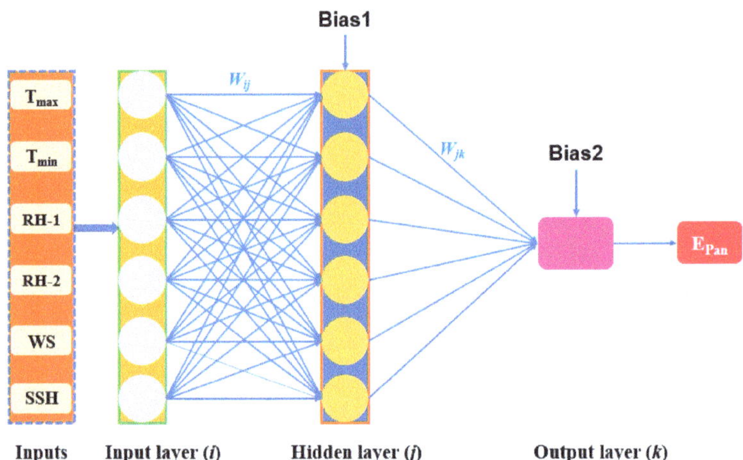

Figure 3. Three-layered structure of the artificial neural network.

2.3.2. Wavelet Artificial Neural Network (WANN)

The wavelet analysis (WA) offers a spectral analysis dependent on the time that explains processes and their relationships in time-frequency space by breaking down time series [40]. WA is an effective method of time-frequency processing, with more benefits than Fourier analysis [41]. WA is an improvement over the Fourier transformation variant used to detect time functionality in data [40]. Wavelet transformation analysis, breaking down time series into essential functions at different frequencies, improves the potential of a predictive model by gathering sufficient information from different resolution levels [25]. There is excellent literature on wavelet transforming theory [42,43]; we will not go into it in depth here. It is vital to choose the base function carefully (called the mother wavelet). The essential functions are generated by translation and dilation [44]. In general, the discrete wavelength transformation (DWT) has been used preferentially in data decomposition, as compared to continuous wavelet transformation (CWT), because CWT is time-consuming [3,18].

The present used the DWT method for daily E_{Pan} (mm) estimation. DWT decomposes the original input time series data of T_{max}, T_{min}, RH-1, RH-2, WS, and SSH into different frequencies (Figure 4), adapted from Rajaee [44].

This analysis used three stages of the Haar à trous decomposition algorithm using Equations (1) and (2):

$$C_r(t) = \sum_{l=0}^{+\infty} h(l) C_{r-1}(t+2^r) \qquad (r = 1, 2, 3, \ldots, n) \tag{1}$$

$$W_r(t) = C_{r-1}(t) - C_r(t) \qquad (r = 1, 2, 3, \ldots, n) \tag{2}$$

where $h(l)$ is the discrete low-pass filter, $C_r(t)$ and $W_r(t)$ (r = 1, 2, 3, ..., n) are scale coefficient and wavelet coefficient at the resolution level. Two sets of filters, including low and high passes, are employed by DWT to decompose the main time series. It is discontinuous and resembles a step feature that is ideal for certain time series of abrupt transitions. The abovementioned wave types were evaluated, and finally, the measured monthly time series, H, were decomposed into multi-frequency time series including details (HD1; HD2; . . . ; HDn) and approximation (Ha) by optimum DWT (Qasem et al., 2019).

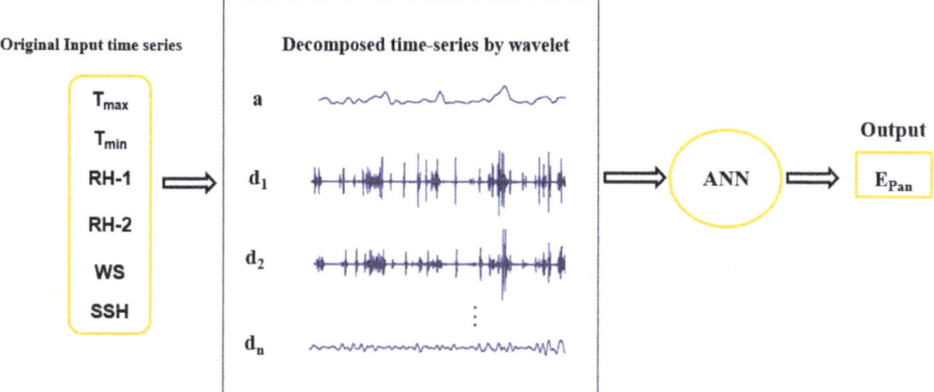

Figure 4. Schematic representation of WANN.

The obtained decomposed frequency values function as an ANN input. Hybridizing the decomposed input time series data of T_{max}, T_{min}, RH-1, RH-2, WS, and SSH with ANN results in a wavelet artificial neural network (WANN) [42]. Three levels of the Haar à trous decomposition algorithm were used in this study. For the model's training, the Levenberg–Marquardt algorithm was used. The hyperbolic tangent sigmoid transfer function was also used to measure a layer's output from its net input.

2.3.3. Support Vector Machine

The support vector machine (SVM) was developed by [33] for classification and regression procedures. The fundamental concept of an SVM is to add a kernel function, map the input data by non-linear mapping into a high-dimensional function space, and then perform a linear regression in the feature space [45]. SVM is a modern classifier focused on two principles (Figure 5) adapted from Lin et al. [46]. First, data transformation into a high-dimensional space can render complicated problems easier, utilizing linear discriminate functions. Secondly, SVM is inspired by the training principle and uses only specific inputs nearest to the decision region since they have the most detail regarding classification [47].

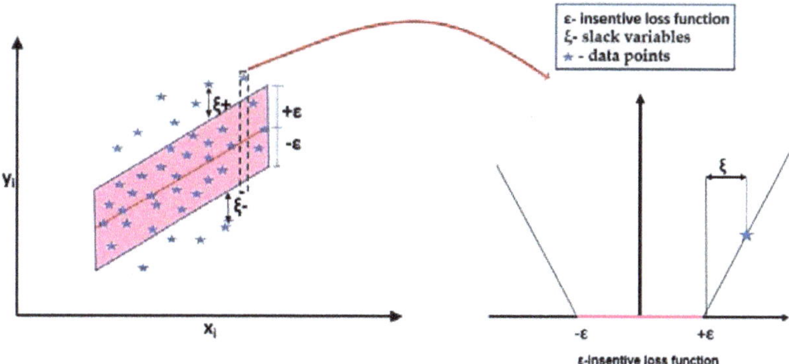

Figure 5. SVM Layout.

We assume a non-linear function $f(x)$ is given by:

$$f(x) = \mathbf{w}^{\mathrm{T}}\boldsymbol{\Phi}(\mathbf{x}_i) + b \tag{3}$$

where **w** is the weight vector, b is the bias, and $\mathbf{\Phi(x_i)}$ is the high dimensional feature space, linearly mapped from the input space x. Equation (3) can be transformed into higher dimensions and gives final expression as:

$$f(x) = \sum_{i=1}^{m} \left(\alpha_i^+ - \alpha_i^-\right) K\left(x_i,\ x_j\right) + b; \tag{4}$$

where, α_i^+,α_i^- are Lagrangian multipliers which are used to eliminate some primal variables, and the term $K\left(x_i,\ x_j\right)$ is the kernel function. The derivation and excellent literature about SVM can be obtained from [48]. The study's kernel function was a linear function (LF) and radial function (RF).

- Linear kernel function (LF): the most basic form of kernel function is written as:

$$K\left(x_i,\ x_j\right) = \left(x_i,\ x_j\right) \tag{5}$$

- Radial basis function (RBF): a mapping of RBF is identically represented as Gaussian bell shapes:

$$K\left(x_i,\ x_j\right) = \exp\left(-\gamma||x_i - x_j||^2\right) \tag{6}$$

  where γ is the Gaussian RBF kernel parameter width; the RBF is widely used among all the kernel functions in the SVM technique.

The efficiency of the SVR technique depends on the environment for an ε-insensitive loss function of three training parameters (kernel, C, γ, and ε). However, the values of C and ε influence the complexity of the final model for every specific type of kernel. The ε value measures the number of support vectors (SV) used for predictions. The best value of ε intuitively results in fewer supporting vectors, leading to less complicated regression estimates. However, C's value is the trade-off between model complexity and the degree of deviations permitted within the optimization formulation. Therefore, a more considerable value of C undermines model complexity [49]. The selection of optimum values for these training parameters (C and ε) guaranteeing fewer complex models is an active research area.

2.3.4. Multiple Linear Regression (MLR)

A linear regression analysis in which more than one independent variable is involved is called MLR. The advantage of MLR is that it is simple, showing how dependent variables interact with independent variables. The overall model of the MLR is:

$$y = c_0 + c_1x_1 + c_2x_2 + \ldots + c_nx_n \tag{7}$$

where y is the dependent variable, and $x_1, x_2, \ldots, x_n$ are independent variables, $c_1, c_2, \ldots,$ c_n are regression coefficients, and c_0 is intercepted. These values are the local behavior calculated using the least square rule or other regression [27].

2.4. Modeling Methodology

In the present study, the daily pan evaporation (E_{Pan}) was estimated based on different input climatic variables (T_{max}, T_{min}, RH-1, RH-2, W.S., and S.S.H.). The five different techniques used for estimation were the artificial neural network (ANN), wavelet-based artificial neural network (WANN), radial function-based support vector machine (SVM-RF), linear function-based support vector machine (SVM-LF), and multi-linear regression (MLR) models. The climatic parameters were collected from 2013 to 2017 and split into three different scenarios, based on the percentage of training and testing datasets for model development (Table 3).

Table 3. Different scenarios of training and testing datasets used in this study.

Scenarios	Training Data Length (%)	Testing Data Length (%)
Scenario 1	60% (2013–2015)	40% (2016–2017)
Scenario 2	70%	30%
Scenario 3	80% (2013–2016)	20% (2017)

Scenario 1 contains 60% (2013–2015) data for training and 40% (2016–2017) data for testing. Scenario 2 contains 70% data for training and 30% data for testing from 2016. Scenario 3 contains 80% (2013–2016) data for training and 20% (2017) data for testing. The training datasets were used for calibration purposes, while the testing dataset was used for validation purposes.

The results of the applied models in three different scenarios were evaluated through different performance evaluators described in Section 2.5.

2.5. Performance Evaluation Criteria

There were four criteria used to measure the performance of the scenarios mentioned above, quantitatively evaluated using root mean square error (RMSE), Nash–Sutcliffe Efficiency (NSE), Pearson's correlation coefficient (PCC), and Willmott index (W.I.), and qualitatively evaluated through graphical interpretation (time-series plot, scatter plot, and Taylor diagram). The RMSE range is zero to infinity (0 < RMSE < ∞); the lower the RMSE, the better the model's performance. The NSE ranges from minus infinity to one (−∞ < NSE < 1). NSE below zero (NSE < 0) indicates that the observed mean only as strong as the average, whereas negative values suggest that the observed mean a more robust indicator than the average [48]. The PCC is also known as the correlation coefficient and is used to calculate the degree of collinearity between observed and estimated values. The PCC varies from minus one to plus one (−1 < PCC < 1) [39]. The WI is also known as the index of agreement. The WI ranges from zero to one (0 < WI < 1); approximately 1 is ideal agreement/fit [3]. The most accurate models were selected based on the highest values of PCC, NSE, and WI, while showing the lowest values of RMSE among all developed models.

$$RMSE = \sqrt{\frac{\sum_{i=1}^{N}\left(E_{p_{obs,i}} - E_{p_{pre,i}}\right)^2}{N}}; \tag{8}$$

$$NSE = 1 - \left[\frac{\sum_{i=1}^{N}(\overline{E}_{p_{obs,i}} - E_{p_{pre,i}})^2}{\sum_{i=1}^{N}(E_{p_{obs,i}} - \overline{E}_{p_{obs,i}})^2}\right]; \tag{9}$$

$$\mathrm{PCC} = \frac{\sum_{i=1}^{N}(E_{p_{obs,i}} - \overline{E}_{p_{obs,i}})(E_{p_{pre,i}} - \overline{E}_{p_{pre,i}})}{\sqrt{\sum_{i=1}^{N}\left(E_{p_{obs,i}} - \overline{E}_{p_{obs,i}}\right)^2 \sum_{i=1}^{N}\left(E_{p_{pre,i}} - \overline{E}_{p_{pre,i}}\right)^2}}; \tag{10}$$

$$\mathrm{WI} = 1 - \frac{\sum_{i=1}^{N}\left(E_{p_{obs,i}} - E_{p_{pre,i}}\right)^2}{\sum_{i=1}^{N}\left(\left|E_{p_{pre,i}} - \overline{E}_{p_{obs,i}}\right| + \left|E_{p_{obs,i}} - \overline{E}_{p_{obs,i}}\right|\right)^2}. \tag{11}$$

where $E_{p_{obs,i}}$, $E_{p_{pre,i}}$ observed and predicted pan evaporation values on the ith day. $\overline{E}_{p_{obs,i}}$, $\overline{E}_{p_{pre,i}}$ are average of observed and predicted values, respectively.

3. Results

3.1. Quantitative and Qualitative Evaluation of Results

This section deals with quantitative and qualitative results obtained for the developed models. ANN and WANN trials were conducted depending on the different number of neurons in hidden layers. In contrast, SVM-LF and SVM-RF trials were performed by taking several values of SVM-g, SVM-c, and SVM-e parameters. These were represented in Tables 4–6 as a structure for the model.

3.2. Comparison of Training and Testing Datasets for Scenario 1

The training results obtained by ANN, Wavelet, and SVM have been shown in Table 4. As depicted in Table 4, for three developed ANN models, namely ANN-1, ANN-2, and ANN-3, ANN-1 has the highest PCC value of 0.832, the lowest RMSE value of 0.993, the highest NSE value of 0.685, and the highest WI value of 0.904.

Similarly, for the developed WANN model, WANN-1 has shown better performance, with a PCC value of 0.773. Furthermore, the WANN model also has the lowest RMSE value of 1.123, the highest NSE value of 0.597, and the highest WI value of 0.860. Furthermore, among developed SVM-RF and SVM-LF models, SVM-RF-3 has shown better performance than other developed models. The SVM-RF-3 model has the highest PCC value of 0.857; it has the lowest RMSE value of 0.956, the highest NSE value of 0.708, and the highest WI value of 0.895 during training datasets. The value of PCC, RMSE, NSE, and WI for MLR techniques was 0.695, 1.274, 0.483, and 0.800. Thus, it can be stated that SVM-RF has modeled the E_{pan} most efficiently of all the machine learning algorithms developed for training.

Table 4. Results for ANN, WANN, SVM-RF, SVM-LF, and M.L.R. during the training and testing period for Scenario 1 (60–40: Training–Testing).

Model	Structure	Dataset	PCC	RMSE	NSE	WI
ANN-1	6-5-1	Training	0.832	0.993	0.685	0.904
		Testing	0.589	1.387	0.136	0.708
ANN-2	6-8-1	Training	0.739	1.254	0.498	0.840
		Testing	0.585	1.486	0.010	0.732
ANN-3	6-12-1	Training	0.769	1.157	0.573	0.846
		Testing	0.531	1.529	−0.048	0.705
WANN-1	24-6-1	Training	0.773	1.123	0.597	0.860
		Testing	0.505	1.394	0.129	0.676
WANN-2	24-11-1	Training	0.694	1.286	0.472	0.813
		Testing	0.428	1.491	0.003	0.614
WANN-3	24-16-1	Training	0.634	1.502	0.281	0.766
		Testing	0.477	1.643	−0.211	0.681
SVM-RF-1	$c = 1, \varepsilon = 0.001, \gamma = 0.16$	Training	0.777	1.122	0.599	0.856
		Testing	0.595	1.369	0.159	0.746
SVM-RF-2	$c = 1, \varepsilon = 0.01, \gamma = 0.16$	Training	0.794	1.088	0.622	0.864
		Testing	0.604	1.344	0.190	0.749
SVM-RF-3	$c = 1, \varepsilon = 0.1, \gamma = 0.16$	Training	0.857	0.956	0.708	0.895
		Testing	0.607	1.349	0.183	0.749
SVM-LF-1	$c = 1, \varepsilon = 0.1, \gamma = 0.5$	Training	0.687	1.297	0.463	0.804
		Testing	0.592	1.406	0.113	0.731
SVM-LF-2	$c = 1, \varepsilon = 0.1, \gamma = 0.8$	Training	0.687	1.297	0.463	0.804
		Testing	0.592	1.406	0.113	0.731
SVM-LF-3	$c = 1, \varepsilon = 0.1, \gamma = 0.16$	Training	0.687	1.297	0.463	0.807
		Testing	0.592	1.406	0.113	0.731
MLR		Training	0.695	1.274	0.483	0.800
		Testing	0.587	1.345	0.188	0.725

Among developed ANN models, ANN-1 has the highest PCC value of 0.589; it has the lowest RMSE value of 1.387 and the highest NSE value of 0.136. Similarly, for the WANN model, WANN-1 has shown better performance with a PCC value of 0.505, the lowest RMSE value of 1.394, the highest NSE value of 0.129, and a WI value of 0.676.

Furthermore, among developed SVM-RF and SVM-LF models, SVM-RF-3 has shown better performance than other developed models. The SVM-RF-3 model has the highest PCC value of 0.607, RMSE value of 1.349, NSE value of 0.183, and the highest WI value of 0.749 training datasets. The values of PCC, RMSE, NSE, and WI for MLR techniques were 0.587, 1.345, 0.188, and 0.725, respectively. The scatter plot and line diagram for the testing data set has been shown in Figure 6. From the line diagram, it can be observed that the obtained results were under-predicted for all models. The scatter plot shows that the highest value of the determination (R^2) coefficients was obtained for the SVM-RF model. Thus, it can be suggested that SVM-RF has modeled the E_{pan} most efficiently among all the machine learning algorithms developed for testing.

3.3. Comparison of Training and Testing Datasets for Scenario 2

In Scenario 2, 70% of the entire data set has been used for training, and the rest of the data has been used for testing the developed model. The training results obtained by ANN, Wavelet, and SVM have been shown in Table 5.

As shown in Table 5, among three developed ANN models, the ANN-1 has the highest PCC value of 0.760, the lowest RMSE value of 1.180, the highest NSE value of 0.577, and the highest WI value of 0.854. Similarly, for the WANN model, WANN-2 has shown better performance with a PCC value of 0.725, a lowest RMSE value of 1.264, a highest NSE value of 0.515, and a highest WI value of 0.831. Furthermore, among developed SVM-RF and SVM-LF models, SVM-RF-3 has shown better performance than other developed models. The SVM-RF-3 model has the highest PCC value of 0.812, the lowest RMSE value of 1.262, the highest NSE value of 0.650, and the highest WI value of 0.714 during training datasets. The values of PCC, RMSE, NSE, and WI for MLR techniques were 0.693, 1.308, 0.481, and 0.799, respectively, during training processes. Thus, it can be stated that SVM-RF has modeled the E_{pan} most efficiently among all the machine-learning algorithms developed for training.

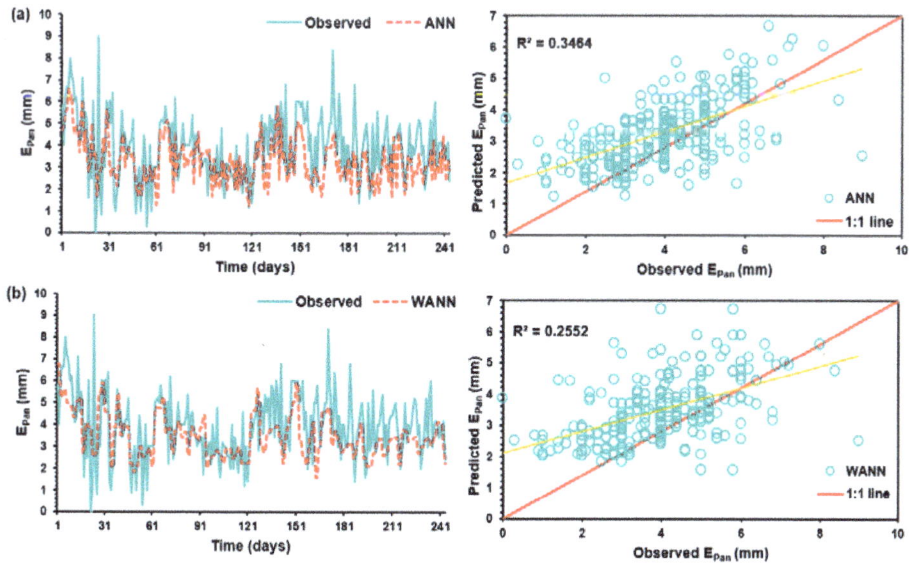

Figure 6. *Cont.*

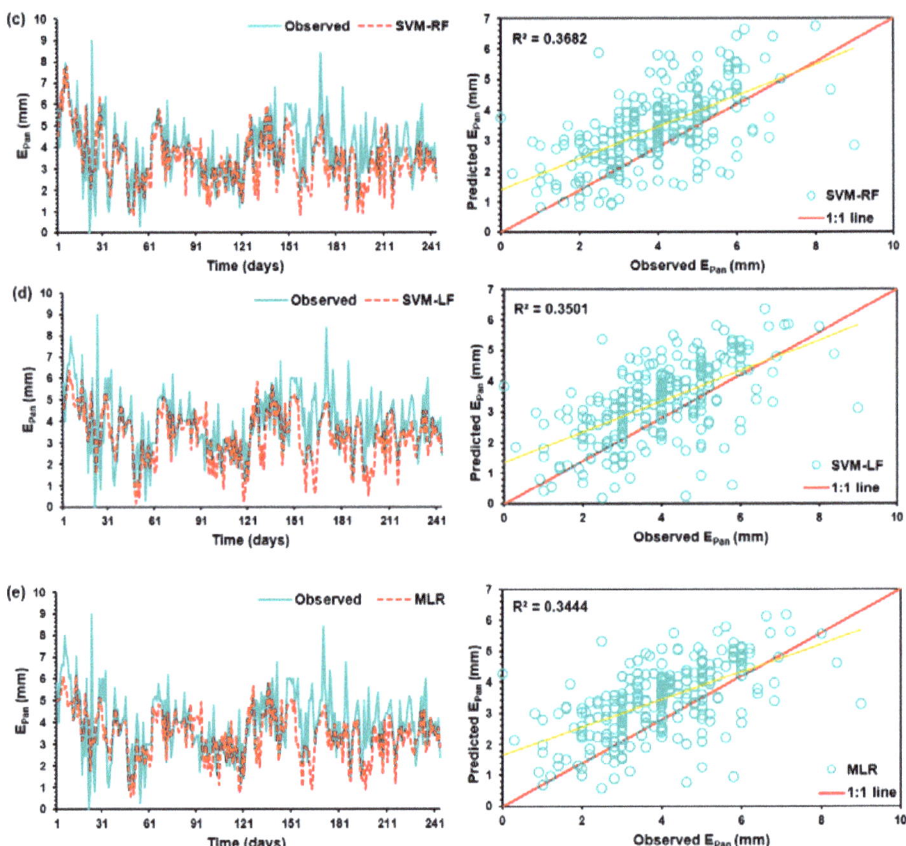

Figure 6. Line and scatter plot between observed and predicted data at Scenario 1 for (**a**) ANN, (**b**) WANN (**c**) SVM-RF, (**d**) SVM-LF, and (**e**) MLR for the study area.

For Scenario 2, where 30% of the data set has been used for testing, model ANN-1 has the highest PCC value of 0.547, the lowest RMSE value of 1.222, the highest NSE value of 0.046, and a WI value of 0.704 among ANN models. Similarly, WANN-1 has shown better performance, with a PCC value of 0.457, the lowest RMSE value of 1.252, the highest NSE value of −0.002, and the highest WI value of 0.639 WANN models. Furthermore, SVM-RF-3 has shown better performance as compared to other developed models among SVM-RF and SVM-LF models. The SVM-RF-3 model has the highest PCC value of 0.568, the lowest RMSE value of 1.262, and the highest WI value of 0.714 during training datasets. The values of PCC, RMSE, NSE, and WI for MLR techniques were 0.531, 1.262, −0.017, and 0.700, respectively. The scatter plot and line diagram for testing have been shown in Figure 7. It can be seen from the line diagram that the obtained results were under-predicted for all models. The scatter plot showed that the highest value of the coefficient of determination (R^2) was obtained for SVM-RF models of 0.3221. Thus, it can be shown that SVM-RF has modeled the E_{pan} most efficiently among all the machine learning algorithms developed for testing.

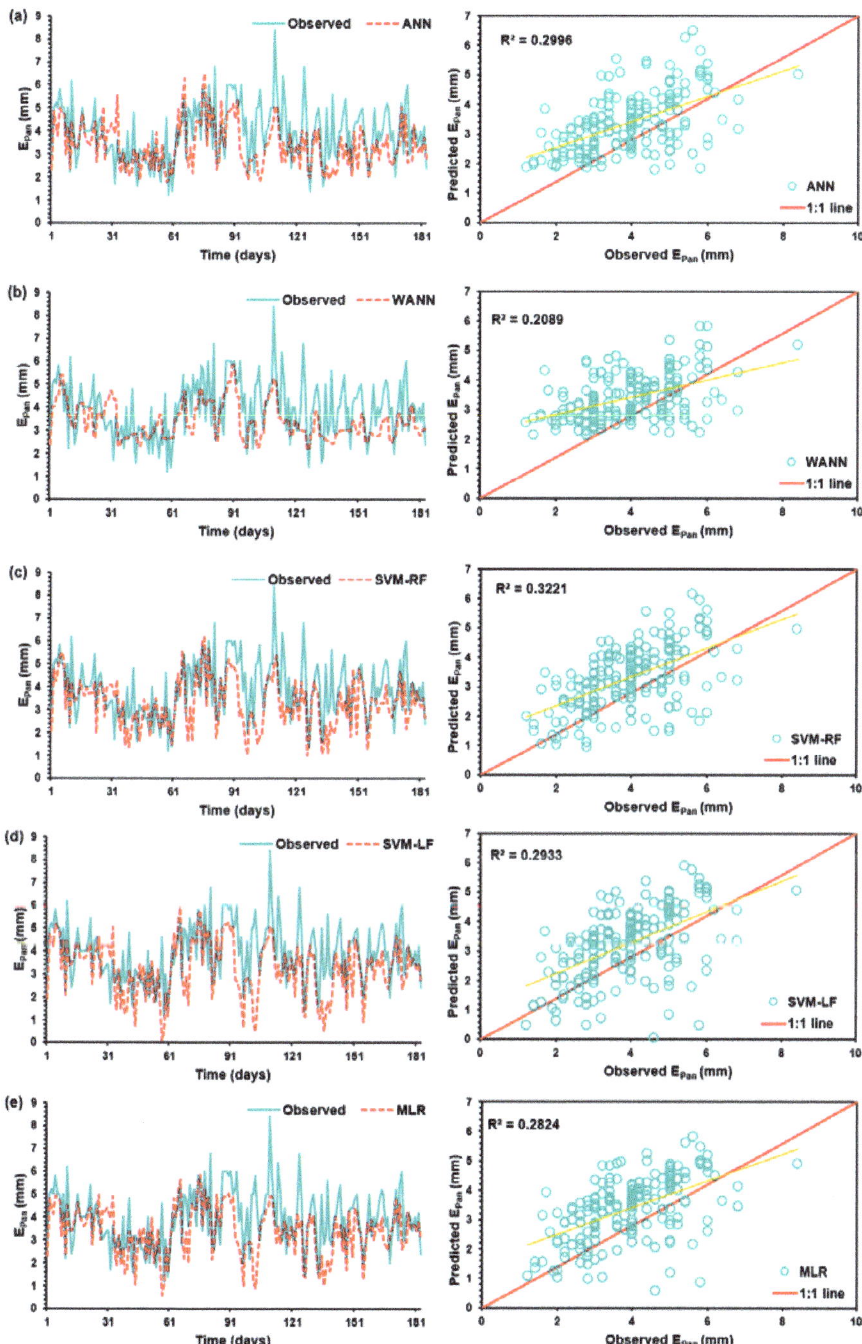

Figure 7. Line and scatter plots between observed and predicted data at Scenario 2 for (**a**) ANN, (**b**) WANN (**c**) SVM-RF, (**d**) SVM-LF, and (**e**) MLR, for the study area.

Table 5. Results for ANN, WANN, SVM-RF, SVM-LF, and MLR during training and testing period for Scenario 2 (70–30: Training–Testing).

Model	Structure	Dataset	PCC	RMSE	NSE	WI
ANN-1	6-1-1	Training	0.760	1.180	0.577	0.854
		Testing	0.547	1.222	0.046	0.704
ANN-2	6-4-1	Training	0.749	1.209	0.557	0.842
		Testing	0.535	1.333	−0.135	0.691
ANN-3	6-10-1	Training	0.716	1.278	0.504	0.824
		Testing	0.546	1.235	0.026	0.727
WANN-1	24-1-1	Training	0.672	1.344	0.452	0.781
		Testing	0.439	1.316	−0.106	0.602
WANN-2	24-6-1	Training	0.725	1.264	0.515	0.831
		Testing	0.457	1.252	−0.002	0.639
WANN-3	24-9-1	Training	0.716	1.281	0.502	0.802
		Testing	0.413	1.275	−0.039	0.604
SVM-RF-1	$c = 1, \varepsilon = 0.001, \gamma = 0.16$	Training	0.764	1.178	0.579	0.847
		Testing	0.560	1.285	−0.055	0.704
SVM-RF-2	$c = 1, \varepsilon = 0.01, \gamma = 0.16$	Training	0.765	1.177	0.579	0.848
		Testing	0.561	1.286	−0.056	0.705
SVM-RF-3	$c = 1, \varepsilon = 0.1, \gamma = 0.16$	Training	0.812	1.073	0.650	0.875
		Testing	0.568	1.262	−0.018	0.714
SVM-LF-1	$c = 1, \varepsilon = 0.1, \gamma = 0.9$	Training	0.689	1.326	0.466	0.805
		Testing	0.539	1.356	−0.175	0.696
SVM-LF-2	$c = 1, \varepsilon = 0.01, \gamma = 0.16$	Training	0.688	1.330	0.463	0.807
		Testing	0.542	1.360	−0.182	0.700
SVM-LF-3	$c = 1, \varepsilon = 0.1, \gamma = 0.16$	Training	0.689	1.326	0.466	0.805
		Testing	0.539	1.356	−0.175	0.696
MLR		Training	0.693	1.308	0.481	0.799
		Testing	0.531	1.262	−0.017	0.700

3.4. Comparison of Training and Testing Datasets for Scenario 3

In Scenario 3, 80% of the total dataset was used for training periods, while the rest, 20%, was used to test the models. The training results obtained by ANN, wavelet analysis, and SVM have been shown in Table 6.

As depicted from Table 6, for developed ANN models, model ANN-3 has the highest PCC value of 0.520; it has an RMSE value of 1.333 and a W.I. value of 0.688. Similarly, for the WANN model, WANN-1 has shown better performance with a PCC value of 0.725, the lowest RMSE value of 1.213, the highest NSE value of 0.519, and the highest WI value of 0.812. Further, SVM-RF-3 has shown better performance compared to other developed models. The SVM-RF-3 model has the highest PCC value of 0.893, the lowest RMSE value of 0.858, the highest NSE value of 0.760, and the highest WI value of 0.913 during training datasets. The values of PCC, RMSE, NSE, and WI for MLR techniques were 0.688, 1.269, 0.474, and 0.795, respectively. Thus, it can be depicted that SVM-RF has modeled the E_{pan} most efficiently among all the machine learning algorithms developed for training.

For testing datasets, for developed ANN models, ANN-3 has the highest PCC value of 0.520, an RMSE value of 1.333, and the highest W.I. value of 0.688. Similarly, for the WANN model, WANN-1 has shown better performance with a PCC value of 0.467, an RMSE value of 1.447, and WI value of 0.639. Furthermore, among developed SVM-RF and SVM-LF models, SVM-RF-1 has shown better performance than other developed models. The SVM-RF-1 model has the highest PCC value of 0.528, the lowest RMSE value of 1.411, and the highest WI value of 0.665 during the testing of datasets.

Table 6. Results for ANN, WANN, SVM-RF, SVM-LF, and M.L.R. during the training and testing period for Scenario 3 (80–20: Training–Testing).

Model	Structure	Dataset	PCC	RMSE	NSE	WI
ANN-1	6-1-1	Training	0.701	1.250	0.490	0.809
		Testing	0.512	1.321	−0.152	0.681
ANN-2	6-9-1	Training	0.764	1.136	0.578	0.847
		Testing	0.514	1.260	−0.049	0.695
ANN-3	6-13-1	Training	0.789	1.079	0.620	0.879
		Testing	0.520	1.333	−0.172	0.688
WANN-1	24-2-1	Training	0.725	1.213	0.519	0.812
		Testing	0.467	1.447	−0.382	0.608
WANN-2	24-7-1	Training	0.693	1.267	0.476	0.813
		Testing	0.369	1.434	−0.357	0.586
WANN-3	24-11-1	Training	0.721	1.221	0.513	0.812
		Testing	0.439	1.334	−0.175	0.603
SVM-RF-1	$c = 1$, $\varepsilon = 0.001$, $\gamma = 0.16$	Training	0.768	1.128	0.584	0.849
		Testing	0.527	1.415	−0.322	0.660
SVM-RF-2	$c = 1$, $\varepsilon = 0.1$, $\gamma = 0.2$	Training	0.850	0.951	0.705	0.894
		Testing	0.526	1.413	−0.318	0.664
SVM-RF-3	$c = 1$, $\varepsilon = 0.1$, $\gamma = 0.16$	Training	0.893	0.858	0.760	0.913
		Testing	0.528	1.411	−0.315	0.665
SVM-LF-1	$c = 1$, $\varepsilon = 0.1$, $\gamma = 0.3$	Training	0.684	1.286	0.460	0.802
		Testing	0.496	1.453	−0.394	0.658
SVM-LF-2	$c = 1$, $\varepsilon = 0.1$, $\gamma = 0.6$	Training	0.684	1.286	0.460	0.802
		Testing	0.496	1.453	−0.394	0.658
SVM-LF-3	$c = 1$, $\varepsilon = 0.001$, $\gamma = 0.16$	Training	0.683	1.286	0.460	0.803
		Testing	0.490	1.465	−0.417	0.654
MLR		Training	0.688	1.269	0.474	0.795
		Testing	0.506	1.363	−0.227	0.665

The values of PCC, RMSE, NSE, and WI for MLR techniques were 0.506, 1.363, −0.227, and 0.665. The scatter plot and line diagram for testing have been shown in Figure 8. From the line diagram, it has been observed that obtained results were under-predicted and over-predicted for all models. The scatter plot showed that the highest value of the coefficient of determination (R^2) was obtained for SVM-RF models of 0.2791. Thus, it can be seen that SVM-RF has modeled the daily E_{pan} most efficiently among all the machine learning algorithms developed for testing.

The comparative results of training and testing data results have been shown in Table 7. This table could suggest that training and testing data using the SVM-RF model, E_{pan}, can be modeled more accurately than ANN and WANN.

The performance of models from best to lowest is SVM > ANN > MLR > WANN for all three scenarios. Table 7 also showed that the WANN model performed poorly compared to other models. This is because wavelet transformation does not reveal the hidden information present in the primary time-series data through different sub-series. It is also observed that, with an increase in the sample set for training, the testing data will show a less accurate modeled result.

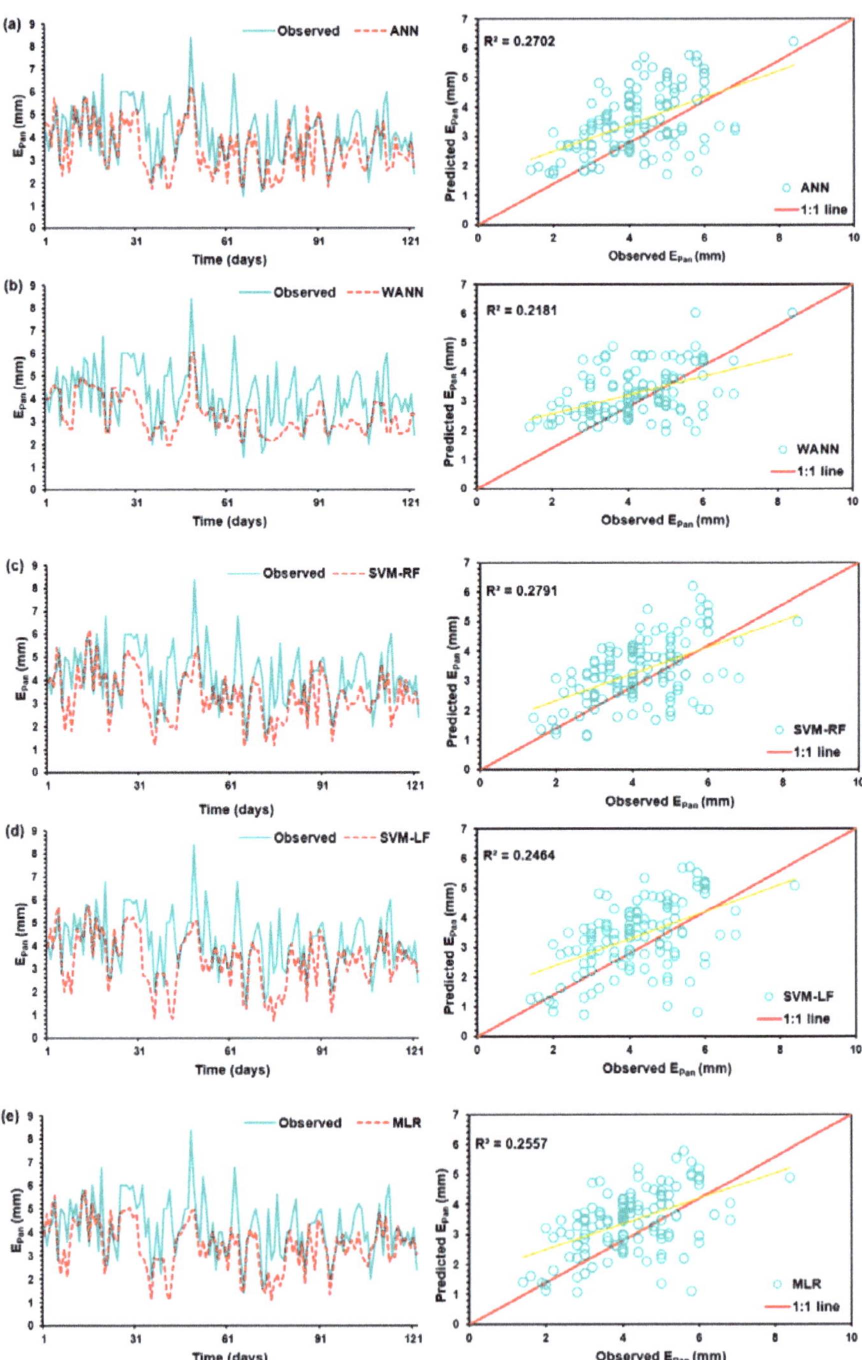

Figure 8. Line and scatter plot between observed and predicted data at scenario 3 for (**a**) ANN, (**b**) WANN (**c**) SVM-RF, (**d**) SVM-LF, and (**e**) MLR, for the study area.

The comparative result of all three scenarios of all developed models has also been shown through Taylor's diagram [50] in Figure 9a–c, which acquires information based on correlation coefficient, standard deviation, and root mean square difference [27]. Figure 9a–c indicates that the SVM-RF model predictions in all three scenarios are very close to the daily values of E_{pan}, which are tending more toward observed point values at abscissa. The performance-based correlation coefficient, standard deviation, and root mean square difference are also superior compared to others. Therefore, the SVM-RF model with T_{max}, T_{min}, RH-1, RH-2, WS, and SSH climate variables can be used for daily E_{pan} estimation at the Pusa station.

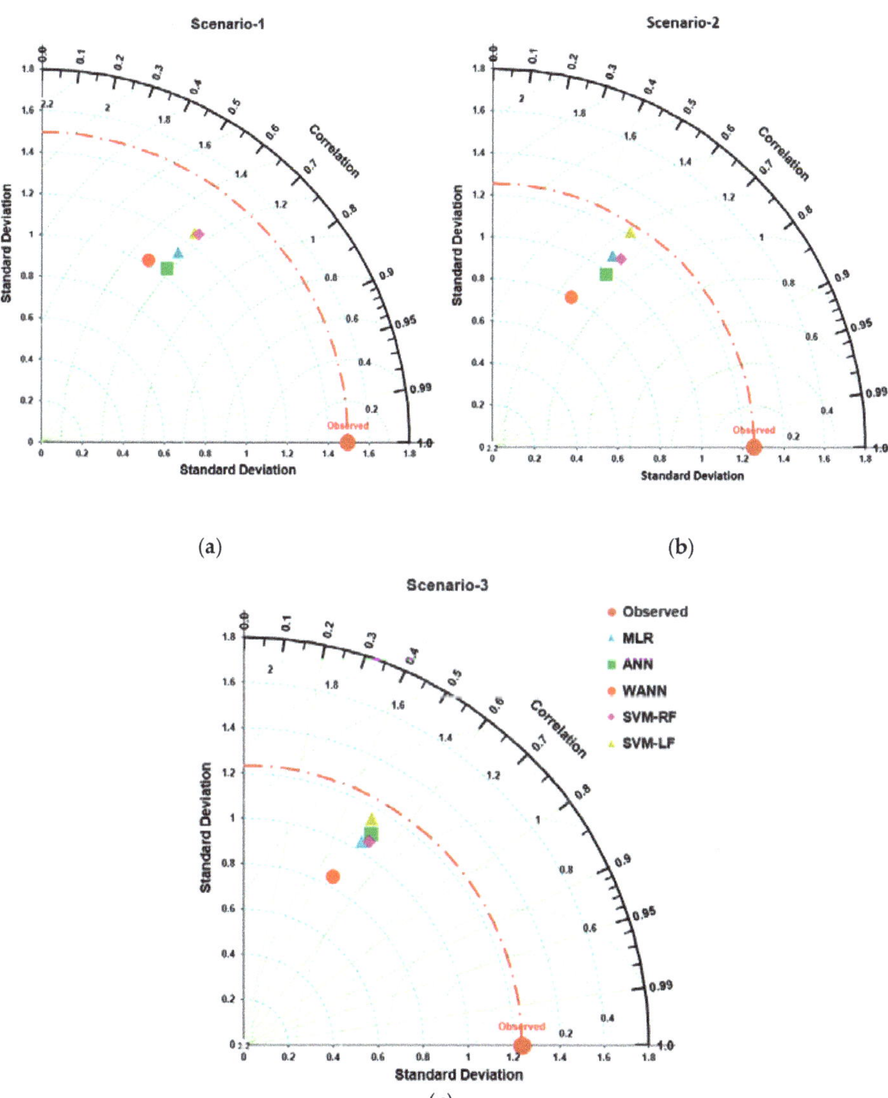

Figure 9. Taylor diagrams of ANN, WANN, SVM-RF, SVM-LF, and MLR corresponding to (**a**) Scenario 1, (**b**) Scenario 2, (**c**) Scenario 3 during the testing period at the study site.

Table 7. Results for best ANN, WANN, SVM-RF, and MLR during the training and testing period for all scenarios.

Scenario	Model	Dataset	PCC	RMSE	NSE	WI
1	ANN-1	Training	0.832	0.993	0.685	0.904
		Testing	0.589	1.387	0.136	0.708
	WANN-1	Training	0.773	1.123	0.597	0.860
		Testing	0.505	1.394	0.129	0.676
	SVM-RF-3	Training	0.857	0.956	0.708	0.895
		Testing	0.607	1.349	0.183	0.749
	MLR	Training	0.695	1.274	0.483	0.800
		Testing	0.587	1.345	0.188	0.725
2	ANN-1	Training	0.760	1.180	0.577	0.854
		Testing	0.547	1.222	0.046	0.704
	WANN-2	Training	0.725	1.264	0.515	0.831
		Testing	0.457	1.252	−0.002	0.639
	SVM-RF-3	Training	0.812	1.073	0.650	0.875
		Testing	0.568	1.262	−0.018	0.714
	MLR	Training	0.693	1.308	0.481	0.799
		Testing	0.531	1.262	−0.017	0.700
3	ANN-3	Training	0.789	1.079	0.620	0.879
		Testing	0.520	1.333	−0.172	0.688
	WANN-1	Training	0.725	1.213	0.519	0.812
		Testing	0.467	1.447	−0.382	0.608
	SVM-RF-3	Training	0.893	0.858	0.760	0.913
		Testing	0.528	1.411	−0.315	0.665
	MLR	Training	0.688	1.269	0.474	0.795
		Testing	0.506	1.363	−0.227	0.665

4. Discussion

Our results as obtained are similar to the results of [17,39]. They modeled pan evaporation and found that the ANN and SVR models achieved high correlation coefficients ranging from 0.81 to 0.90. In addition, our findings are in agreement with Cobaner [15], who observed that the ANN model with Bayesian Regularization (BR) and algorithm during training, validation, and testing generated 0.76, 0.67, and 0.72, respectively. Applying Levenberg–Marquardt (LM) algorithm, the corresponding values were 0.77, 0.69, and 0.71, respectively. Furthermore, for SVR, this model's findings are close to those of Tezel and Buyukyildiz [51]. They concluded that the SVR gave high correlations, ranging from 0.86 to 0.90, for evaporation forecasting. Moreover, the results obtained with SVR are in line with Pammar and Deka [52]. They stated that the correlation coefficients and RMSE ranged from 0.79 to 0.84 and from 0.90 to 1.03 under the different kernels. The values of RMSE conducted by Alizamir et al. [17] were 0.836 and 0.882 for ANN 4-6-6-1 and 1.028 and 1.106 for MLR models through the training and testing period. Their results found that ANN's evaporation estimation was better than the estimation through MLR and agreed with the present study results. The ANN model of pan evaporation, with all available variables as inputs, proposed by Rahimi Khoob [21] was the most accurate, delivering an R^2 of 0.717 and an RMSE of 1.11 mm independent evaluation data set, which correlates with our outcomes. As reported by Keskin and Terzi [25], the R^2 values of the ANN 3, 6, 1, ANN 6, 2, 1, and ANN 7, 2, 1 model equaling 0.770, 0.787, and 0.788 for modeling E_{pan} are also acceptable and agree with our results. These developed models produced a more acceptable outcome than Kim et al. [53]. The latter stated that the ANN and MLR generated R^2 values ranging from 0.69 to 0.74 and from 0.61 to 0.64. The RMSE for these models varied from 1.38 to 1.48 and from 1.56 to 1.60, respectively. However, all developed models in this manuscript could not capture the variability of extreme values present in the input and output parameters at the given study location. The models' efficiency might be improved if the extreme values are removed. This is one of the limitations of the study outlined in this paper.

5. Conclusions

Evaporation processes are strongly non-linear and stochastic phenomena affected by relative humidity, temperature, vapor pressure deficit, and wind speed. In the present study, daily pan evaporation (E_{pan}) estimation was evaluated using ANN, WANN, SVM-RF, SVM-LF, and MLR models. The input climatic variables for the estimation of daily E_{pan} were: maximum and minimum temperatures (T_{max} and T_{min}), relative humidity (RH-1 and RH-2), wind speed (W.S.), and bright sunshine hours (SSH). The free availability of these meteorological parameters for other stations in Bihar, India, is a significant concern and limitation of this research. The proposed models were trained and tested in three separate scenarios, i.e., Scenario 1, Scenario 2, and Scenario 3, utilizing different percentages of data points. The models above were evaluated using statistical tools, namely, PCC, RMSE, NSE, and WI, through visual inspection using a line diagram, scatter plot, and Taylor diagram. Research results evidenced the SVM-RF model's ability to estimate daily $E_{pan,}$ integrating all weather details like T_{max}, T_{min}, RH-1, RH-2, WS, and SSH The SVM-RF model's dominance was found at Pusa station for all scenarios investigated. It is also clear that, with an increase in the sample set for training, the testing data will show a less accurate modeled result. Since the Pusa dataset has many extreme values, the developed model could not capture extreme values very efficiently; this is one of the limitations of this paper. Overall, the current research outcome showed the SVM-RF model's viability as a newly established data-intelligent method to simulate pan evaporation in the Indian area. It can be extended to many water resource engineering applications. It is also recommended that SVM-RF models can be applied under the same climatic conditions and the availability of the same meteorological parameters.

Author Contributions: Conceptualization, M.K., A.K. (Anuradha Kumari), D.K. and A.K. (Ambrish Kumar); methodology, M.K. and D.K.; software, M.K., A.K. (Anuradha Kumari) and R.K.; validation, M.K., A.K. (Anuradha Kumari), D.K. and A.K. (Ambrish Kumar); formal analysis, M.K., D.K. and A.K. (Alban Kuriqi); investigation, M.K.; resources, M.K., D.K. and A.K. (Ambrish Kumar); data curation, M.K. and A.K. (Anuradha Kumari); writing—original draft preparation, M.K., A.K. (Anuradha Kumari), R.A. and R.K.; writing—review and editing, M.K., D.K., R.A., A.E. and A.K. (Alban Kuriqi); visualization, D.K., N.A.-A., R.A., A.E. and A.K. (Alban Kuriqi); supervision, D.K., N.A.-A., A.K. (Ambrish Kumar), A.E. and A.K. (Alban Kuriqi); project administration, A.K. (Alban Kuriqi); funding acquisition, N.A-A. Please refer to the CRediT taxonomy for the term explanation. Authorship must be limited to those who have contributed substantially to work reported. All authors have read and agreed to the published version of the manuscript.

Funding: This research received no external funding.

Institutional Review Board Statement: Not applicable.

Informed Consent Statement: Not applicable.

Data Availability Statement: Not available.

Acknowledgments: The authors would like to thank the anonymous reviewers for their valuable comments and suggestions to improve this manuscript further.

Conflicts of Interest: The authors declare no conflict of interest.

References

1. Alizadeh, M.J.; Kavianpour, M.R.; Kisi, O.; Nourani, V. A new approach for simulating and forecasting the rainfall-runoff process within the next two months. *J. Hydrol.* **2017**, *548*, 588–597. [CrossRef]
2. Adnan, R.M.; Liang, Z.; Parmar, K.S.; Soni, K.; Kisi, O. Modeling monthly streamflow in mountainous basin by MARS, GMDH-NN and DENFIS using hydroclimatic data. *Neural Comput. Appl.* **2021**, *33*, 2853–2871. [CrossRef]

3. Mbangiwa, N.C.; Savage, M.J.; Mabhaudhi, T. Modelling and measurement of water productivity and total evaporation in a dryland soybean crop. *Agric. For. Meteorol.* **2019**, *266–267*, 65–72. [CrossRef]
4. Sayl, K.N.; Muhammad, N.S.; Yaseen, Z.M.; El-shafie, A. Estimation the Physical Variables of Rainwater Harvesting System Using Integrated GIS-Based Remote Sensing Approach. *Water Resour. Manag.* **2016**, *30*, 3299–3313. [CrossRef]
5. Sanikhani, H.; Kisi, O.; Maroufpoor, E.; Yaseen, Z.M. Temperature-based modeling of reference evapotranspiration using several artificial intelligence models: Application of different modeling scenarios. *Theor. Appl. Climatol.* **2019**, *135*, 449–462. [CrossRef]
6. Rajaee, T.; Nourani, V.; Zounemat-Kermani, M.; Kisi, O. River suspended sediment load prediction: Application of ANN and wavelet conjunction model. *J. Hydrol. Eng.* **2011**, *16*, 613–627. [CrossRef]
7. Aytek, A. Co-active neurofuzzy inference system for evapotranspiration modeling. *Soft Comput.* **2008**, *13*, 691. [CrossRef]
8. Wang, K.; Liu, X.; Tian, W.; Li, Y.; Liang, K.; Liu, C.; Li, Y.; Yang, X. Pan coefficient sensitivity to environment variables across China. *J. Hydrol.* **2019**, *572*, 582–591. [CrossRef]
9. Adnan, R.M.; Liang, Z.; Heddam, S.; Zounemat-Kermani, M.; Kisi, O.; Li, B. Least square support vector machine and multivariate adaptive regression splines for streamflow prediction in mountainous basin using hydro-meteorological data as inputs. *J. Hydrol.* **2020**, *586*, 124371. [CrossRef]
10. Snyder, R.L. Equation for Evaporation Pan to Evapotranspiration Conversions. *J. Irrig. Drain. Eng.* **1992**, *118*, 977–980. [CrossRef]
11. Adnan, R.M.; Liang, Z.; Trajkovic, S.; Zounemat-Kermani, M.; Li, B.; Kisi, O. Daily streamflow prediction using optimally pruned extreme learning machine. *J. Hydrol.* **2019**, *577*, 123981. [CrossRef]
12. Yuan, X.; Chen, C.; Lei, X.; Yuan, Y.; Muhammad Adnan, R. Monthly runoff forecasting based on LSTM–ALO model. *Stoch. Environ. Res. Risk Assess.* **2018**, *32*, 2199–2212. [CrossRef]
13. Zerouali, B.; Al-Ansari, N.; Chettih, M.; Mohamed, M.; Abda, Z.; Santos, C.A.G.; Zerouali, B.; Elbeltagi, A. An Enhanced Innovative Triangular Trend Analysis of Rainfall Based on a Spectral Approach. *Water* **2021**, *13*, 727. [CrossRef]
14. Malik, A.; Rai, P.; Heddam, S.; Kisi, O.; Sharafati, A.; Salih, S.Q.; Al-Ansari, N.; Yaseen, Z.M. Pan Evaporation Estimation in Uttarakhand and Uttar Pradesh States, India: Validity of an Integrative Data Intelligence Model. *Atmosphere* **2020**, *11*, 553. [CrossRef]
15. Cobaner, M. Evapotranspiration estimation by two different neuro-fuzzy inference systems. *J. Hydrol.* **2011**, *398*, 292–302. [CrossRef]
16. Muhammad Adnan, R.; Chen, Z.; Yuan, X.; Kisi, O.; El-Shafie, A.; Kuriqi, A.; Ikram, M. Reference Evapotranspiration Modeling Using New Heuristic Methods. *Entropy* **2020**, *22*, 547. [CrossRef]
17. Alizamir, M.; Kisi, O.; Muhammad Adnan, R.; Kuriqi, A. Modelling reference evapotranspiration by combining neuro-fuzzy and evolutionary strategies. *Acta Geophys.* **2020**, *68*, 1113–1126. [CrossRef]
18. Vallet-Coulomb, C.; Legesse, D.; Gasse, F.; Travi, Y.; Chernet, T. Lake evaporation estimates in tropical Africa (Lake Ziway, Ethiopia). *J. Hydrol.* **2001**, *245*, 1–18. [CrossRef]
19. Moghaddamnia, A.; Ghafari Gousheh, M.; Piri, J.; Amin, S.; Han, D. Evaporation estimation using artificial neural networks and adaptive neuro-fuzzy inference system techniques. *Adv. Water Resour.* **2009**, *32*, 88–97. [CrossRef]
20. Guven, A.; Kişi, Ö. Daily pan evaporation modeling using linear genetic programming technique. *Irrig. Sci.* **2011**, *29*, 135–145. [CrossRef]
21. Rahimi Khoob, A. Artificial neural network estimation of reference evapotranspiration from pan evaporation in a semi-arid environment. *Irrig. Sci.* **2008**, *27*, 35–39. [CrossRef]
22. Trajkovic, S. Testing hourly reference evapotranspiration approaches using lysimeter measurements in a semiarid climate. *Hydrol. Res.* **2009**, *41*, 38–49. [CrossRef]
23. Sudheer, K.P.; Gosain, A.K.; Mohana Rangan, D.; Saheb, S.M. Modelling evaporation using an artificial neural network algorithm. *Hydrol. Process.* **2002**, *16*, 3189–3202. [CrossRef]
24. Keskin, M.E.; Terzi, Ö.; Taylan, D. Fuzzy logic model approaches to daily pan evaporation estimation in western Turkey/Estimation de l'évaporation journalière du bac dans l'Ouest de la Turquie par des modèles à base de logique floue. *Hydrol. Sci. J.* **2004**, *49*, 1010. [CrossRef]
25. Keskin, M.E.; Terzi, Ö. Artificial Neural Network Models of Daily Pan Evaporation. *J. Hydrol. Eng.* **2006**, *11*, 65–70. [CrossRef]
26. Tan, S.B.K.; Shuy, E.B.; Chua, L.H.C. Modelling hourly and daily open-water evaporation rates in areas with an equatorial climate. *Hydrol. Process. Int. J.* **2007**, *21*, 486–499. [CrossRef]
27. Kisi, Ö.; Çobaner, M. Modeling River Stage-Discharge Relationships Using Different Neural Network Computing Techniques. *CLEAN Soil Air Water* **2009**, *37*, 160–169. [CrossRef]
28. Piri, J.; Amin, S.; Moghaddamnia, A.; Keshavarz, A.; Han, D.; Remesan, R. Daily Pan Evaporation Modeling in a Hot and Dry Climate. *J. Hydrol. Eng.* **2009**, *14*, 803–811. [CrossRef]
29. Keskin, M.E.; Terzi, Ö.; Taylan, D. Estimating daily pan evaporation using adaptive neural-based fuzzy inference system. *Theor. Appl. Climatol.* **2009**, *98*, 79–87. [CrossRef]
30. Dogan, E.; Gumrukcuoglu, M.; Sandalci, M.; Opan, M. Modelling of evaporation from the reservoir of Yuvacik dam using adaptive neuro-fuzzy inference systems. *Eng. Appl. Artif. Intell.* **2010**, *23*, 961–967. [CrossRef]
31. Tabari, H.; Marofi, S.; Sabziparvar, A.-A. Estimation of daily pan evaporation using artificial neural network and multivariate non-linear regression. *Irrig. Sci.* **2010**, *28*, 399–406. [CrossRef]

32. Chu, H.-J.; Chang, L.-C. Application of Optimal Control and Fuzzy Theory for Dynamic Groundwater Remediation Design. *Water Resour. Manag.* **2009**, *23*, 647–660. [CrossRef]
33. Vapnik, V. *The Nature of Statistical Learning Theory*; Springer: New York, NY, USA, 1995.
34. Kim, S.; Shiri, J.; Kisi, O. Pan Evaporation Modeling Using Neural Computing Approach for Different Climatic Zones. *Water Resour. Manag.* **2012**, *26*, 3231–3249. [CrossRef]
35. Tikhamarine, Y.; Malik, A.; Pandey, K.; Sammen, S.S.; Souag-Gamane, D.; Heddam, S.; Kisi, O. Monthly evapotranspiration estimation using optimal climatic parameters: Efficacy of hybrid support vector regression integrated with whale optimization algorithm. *Environ. Monit. Assess.* **2020**, *192*, 696. [CrossRef]
36. Elbeltagi, A.; Deng, J.; Wang, K.; Hong, Y. Crop Water footprint estimation and modeling using an artificial neural network approach in the Nile Delta, Egypt. *Agric. Water Manag.* **2020**, *235*, 106080. [CrossRef]
37. Elbeltagi, A.; Aslam, M.R.; Malik, A.; Mehdinejadiani, B.; Srivastava, A.; Bhatia, A.S.; Deng, J. The impact of climate changes on the water footprint of wheat and maize production in the Nile Delta, Egypt. *Sci. Total Environ.* **2020**, *743*, 140770. [CrossRef] [PubMed]
38. Elbeltagi, A.; Deng, J.; Wang, K.; Malik, A.; Maroufpoor, S. Modeling long-term dynamics of crop evapotranspiration using deep learning in a semi-arid environment. *Agric. Water Manag.* **2020**, *241*, 106334. [CrossRef]
39. Elbeltagi, A.; Aslam, M.R.; Mokhtar, A.; Deb, P.; Abubakar, G.A.; Kushwaha, N.L.; Venancio, L.P.; Malik, A.; Kumar, N.; Deng, J. Spatial and temporal variability analysis of green and blue evapotranspiration of wheat in the Egyptian Nile Delta from 1997 to 2017. *J. Hydrol.* **2020**, 125662. [CrossRef]
40. Kim, T.-W.; Valdés, J.B. Nonlinear Model for Drought Forecasting Based on a Conjunction of Wavelet Transforms and Neural Networks. *J. Hydrol. Eng.* **2003**, *8*, 319–328. [CrossRef]
41. Adamowski, J.; Fung Chan, H.; Prasher, S.O.; Ozga-Zielinski, B.; Sliusarieva, A. Comparison of multiple linear and nonlinear regression, autoregressive integrated moving average, artificial neural network, and wavelet artificial neural network methods for urban water demand forecasting in Montreal, Canada. *Water Resour. Res.* **2012**, 48. [CrossRef]
42. Labat, D.; Ababou, R.; Mangin, A. Rainfall–runoff relations for karstic springs. Part II: Continuous wavelet and discrete orthogonal multiresolution analyses. *J. Hydrol.* **2000**, *238*, 149–178. [CrossRef]
43. Kişi, Ö. Daily suspended sediment estimation using neuro-wavelet models. *Int. J. Earth Sci.* **2010**, *99*, 1471–1482. [CrossRef]
44. Rajaee, T. Wavelet and ANN combination model for prediction of daily suspended sediment load in rivers. *Sci. Total Environ.* **2011**, *409*, 2917–2928. [CrossRef]
45. Adnan, R.M.; Khosravinia, P.; Karimi, B.; Kisi, O. Prediction of hydraulics performance in drain envelopes using Kmeans based multivariate adaptive regression spline. *Appl. Soft Comput.* **2021**, *100*, 107008. [CrossRef]
46. Lin, J.-Y.; Cheng, C.-T.; Chau, K.-W. Using support vector machines for long-term discharge prediction. *Hydrol. Sci. J.* **2006**, *51*, 599–612. [CrossRef]
47. Tripathi, S.; Srinivas, V.V.; Nanjundiah, R.S. Downscaling of precipitation for climate change scenarios: A support vector machine approach. *J. Hydrol.* **2006**, *330*, 621–640. [CrossRef]
48. Liu, Q.-J.; Shi, Z.-H.; Fang, N.-F.; Zhu, H.-D.; Ai, L. Modeling the daily suspended sediment concentration in a hyperconcentrated river on the Loess Plateau, China, using the Wavelet–ANN approach. *Geomorphology* **2013**, *186*, 181–190. [CrossRef]
49. Cherkassky, V.; Ma, Y. Practical selection of SVM parameters and noise estimation for SVM regression. *Neural Netw.* **2004**, *17*, 113–126. [CrossRef]
50. Taylor, K.E. Summarizing multiple aspects of model performance in a single diagram. *J. Geophys. Res. Atmos.* **2001**, *106*, 7183–7192. [CrossRef]
51. Tezel, G.; Buyukyildiz, M. Monthly evaporation forecasting using artificial neural networks and support vector machines. *Theor. Appl. Climatol.* **2016**, *124*, 69–80. [CrossRef]
52. Pammar, L.; Deka, P.C. Daily pan evaporation modeling in climatically contrasting zones with hybridization of wavelet transform and support vector machines. *Paddy Water Environ.* **2017**, *15*, 711–722. [CrossRef]
53. Kim, S.; Singh, V.P.; Seo, Y. Evaluation of pan evaporation modeling with two different neural networks and weather station data. *Theor. Appl. Climatol.* **2014**, *117*, 1–13. [CrossRef]

MDPI

Article

Spatial and Temporal Analysis of Dry and Wet Spells in the Wadi Cheliff Basin, Algeria

Mohammed Achite [1,2], Nir Y. Krakauer [3], Andrzej Wałęga [4,*] and Tommaso Caloiero [5]

1 Laboratory of Water & Environment, Faculty of Nature and Life Sciences, University Hassiba Benbouali of Chlef, Chlef 02180, Algeria; achitemohammed@gmail.com
2 National Higher School of Agronomy, ENSA, Hassan Badi, El Harrach, Algiers 16200, Algeria
3 Department of Civil Engineering and NOAA-CREST, The City College of New York, New York, NY 10031, USA; nkrakauer@ccny.cuny.edu
4 Department of Sanitary Engineering and Water Management, University of Agriculture in Krakow, Mickiewicza 24/28 Street, 30-059 Krakow, Poland
5 National Research Council of Italy, Institute for Agriculture and Forest Systems in the Mediterranean (CNR-ISAFOM), 87036 Rende (CS), Italy; tommaso.caloiero@isafom.cnr.it
* Correspondence: andrzej.walega@urk.edu.pl

Abstract: The Mediterranean Basin, located in a transition zone between the temperate and rainy climate of central Europe and the arid climate of North Africa, is considered a major hotspot of climate change, subject to water scarcity and drought. In this work, dry and wet spells have been analyzed in the Wadi Cheliff basin (Algeria) by means of annual precipitation observed at 150 rain gauges in the period 1970–2018. In particular, the characteristics of dry and wet spells (frequency, duration, severity, and intensity) have been evaluated by means of the run theory applied to the 12-month standardized precipitation index (SPI) values. Moreover, in order to detect possible tendencies in the SPI values, a trend analysis has been performed by means of two non-parametric tests, the Theil–Sen and Mann–Kendall test. The results indicated similar values of frequency, severity, duration, and intensity between the dry and the wet spells, although wet events showed higher values in the extreme. Moreover, the results of the trend analysis evidenced a different behavior between the northern side of the basin, characterized by a negative trend in the 12-month SPI values, and the southern side, in which positive trends were detected.

Keywords: drought; SPI; run theory; Sen's estimator; Mann–Kendall; Wadi Cheliff Basin

Citation: Achite, M.; Krakauer, N.Y.; Wałęga, A.; Caloiero, T. Spatial and Temporal Analysis of Dry and Wet Spells in the Wadi Cheliff Basin, Algeria. *Atmosphere* **2021**, *12*, 798. https://doi.org/10.3390/atmos12060798

Academic Editor: Alexander V. Chernokulsky

Received: 14 May 2021
Accepted: 17 June 2021
Published: 21 June 2021

1. Introduction

Drought, similar to floods, is a dangerous natural hazard that can affect almost every region of the world at any time. Its genesis and course depend on many factors, both natural and those resulting from human pressure. Unlike floods, drought develops gradually and exhibits a high temporal inertia, so its symptoms are often underestimated and mistakenly perceived as less of a threat to humans compared to other natural disasters. Long-term droughts affect all sectors of the economy and, as a result, society as a whole. Drought is mainly related to a rainfall deficit leading to a decrease in water supplies affecting the flora and fauna of a given region [1,2]. Meteorological drought is characterized by a deficit of precipitation, an elevated temperature, and low humidity. These anomalies then propagate to impact the surface water and groundwater sources, ecosystems, and human activities. The impact of drought on society, the environment, and the economy depends on its duration and spatial extent. Water stress or water deficit caused by drought has a substantial influence on low production in major agricultural crops [3,4].

The extent of the water deficit on the land surface can be quantified by various indices based on meteorological variables. These include the Palmer Drought Severity Index (PDSI), the Crop Moisture Index (CMI), the Surface Water Supply Index (SWSI), the Rainfall Anomaly Index (RAI), the Standardized Precipitation Index (SPI), and the

Standardized Precipitation Evapotranspiration Index (SPEI) [5,6]. The World Meteorological Organization has recommended that the standardized precipitation index (SPI) developed by McKee et al. [7] is used as a universal meteorological drought index because of its standardized form and the lower requirement of available data that is needed [8–11]. For example, a study described by Ekewzuo and Ezeh [12] and performed in West Africa used a 3-month SPI and showed that the most exposed area to extreme drought conditions occurs over the northern Sahel domain though the frequency of occurrence is very low.

In general, climate change has worsened the extremes of high temperature and both low and high precipitation, and thus has increased the risk of drought [13,14]. For example, Vilaj et al. [15] reported that the SPI reveals an increasing occurrence of droughts in Kerala, India, caused by a decreasing trend of extreme precipitation indexes and an increasing trend of extreme temperature indexes. Algeria is a good example of the worrying manifestations of climate change. As Hadour et al. [16] reported for the RCP8.5 scenario, a decrease in winter rains for the 2039, 2069, and 2099 horizons is projected, while the temperatures will increase. To help with water management under this increasing risk of drought, spatially detailed long-term meteorological data are needed. These data can inform an analysis of the tendency of meteorological drought indicators and better represent their spatiotemporal complexity. The western part of Algeria has experienced several droughts over the last century [16–18]. Drought effects within the country are modulated by the high heterogeneity of the spatial distribution of the rainfall [19].

The variability of precipitation and thus the variability of the drought intensity is linked with many physical mechanisms. For example, in the Iberian Peninsula, Vicente-Serrano et al. [20] linked the increasing drought tendency with greater atmospheric evaporative demand associated with temperature rises. Markonis et al. [21] showed that drier conditions over the Mediterranean are in accordance with a recent north–south polarization of drought patterns over Europe. Parry et al. [22] analyzed three major pan-European droughts in the second half of the twentieth century through synoptic conditions and large-scale circulation patterns, emphasizing that each major drought episode had its own unique spatiotemporal signature. Studies performed by Littman [23] evidenced the influence of the phase pace of the NAO (North Atlantic Oscillation) teleconnection pattern on the precipitation and temperature variability in Turkey. Kingston et al. [24] found that the combination of the NAO and the EA/WR circulation patterns was the most important driver of drought that the European land area was experiencing on a monthly time scale. López-Moreno and Vicente-Serrano [25] found opposing NAO–SPI relationships between northern and southern Europe. Precipitation and drought are also linked to sea surface temperature (SST) anomalies [26,27].

The Wadi Cheliff is the longest river in Algeria and plays a vital role in its socio-economic development. The Wadi originates from the Saharan Atlas, near Aflou in the mountains of the Jebel Amour, and has a length of approximately 750 km, flowing into the Mediterranean Sea. Accordingly, the present paper shows the results from an analysis of meteorological drought performed on a large number of (150) rainfall stations with long-term precipitation records, which is a unique contribution for the Wadi Cheliff basin.

The aim of the paper is the spatiotemporal analysis of the Standardized Precipitation Index (SPI) variability on the Wadi Cheliff basin in the period 1970–2018. Run theory has been applied on the 12-month SPI series and some characteristics of the drought and wet spells have been identified. Additionally, trends of annual 12-month SPI are shown.

2. Materials and Methods

2.1. Study Area and Data

The Wadi Cheliff Basin (WCB) covers an area of 43,750 km^2 and lies between 00°07′44″ E and 03°31′07″ E and between 33°53′13″ N and 36°26′34″ N (Figure 1). The topography of the basin is complex and rugged. The altitude varies from −4 m to 1969 m.

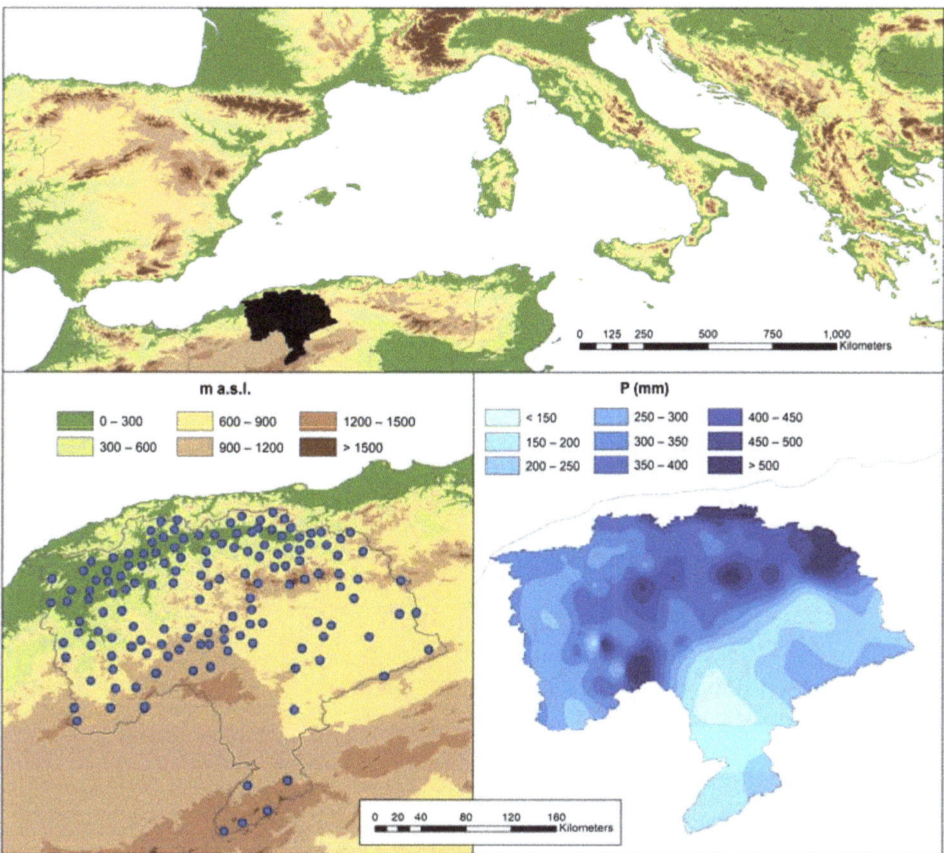

Figure 1. Location of the study area and of the selected 150 rain gauges on an elevation map, along with the distribution of the average annual precipitation evaluated from the 150 rain gauges with a spline interpolation.

Climatically, the basin is arid and semi-arid. The mean annual temperature decreases gradually from the north to south as the elevation increases going upstream [28]. The mean annual precipitation recorded at different stations (1970–2018) ranged from 161 mm to 662 mm, 80% of which fell between November and March. For this study, datasets of several rainfall stations (Figure 1) with long-term annual precipitation records from 1970 to 2018 across the WCB were taken from the National Agency of the Water Resources (ANRH). However, the period of the records for these stations varies and some have missing records. To improve the data quality, only the observing stations with data series accounting for 70% or more of the overall period were chosen for our study. After excluding the stations with too many missing values, the double mass curve technique was used to analyze the remaining missing data. The data was subjected to quality control and data gap filling using the linear regression method. The period of study was chosen as 1970–2018, which is as long as possible based on the availability of recorded data for the majority of the stations in the region [29].

2.2. The Standardized Precipitation Index (SPI)

The SPI is an index by which we can evaluate wet and the dry spells for any region in the world. According to McKee et al. [7], drought has a beginning date, an end date, a drought intensity and a drought magnitude. The SPI quantifies the intensity of a drought or wet spell and is mathematically based on the cumulative probability of the precipitation amount recorded at each station.

A period of observation at one meteorological station was used to determine the parameters of scaling and the forms of precipitation probability density function:

$$g(x) = \frac{1}{\beta^{\alpha}\Gamma(\alpha)} x^{\alpha-1} e^{-x/\beta} \text{ for } x > 0 \tag{1}$$

where α and β are the shape and scale parameters respectively, x is the precipitation amount and $\Gamma(\alpha)$ is the gamma function. The gamma function is defined as follows:

$$\Gamma(\alpha) = \int_0^{\infty} y^{\alpha-1} e^{-y} \mathrm{d}y. \tag{2}$$

The shape and scale parameters can be estimated using the approximation of Thom [30]:

$$\alpha = \frac{1}{4A}\left(1 + \sqrt{1 + \frac{4A}{3}}\right), \tag{3}$$

and

$$\beta = \frac{\overline{x}}{\alpha}, \tag{4}$$

with

$$A = \ln(\overline{x}) - \frac{\sum \ln(x)}{n}, \tag{5}$$

where $\overline{x}$ is the mean value of the precipitation quantity; n is the precipitation measurement number; x is the quantity of the precipitation in a sequence of data.

The acquired parameters were further applied to determine the cumulative probability of a certain precipitation for a specific time period in a time scale of all the recorded precipitation. The cumulative probability can be presented as:

$$G(x) = \int_0^x g(x)\mathrm{d}x = \frac{1}{\beta^{\alpha}\Gamma(\alpha)} \int_0^x x^{\alpha-1} e^{-x/\beta} \mathrm{d}x, \tag{6}$$

Since the gamma distribution is undefined for a rainfall amount $x = 0$, in order to take into account the zero values that occur in a sample set, a modified cumulative distribution function (CDF) must be considered.

$$H(x) = q + (1-q)G(x), \tag{7}$$

with $G(x)$ the CDF and q the probability of zero precipitation, given by the ratio between the number of zeros in the rainfall series (m) and the number of observations (n).

The calculation of the SPI is presented on the basis of the following equation [31,32]

$$\mathrm{SPI} = \begin{cases} -\left(t - \frac{c_0 + c_1 t + c_2 t^2}{1 + d_1 t + d_2 t^2 + d_3 t^3}\right) 0 < \mathrm{H}(x) \leq 0.5 \\ +\left(t - \frac{c_0 + c_1 t + c_2 t^2}{1 + d_1 t + d_2 t^2 + d_3 t^3}\right) 0.5 < \mathrm{H}(x) \leq 1.0 \end{cases}, \tag{8}$$

where t is determined as:

$$t = \begin{cases} \sqrt{\ln\left(\frac{1}{(H(x))^2}\right)} 0 < \mathrm{H}(x) \leq 0.5 \\ \sqrt{\ln\left(\frac{1}{(1-H(x))^2}\right)} 0.5 < \mathrm{H}(x) \leq 1.0 \end{cases}, \tag{9}$$

and c_0, c_1, c_2, d_1, d_2 and d_3 are coefficients whose values are:

$c_0 = 2.515517$, $c_1 = 0.802853$, $c_2 = 0.010328$

$d_1 = 1.432788, d_2 = 0.189269\ d_3 = 0.001308$

According to the criteria of McKee et al. [7], severe and extreme droughts correspond to categories of negative SPI (below-average precipitation amount), as shown in Table 1.

Table 1. Climate classification according to the SPI values.

SPI Value	Class	Probability (%)
$SPI \geq 2.00$	Extremely wet	2.3
$1.50 \leq SPI < 2.00$	Severely wet	4.4
$1.00 \leq SPI < 1.50$	Moderately wet	9.2
$0.00 \leq SPI < 1.00$	Mildly wet	34.1
$-1.00 \leq SPI < 0.00$	Mild drought	34.1
$-1.50 \leq SPI < -1.00$	Moderate drought	9.2
$-2.00 \leq SPI < -1.50$	Severe drought	4.4
$SPI < -2.00$	Extreme drought	2.3

2.3. Run Theory

The run theory proposed by Yevjevich [33] refers to the occurrence of consecutive comparable conditions, such as wet or dry periods, allowing the characterization of each spell by assessing some characteristics such as duration, frequency, severity, and intensity. Figure 2 shows an example of the run theory for a fixed threshold.

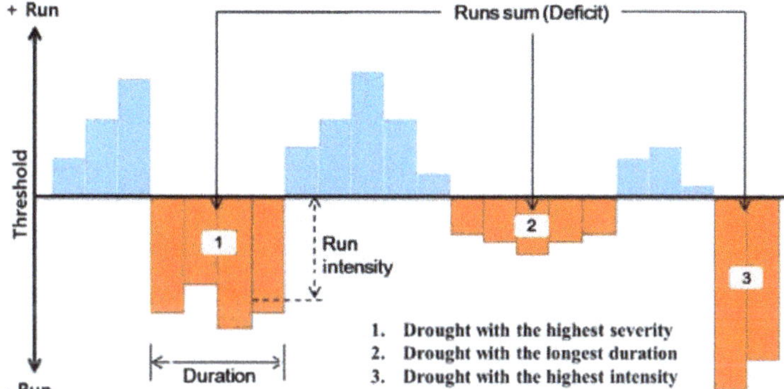

Figure 2. Example of drought characteristics evaluated using the run theory for a given threshold level [34].

A "run" is defined as an interval in which the values are all above (positive run) or below (negative run) the threshold [33]. Once the runs, and thus the dry (SPI below the threshold) or wet (SPI over the threshold) spells, have been identified, it is possible to extract some characteristics. The percentage of dry or wet spells over the study period constitute the dry (*DF*) and wet (*WF*) frequencies. The dry and wet durations (*DD* and *WD*) are the time period lengths in which SPI values are constantly below or above the threshold. Durations can be expressed in weeks, months, years or any other time period. The average drought and wet durations (*ADD* and *AWD*) are the ratio between the sum of the durations of all the drought and wet spells and the number of drought and wet spells (*ND* and *NW*). The cumulated drought and wet values during each spell represent the drought and wet severities (*DS* and *WS*). The average drought and wet severities (*ADS* and *AWS*) are the ratio between the sum of the *DS* and *WS* of all the spells and *ND* and *NW*, respectively. The drought and wet intensities (*DI* and *WI*) were evaluated, for each event, as the ratio between the *DS* and *DD* and the *WS* and *WD*, respectively, thus the average drought and wet intensities (*ADI* and *AWI*) are the ratios between the sum of the *DI* and *WI* of all the spells and *ND* and *NW*, respectively.

In this study, drought and wet periods were evaluated considering the SPI thresholds of −1 and 1, respectively. Moreover, duration, severity, and intensity were estimated using the run theory applied to the 12-month SPI series, and thus *DD* and *WD* are expressed in years. This methodology has been applied in past studies performed for several areas of the world [35].

2.4. Theil–Sen Estimator

Considering that precipitation is often non normally distributed, the Theil–Sen estimator is generally considered more powerful than the linear regression methods in trend magnitude evaluation, because it is not subject to the influence of extreme values [36]. Given $x_1, x_2, \ldots,$ x_n precipitation observations at times $t_1, t_2, \ldots, t_n$ (with $t_1 < t_2 < \ldots < t_n$), for each N pairs of observations x_j and x_i taken at times t_j and t_i, the gradient Q_k can be calculated as:

$$Q_k = \frac{x_j - x_i}{t_j - t_i} \text{ for } k = 1, \ldots, N, \quad (10)$$

with $1 < i < j < n$ and $t_j > t_i$.

The estimate of the trend in the data series $x_1, x_2, \ldots, x_n$ can then be calculated as the median Q_{med} of the N values of Q_k, ranked from the smallest to the largest:

$$Q_{med} = \begin{cases} Q_{[(N+1)/2]} if N is odd \\ \frac{Q_{[N/2]} + Q_{[(N+2)/2]}}{2} if N is even \end{cases} \quad (11)$$

The Q_{med} sign reveals the trend behavior, while its value indicates the magnitude of the trend.

2.5. Mann–Kendall Test

As regards the MK test [37,38], in order to evaluate the trend significance, the statistic S based on the rank sums is calculated as:

$$S = \sum_{i=1}^{n-1} \sum_{j=i+1}^{n} \text{sgn}(x_j - x_i) \text{ where } \text{sgn}(x_j - x_i) = \begin{cases} 1 \text{ if } (x_j - x_i) > 0 \\ 0 \text{ if } (x_j - x_i) = 0 \\ -1 \text{ if } (x_j - x_i) < 0 \end{cases} \quad (12)$$

In which x_j and x_i are the observations taken at times j and i (with $j > i$), respectively, and n is the dimension of the series.

Under the null hypothesis H_0, the distribution of S is symmetrical and is normal in the limit as n becomes large, with zero mean and variance:

$$\text{Var}(S) = \left[n(n-1)(2n+5) - \sum_{i=1}^{m} t_i i(i-1)(2i+5) \right] / 18 \quad (13)$$

in which t_i indicates the number of ties with extend i.

Given the variance of S, it is possible to evaluate the standardized statistic Z_{MK} as:

$$Z_{\text{MK}} = \begin{cases} \frac{S-1}{\sqrt{\text{Var}(S)}} \text{for } S > 0 \\ 0 \text{for } S = 0 \\ \frac{S+1}{\sqrt{\text{Var}(S)}} \text{for } S < 0 \end{cases} \quad (14)$$

By applying a two-tailed test, for a specified significance level α, the significance of the trend can be evaluated.

3. Results

Figure 3 shows the boxplots with the main statistics of *DF, WF, ADD, AWD, ADS, AWS, ADI,* and *AWI*. Generally, similar values were evaluated between the dry and the wet spells although higher frequencies and durations of wet events were detected in the minimum and in the maximum values across the 150 stations. In fact, as regards the frequency, the *DF* ranged between 0 and almost 23% while the *WF* ranged between 8.3 and 25%. Similarly, considering the duration, *ADD* and *AWD* showed almost the same median, but maximum values of 3 and 4 years were obtained for *ADD* and *AWD*, respectively. This behavior has been confirmed also for the severity and the intensity, with *AWS* and *AWI* showing minimum and maximum values higher than the ones obtained for *ADS* and *ADI*.

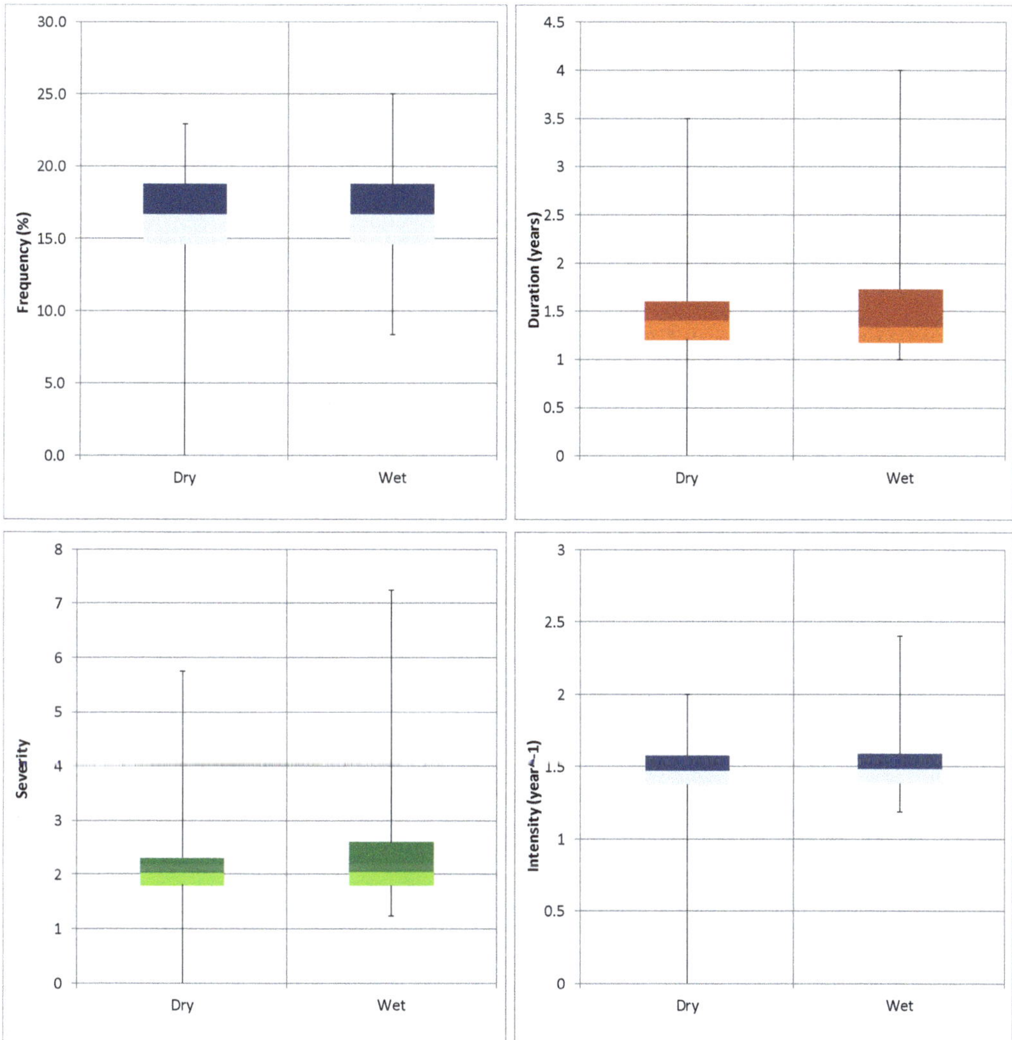

Figure 3. Characterization through boxplots of frequency (%), average duration (in years), average severity, and average intensity (year^{-1}). The top and the bottom of the boxes are the third and the second quartiles, respectively; the band inside the box is the median and the ends of the whiskers represent the minimum and maximum of all of the data.

Figure 4 shows the spatial distribution of *DF*, *ADD*, *ADS* and *ADI*. With respect to the dry frequency, the highest values largely involved the northern side of the basin, although in one station on the southern side a *DF* value higher than 20% was detected. Conversely, the southern side of the basin, which is the area with the highest elevation in which few rain gauges are located, was characterized by the lowest values of *DF*, and two rain gauges did not show any dry event (Figure 4a).

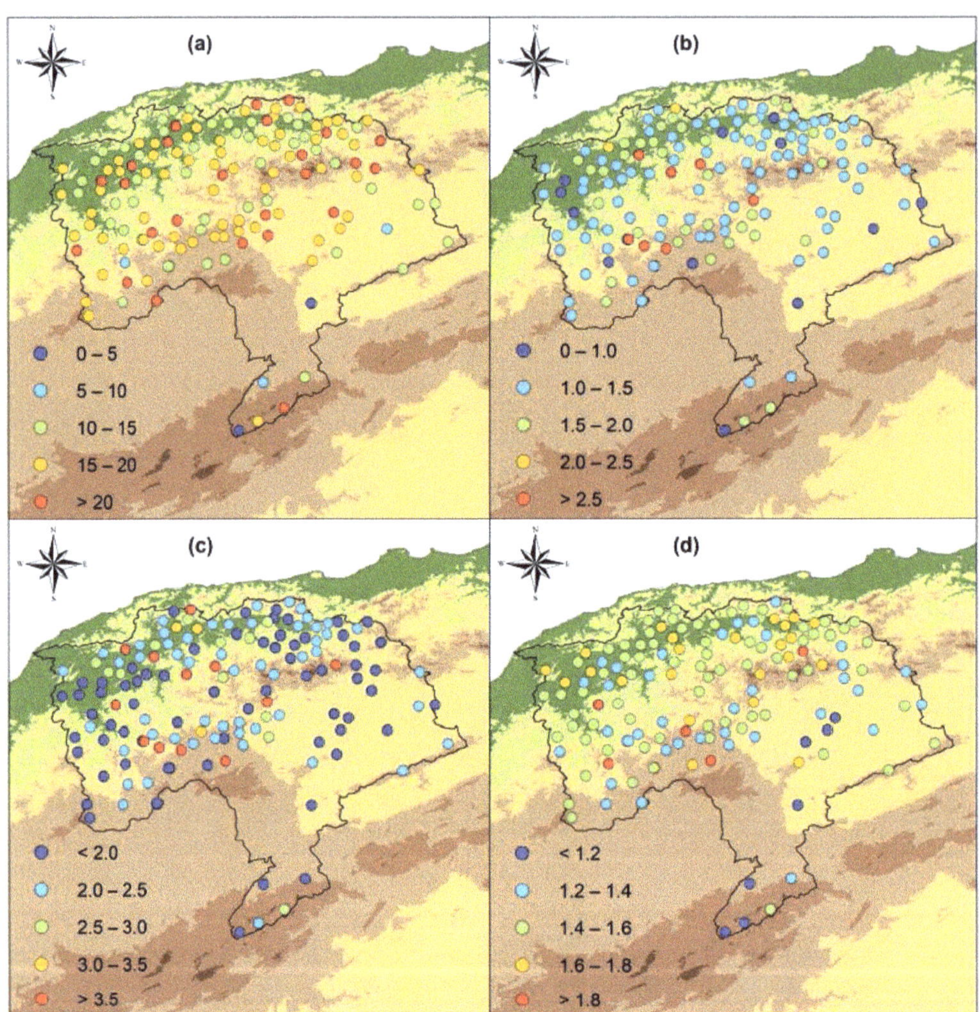

Figure 4. Spatial distribution of the (**a**) dry frequency in %, (**b**) average dry duration in years, (**c**) average dry severity and (**d**) average dry intensity in year^{-1}.

The average dry duration did not present any noteworthy spatial behavior. In fact, the lowest values (≈1 year) were distributed across the basin, while the highest ones (>2.5 years) were localized in the central part of the basin, but without any clear connection with orography (Figure 4b).

A spatial behavior similar to *ADD* was detected for *ADS* (Figure 4c). Indeed, from the spatial distribution of the average dry severity it is not possible to identify definite areas characterized by the highest values (>3.5 year), which were spread across the central part of the basin.

Finally, the spatial distribution of the average dry intensity evidenced some differences between the northern and the southern side of the basin (Figure 4d). In fact, the *ADI* values showed a distribution quite similar to the one obtained for the *DF* values, with intensities lower than 1.2 localized in the southeastern part of the basin, especially in the stations characterized by the highest elevations.

Figure 5 is similar to Figure 4 but for the wet events and, thus, it shows the spatial distribution of *WF*, *AWD*, *AWS*, and *AWI*.

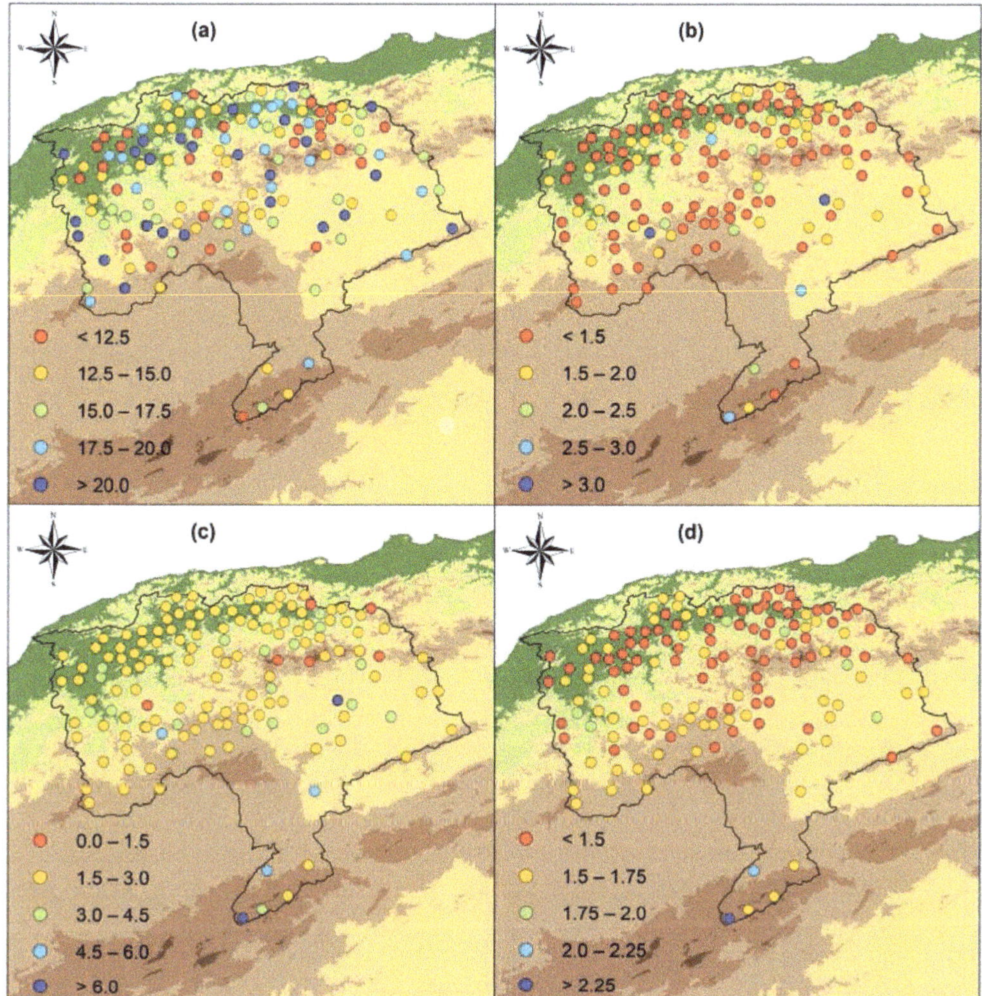

Figure 5. Spatial distribution of the (**a**) wet frequency in %, (**b**) average wet duration in years, (**c**) average wet severity and (**d**) average wet intensity in year^{-1}.

Regarding the frequency of the wet events, values higher than 20% were detected in several areas of the basin, without any particular geographical difference and without any connection with the orography (Figure 5a). Similarly, the average wet duration did not show any significant spatial behavior with the lowest values (<1.5 years) distributed across the basin and only two stations showed *AWD* values higher than 3 years (Figure 5b).

Differently from *DF* and *ADD*, the spatial distributions of the average wet severity and intensity evidenced some differences between the northern and the southern side of the basin (Figure 5c,d). In fact, for both *AWS* and *AWI* the highest values were localized in the southern part of the basin, especially in the stations characterized by the highest elevations.

In order to detect the temporal evolution of drought in the period 1970–2018, the 12-month SPI series were tested for trends through the Theil–Sen estimator and the Mann–Kendall test. As a result, for a SL = 95%, 26 out of 150 stations (i.e., 39%) showed a negative trend while an opposite behavior was detected in 13 out of 150 stations (i.e., 19.5%). Spatially, the negative trend mainly involved the northern areas of the basin, with a maximum decrease of more than −0.3/10 years (Figure 6). On the contrary, a positive trend was evidenced in the southern part of the basin reaching values between 0.2 and 0.3/10 years.

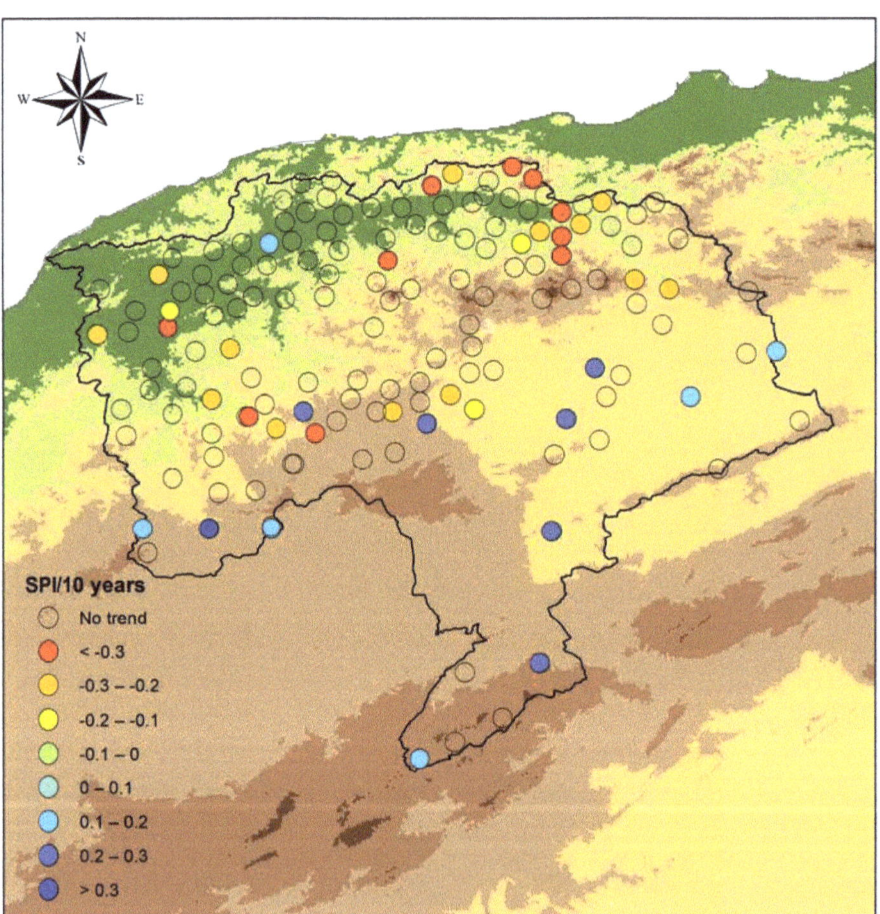

Figure 6. Spatial results of the trend analysis performed on the SPI values. The trend magnitude has been evaluated with the Theil–Sen estimator; the statistical significance of the trends, with a SL = 95%, has been assessed with the Mann–Kendall test. Colored points indicate significant trends.

Finally, with the aim to assess the drought response to the global circulation variability, a correlation analysis between the SPI values and the NAO was computed for each station. As a result, a clear link between drought and the NAO was evidenced. In particular, negative correlation values (<−0.4) mainly involved the northern areas of the basin, while a positive correlation was evidenced in the southern part of the basin (Figure 7).

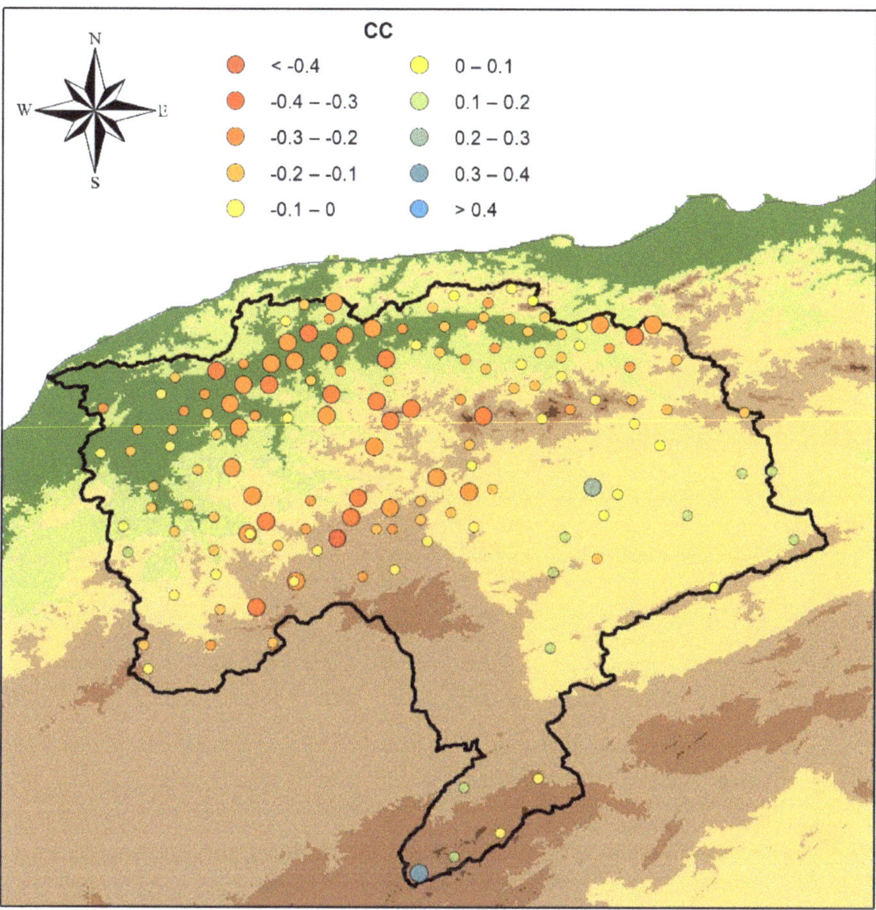

Figure 7. Spatial results of the correlation analysis computed, with the Pearson method and for each station, between the SPI values and the NAO. Large sized points reveal a significant correlation, while small points show non-significant correlations.

4. Discussion

Although a lack of precipitation can be considered an ordinary part of the seasonal climatic cycle within the Mediterranean basin, future climate projections have identified the Mediterranean region as an area particularly subject to water scarcity and drought [39]. In order to monitor the drought phenomenon and to assess the climate anomalies quantitatively in terms of intensity, duration, frequency, recurrence probability, and spatial extent, several indices have been developed [40]. Among the several indices, the SPI has been applied in the trend detection of drought in many areas of the world [41]. While a trend analysis on the SPI values can only point out the possible changes in the expected precipitation amount over the years, the study of some of the main drought characteristics such as frequency, duration, severity, and intensity is essential for politicians, local communities, and stakeholders to identify the most vulnerable areas and their drought features.

The detection of these characteristics is particularly relevant for Algeria, which some studies have identified as one of the countries in North Africa that will become a global hot spot for drought by the end of the twenty-first century [42]. In fact, the results of this study evidenced a great spatiotemporal variability in wet and dry episodes in the Cheliff Basin, with different results obtained between the northern side of the basin, characterized by the highest annual rainfall values, and the southern part, where annual rainfall values below 250 mm have been registered. In particular, the results in this area could be influenced by the low total amount of precipitation, which makes it more difficult to accurately detect drought development. The main outcomes of this study allow an identification of the areas, in the northern side of the basin, that could also face water stress conditions in the future, thus requiring drought monitoring and adequate adaptation strategies. At the same time, in the area currently suffering from a water deficit, an increase in the wet episodes has been detected, which is an important result that can inform people involved in the water resource management of the changing potential for flooding and water storage. The different behavior between the northern and the southern part of the basin has been pointed out in previous works e.g., in [29], the authors of which evidenced an annual rainfall decrease of more than 20 mm/10 years on the northern side and an increase of about 5 mm/10 years on the southern side. These results can be influenced by both global and local factors. Global factors which impact on the intra-annual precipitation distribution include teleconnection patterns [43,44]. In fact, as evidenced in Figure 7, a clear link exists between drought and the NAO. This result agrees with the results obtained by several authors e.g., [29], which evidenced that the Mediterranean rainfall regime is strongly linked to general atmospheric circulation patterns such as the El Niño Southern Oscillation (ENSO), the Mediterranean Oscillation (MO), the Western Mediterranean Oscillation (WeMO), and especially the North Atlantic Oscillation indices that are negatively correlated with precipitation in Algeria [45]. Indeed, positive NAO phases can cause dry conditions across large parts of the Mediterranean Basin, from Spain and Morocco across to the Balkans and western Turkey [46]. Specifically, a predominant negative phase of the NAO occurred between 1940 and 1980, corresponding to a period when precipitation was above normal. This was followed by a predominant positive phase, which significantly contributed to the rainfall reduction observed from the beginning of the 1980s in the Mediterranean basin, and also in Algeria [29]. Moreover, another regional-scale system that could influence the rainfall conditions in North Africa is the well-documented Sahelian drought and its multidecadal variability, which resulted from the response of the African summer monsoon to oceanic forcing and was amplified by land–atmosphere interactions [47].

In order to better appreciate the results of this study, an important final remark must be made concerning the database. In this study a high-quality database in the period 1970–2018 has been used. Unfortunately, only annual data (evaluated for water year) were available and thus it was not possible to perform a detailed analysis at a monthly scale. This study is nevertheless a valuable contribution not only to the quality of the database but also to its spatial resolution, which covers almost all the basin, with some gaps only in the central area. Continuous and spatially distributed data allowed a reliable analysis of the drought characteristics in the basin. Although globally gridded satellite-based precipitation products have become available in recent years as potential sources of data, ground-based precipitation measurements still remain the main and most accurate source in any climatological analysis [48].

5. Conclusions

This study aimed to analyze some of the main characteristics of the wet and dry events affecting the Cheliff Basin in the period 1970–2018 using the 12-month SPI, which is a broad proxy for water resource availability, evaluated from 150 rain gauges distributed across the basin. Generally, the analysis of the main statistics of the frequency, severity, duration, and intensity were evaluated for both dry and wet events, and it was found that their median values were similar, although higher minimum and maximum values were detected for the wet events. These results clearly indicate that the Cheliff Basin is at risk for extreme wet events as well as dry events. From the spatial distribution of the wet and dry characteristics, the basin areas facing water stress were identified. Finally, the results of the trend analysis performed on the 12-month SPI values evidenced a decreasing trend on the northern side of the basin and an opposite behavior on the southern side, characterized by the highest elevations and the lowest annual rainfall values, in which positive values were detected. These results, probably linked with the predominant positive NAO phase which significantly contributed to the rainfall reduction observed from the beginning of the 1980s in Algeria, could be highly relevant for people involved in natural resource management decision-making for sustainable long-term water resource management in semi-arid regions.

Author Contributions: Conceptualization, M.A.; methodology, M.A. and T.C.; software, M.A. and T.C.; formal analysis, M.A. and T.C.; validation: M.A., T.C., A.W. and N.Y.K.; investigation, M.A., T.C., A.W. and N.Y.K.; data curation, M.A.; writing—original draft preparation, M.A., T.C., A.W. and N.Y.K.; writing—review and editing, M.A., T.C., A.W. and N.Y.K.; visualization, T.C.; supervision, A.W. and N.Y.K. All authors have read and agreed to the published version of the manuscript.

Funding: This research received no external funding.

Institutional Review Board Statement: Not applicable.

Informed Consent Statement: Not applicable.

Data Availability Statement: The data presented in this study are available on request from the corresponding authors.

Conflicts of Interest: The authors declare no conflict of interest.

References

1. Dracup, J.A.; Lee, K.S.; Edwin, G.; Paulson, J.R. On the definition of droughts. *Water Resour. Res.* **1980**, *16*, 297–302. [CrossRef]
2. Wilhite, D.A.; Glantz, M.H. Understanding: The Drought Phenomenon: The Role of Definitions. *Water Int.* **1985**, *10*, 111–120. [CrossRef]
3. Kováčová, M.; Bárek, V.; Kišš, V. DENDROMETRIC CHANGES AS WATER STRESS INDICATOR FOR SUNFLOWER (HELIANTHUS ANNUS L.) AND MAIZE (ZEA MAYS L.)—BASIC RESEARCH IN LABORATORY CONDITIONS. *Acta Sci. Pol. Form. Circumiectus* **2020**, *19*, 77–85. [CrossRef]
4. Almendra-Martín, L.; Martínez-Fernández, J.; González-Zamora, Á.; Benito-Verdugo, P.; Herrero-Jiménez, C.M. Agricultural Drought Trends on the Iberian Peninsula: An Analysis Using Modeled and Reanalysis Soil Moisture Products. *Atmosphere* **2021**, *12*, 236. [CrossRef]
5. Palmer, W.C. Meteorological droughts. In *U.S. Department of Commerce Weather Bureau Research Paper 45*; US Deptartment of Commerce: Washington, DC, USA, 1965; p. 58.
6. Zhou, J.; Wang, Y.; Su, B.; Wang, A.; Tao, H.; Zhai, J.; Kundzewicz, Z.; Jiang, T. Choice of potential evapotranspiration formulas influences drought assessment: A case study in China. *Atmos. Res.* **2020**, *242*, 104979. [CrossRef]
7. McKee, T.B.; Doesken, N.J.; Kleist, J. The relationship of drought frequency and duration to time scales. In Proceedings of the 8th Conference on Applied Climatology, Boston, MA, USA, 17–22 January 1993; Volume 17, pp. 179–183.
8. WMO. *Drought Monitoring and Early Warning: Concepts, Progress and Future Challenges, WMO-No. 1006*; World Meteorological Organization: Geneva, Switzerland, 2006; 26p.
9. Svoboda, M.; Hayes, M.; Wood, M. (Eds.) *WMO Standardized Precipitation Index User Guide*; World Meteorological Organization Report WMO-No. 1090; World Meteorological Organization: Geneva, Switzerland, 2012.
10. Hayes, M.J.; Svoboda, M.; Wall, N.A.; Widhalm, M. The Lincoln Declaration on Drought Indices: Universal Meteorological Drought Index Recommended. *Bull. Am. Meteorol. Soc.* **2011**, *92*, 485–488. [CrossRef]
11. Fellag, M.; Achite, M.; Wałęga, A. Spatial-temporal characterization of meteorological drought using the Standardized precipitation index. Case study in Algeria. *Acta Sci. Pol. Form. Circumiectus* **2021**, *20*, 19–31.
12. Ekwazuo, C.S.; Ezeh, C.U. Regional characterization of meteorological drought and floods over West Africa. *Sustain. Water Resour. Manag.* **2020**, *6*, 1–15.
13. Ojeda, M.G.-V.; Gámiz-Fortis, S.R.; Romero-Jiménez, E.; Rosa-Cánovas, J.J.; Yeste, P.; Castro-Díez, Y.; Esteban-Parra, M.J. Projected changes in the Iberian Peninsula drought characteristics. *Sci. Total Environ.* **2021**, *757*, 143702. [CrossRef]
14. Zechiel, P.; Chiao, S. Climate Variability of Atmospheric Rivers and Droughts over the West Coast of the United States from 2006 to 2019. *Atmosphere* **2021**, *12*, 201. [CrossRef]
15. Vijay, A.; Sivan, S.D.; Mudbhatkal, A.; Mahesha, A. Long-Term Climate Variability and Drought Characteristics in Tropical Region of India. *J. Hydrol. Eng.* **2021**, *26*, 05021003. [CrossRef]
16. Hadour, A.; Mahé, G.; Meddi, M. Watershed based hydrological evolution under climate change effect: An example from North Western Algeria. *J. Hydrol. Reg. Stud.* **2020**, *28*, 100671. [CrossRef]
17. Taibi, S.; Meddi, M.; Mahé, G.; Assani, A. Relationships between atmospheric circulation indices and rainfall in Northern Algeria and comparison of observed and RCM-generated rainfall. *Theor. Appl. Clim.* **2015**, *127*, 241–257. [CrossRef]

18. Zeroual, A.; Meddi, M.; Bensaad, S. The impact of climate change on river flow in arid and semi-arid rivers in Algeria. In *Climate and Land-Surface Changes in Hydrology, Proceedings of the H01, IAHS-IAPSO-IASPEI Assembly, Gothenburg, Sweden, 22–26 July 2013*; IAHS Publicaions: Wallingford, UK, 2013; Volume 359, pp. 105–110.
19. Meddi, H.; Meddi, M. Variabilité spatiale et temporelle des précipitations du Nord-Ouest de l'Algérie. *Géogr. Tech.* **2007**, *2*, 49–55.
20. Vicente-Serrano, S.M.; Lopez-Moreno, J.I.; Beguería, S.; Lorenzo-Lacruz, J.; Sanchez-Lorenzo, A.; García-Ruiz, J.M.; Azorin-Molina, C.; Morán-Tejeda, E.; Revuelto, J.; Trigo, R.M.; et al. Evidence of increasing drought severity caused by temperature rise in southern Europe. *Environ. Res. Lett.* **2014**, *9*, 044001. [CrossRef]
21. Markonis, Y.; Hanel, M.; Máca, P.; Kyselý, J.; Cook, E.R. Persistent multi-scale fluctuations shift European hydroclimate to its millennial boundaries. *Nat. Commun.* **2018**, *9*, 1767. [CrossRef] [PubMed]
22. Parry, S.; Prudhomme, C.; Hannaford, J.; Lloyd-Hughes, B. Examining the spatio-temporal evolution and characteristics of large-scale European droughts. In Proceedings of the BHS Third International Symposium, Newcastle, UK, 19–23 July 2010; British Hydrological Society: Loindon, UK, 2010; pp. 135–142.
23. Littmann, T. An empirical classification of weather types in the Mediterranean Basin and their interrelation with rainfall. *Theor. Appl. Clim.* **2000**, *66*, 161–171. [CrossRef]
24. Kingston, D.G.; Stagge, J.H.; Tallaksen, L.M.; Hannah, D.M. European-Scale Drought: Understanding Connections between Atmospheric Circulation and Meteorological Drought Indices. *J. Clim.* **2015**, *28*, 505–516. [CrossRef]
25. López-Moreno, J.I.; Vicente-Serrano, S.M. Positive and negative phases of the winter time North Atlantic Oscillation and drought occurrence over Europe: A multi temporal scale approach. *J. Clim.* **2008**, *21*, 1220–1243. [CrossRef]
26. Zhang, Y.; Wu, R. Asian meteorological droughts on three time scales and different roles of sea surface temperature and soil moisture. *Int. J. Climatol.* **2021**, 1–18. [CrossRef]
27. Hu, Y.; Wang, S. Associations between winter atmospheric teleconnections in drought and haze pollution over Southwest China. *Sci. Total Environ.* **2021**, *766*, 142599. [CrossRef] [PubMed]
28. Elmeddahi, Y.; Mahmoudi, H.; Issaadi, A.; Goosen, M.F.; Ragab, R. Evaluating the Effects of Climate Change and Variability on Water Resources: A Case Study of the Cheliff Basin in Algeria. *Am. J. Eng. Appl. Sci.* **2016**, *9*, 835–845. [CrossRef]
29. Achite, M.; Caloiero, T.; Wałęga, A.; Krakauer, N.; Hartani, T. Analysis of the Spatiotemporal Annual Rainfall Variability in the Wadi Cheliff Basin (Algeria) over the Period 1970 to 2018. *Water* **2021**, *13*, 1477. [CrossRef]
30. Thom, H.C.S. A Note on the Gamma Distribution. *Mon. Weather. Rev.* **1958**, *86*, 117–122. [CrossRef]
31. Abramowitz, M.; Stegun, I.A. *Handbook of Mathematical Functions with Formulas, Graphs, and Mathematical Tables*; Dover Publications, Inc.: New York, NY, USA, 1970.
32. Lloyd-Hughes, B.; Saunders, M. Seasonal prediction of European spring precipitation from El Niño-Southern Oscillation and Local sea-surface temperatures. *Int. J. Clim.* **2002**, *22*, 1–14. [CrossRef]
33. Yevjevich, V. An Objective Approach to Definitions and Investigation of Continental Hydrologic Droughts. In *Hydrology Paper 23*; Colorado State University: Fort Collins, CO, USA, 1967.
34. Lee, S.-H.; Yoo, S.-H.; Choi, J.-Y.; Bae, S. Assessment of the Impact of Climate Change on Drought Characteristics in the Hwanghae Plain, North Korea Using Time Series SPI and SPEI: 1981–2100. *Water* **2017**, *9*, 579. [CrossRef]
35. Caloiero, T.; Caroletti, G.; Coscarelli, R. IMERG-Based Meteorological Drought Analysis over Italy. *Climate* **2021**, *9*, 65. [CrossRef]
36. Sen, P.K. Estimates of the regression coefficient based on Kendall's tau. *J. Am. Stat. Assoc.* **1968**, *63*, 1379–1389. [CrossRef]
37. Kendall, M.G. *Rank Correlation Methods*; Hafner Publishing Company: New York, NY, USA, 1962.
38. Mann, H.B. Nonparametric tests against trend. *Econometrica* **1945**, *13*, 245–259. [CrossRef]
39. Caloiero, T.; Veltri, S.; Caloiero, P.; Frustaci, F. Drought Analysis in Europe and in the Mediterranean Basin Using the Standardized Precipitation Index. *Water* **2018**, *10*, 1043. [CrossRef]
40. Tsakiris, G.; Pangalou, D.; Vangelis, H. Regional Drought Assessment Based on the Reconnaissance Drought Index (RDI). *Water Resour. Manag.* **2007**, *21*, 821–833. [CrossRef]
41. Caloiero, T. SPI Trend Analysis of New Zealand Applying the ITA Technique. *Geosciences* **2018**, *8*, 101. [CrossRef]
42. Schilling, J.; Hertig, E.; Tramblay, Y.; Scheffran, J. Climate change vulnerability, water resources and social implications in North Africa. *Reg. Environ. Chang.* **2020**, *20*, 15. [CrossRef]
43. Vicente-Serrano, S.M.; López-Moreno, J.I.; Lorenzo-Lacruz, J.; Kenawy, A.; Azorin-Molina, C.; Morán-Tejeda, E.; Pasho, E.; Zabalza, J.; Beguería, S.; Angulo-Martínez, M. The NAO Impact on Droughts in the Mediterranean Region. In *Advances in Global Change Research*; Springer Science and Business Media LLC: Berlin, Germany, 2011; pp. 23–40.
44. Hoerling, M.P.; Eischeid, J.K.; Perlwitz, J.; Quan, X.; Zhang, T.; Pegion, P.J. On the Increased Frequency of Mediterranean Drought. *J. Clim.* **2012**, *25*, 2146–2161. [CrossRef]
45. Meddi, H.; Meddi, M.; Assani, A.A. Study of Drought in Seven Algerian Plains. *Arab. J. Sci. Eng.* **2013**, *39*, 339–359. [CrossRef]
46. Quadrelli, R.; Pavan, V.; Molteni, F. Wintertime variability of Mediterranean precipitation and its links with large-scale cir-culation anomalies. *Clim. Dyn.* **2001**, *17*, 457–466. [CrossRef]
47. Giannini, A.; Saravanan, R.; Chang, P. Oceanic Forcing of Sahel Rainfall on Interannual to Interdecadal Time Scales. *Science* **2003**, *302*, 1027–1030. [CrossRef] [PubMed]
48. Sun, Q.; Miao, C.; Duan, Q.; Ashouri, H.; Sorooshian, S.; Hsu, K.L. A review of global precipitation data sets: Data sources, estimation, and intercomparisons. *Rev. Geophys.* **2018**, *56*, 79–107. [CrossRef]

Article

Assessing the Contrasting Effects of the Exceptional 2015 Drought on the Carbon Dynamics in Two Norway Spruce Forest Ecosystems

Caleb Mensah [1,2,3,*], Ladislav Šigut [1], Milan Fischer [1,2], Lenka Foltýnová [1], Georg Jocher [1], Manuel Acosta [1], Natalia Kowalska [1], Lukáš Kokrda [1], Marian Pavelka [1], John David Marshall [1], Emmanuel K. Nyantakyi [3] and Michal V. Marek [1,4]

1 Department of Matter and Energy and Fluxes, Global Change Research Institute CAS, Bělidla 986/4a, 603 00 Brno, Czech Republic; sigut.l@czechglobe.cz (L.Š.); fischer.m@czechglobe.cz (M.F.); foltynova.l@czechglobe.cz (L.F.); jocher.g@czechglobe.cz (G.J.); acosta.m@czechglobe.cz (M.A.); kowalsa.n@czechglobe.cz (N.K.); kokrda.l@czechglobe.cz (L.K.); pavelka.m@czechglobe.cz (M.P.); marshall.j@czechglobe.cz (J.D.M.); marek.mv@czechglobe.cz (M.V.M.)
2 Department of Agrosystems and Bioclimatology, Faculty of Agronomy, Mendel University in Brno, Zěmědelská 1, 613 00 Brno, Czech Republic
3 Earth Observation Research and Innovation Centre, University of Energy and Natural Resources, Sunyani P.O. Box 214, Ghana; emmanuel.nyantakyi@uenr.edu.gh
4 Department of Spatial Planning, Institute of Management, Slovak University of Technology in Bratislava, 811 07 Bratislava, Slovakia
* Correspondence: mensah.c@czechglobe.cz; Tel.: +420-608-294-420

Citation: Mensah, C.; Šigut, L.; Fischer, M.; Foltýnová, L.; Jocher, G.; Acosta, M.; Kowalska, N.; Kokrda, L.; Pavelka, M.; Marshall, J.D.; et al. Assessing the Contrasting Effects of the Exceptional 2015 Drought on the Carbon Dynamics in Two Norway Spruce Forest Ecosystems. *Atmosphere* **2021**, *12*, 988. https://doi.org/10.3390/atmos12080988

Academic Editors: Agnieszka Ziernicka-Wojtaszek and Andrzej Walega

Received: 21 June 2021
Accepted: 24 July 2021
Published: 31 July 2021

Publisher's Note: MDPI stays neutral with regard to jurisdictional claims in published maps and institutional affiliations.

Abstract: The occurrence of extreme drought poses a severe threat to forest ecosystems and reduces their capability to sequester carbon dioxide. This study analysed the impacts of a central European summer drought in 2015 on gross primary productivity (GPP) at two Norway spruce forest sites representing two contrasting climatic conditions—cold and humid climate at Bílý Kříž (CZ-BK1) vs. moderately warm and dry climate at Rájec (CZ-RAJ). The comparative analyses of GPP was based on a three-year eddy covariance dataset, where 2014 and 2016 represented years with normal conditions, while 2015 was characterized by dry conditions. A significant decline in the forest GPP was found during the dry year of 2015, reaching 14% and 6% at CZ-BK1 and CZ-RAJ, respectively. The reduction in GPP coincided with high ecosystem respiration (R_{eco}) during the dry year period, especially during July and August, when several heat waves hit the region. Additional analyses of GPP decline during the dry year period suggested that a vapour pressure deficit played a more important role than the soil volumetric water content at both investigated sites, highlighting the often neglected importance of considering the species hydraulic strategy (isohydric vs. anisohydric) in drought impact assessments. The study indicates the high vulnerability of the Norway spruce forest to drought stress, especially at sites with precipitation equal or smaller than the atmospheric evaporative demand. Since central Europe is currently experiencing large-scale dieback of Norway spruce forests in lowlands and uplands (such as for CZ-RAJ conditions), the findings of this study may help to quantitatively assess the fate of these widespread cultures under future climate projections, and may help to delimitate the areas of their sustainable production.

Keywords: climate change; water stress; soil moisture; atmospheric evaporative demand; eddy covariance; gross primary productivity

1. Introduction

Terrestrial ecosystems, such as forests, play a significant role in the global carbon cycle by sequestering carbon dioxide (CO_2) through photosynthesis and the conversion of biomass into stable soil organic compounds [1–6]. However, carbon uptake and carbon allocation can be reduced by high vapour pressure deficits (VPDs) and soil water deficits, commonly observed during periods with increased temperature and lack of precipitation [7,8]. These physiological responses to drought in plants vary at both local and global scales, depending on the type of plant species (based on the different levels of resilience to water stress), the local climatic condition, and other additional factors [9–13]. It has also been widely demonstrated that climate change increases the likelihood and severity of such drought events, especially within the temperate regions [14–18]. Thus, despite the resilience of some tree species, the projected increase in the occurrence and severity of extreme drought events will further reduce forest ecosystem photosynthesis in many parts of Europe, especially at unsuitable locations with sub-optimal climatic conditions [19].

Drought does not have a single definition and can be used in different contexts. However, typically, it is used to describe the periods with high VPD (meteorological drought) and low soil moisture (edaphic drought) [20]. The severity of the drought impact on carbon exchange depends on the site characteristics, duration and intensity of the drought periods [18,21,22]. In Europe, the 2003 summer drought period resulted in an estimated loss of about 30% in GPP (0.5 P gC y^{-1}) over both the northern Mediterranean forests and temperate deciduous beech forests [23,24]. This was the result of the prolonged abnormal reduction in the soil moisture and high air temperatures below the wilting point and high air temperatures recorded across Europe during the summer period in 2003. The summer drought of 2015 was considered one of the most severe drought events in Europe after the 2003 summer drought [25]. During this period, much of the European continent (Poland, the Czech Republic, Slovakia, Western Ukraine, and Belarus) was severely affected in June and July 2015, due to the exceptionally high maximum daily air temperatures and the prolonged rainfall shortage by 31% since April [26–29]. Consequently, this had a significant effect on forest growth, as observed in both spruce and beech forests, within the Czech Republic. Nonetheless, spruce forest showed a higher sensitivity to drought as compared to beech forests [30,31].

The Norway spruce (*Picea abies* (L.) H. Karst) has significant economic and ecological importance within Europe [32]. However, its shallow root system makes it vulnerable to drought stress, especially at lower altitudes [33]. Since spruce thrives well in cold and humid regions, a significant decrease in soil moisture coupled with high air temperatures could have severe consequences on the spruce forest growth [34]. Therefore, the 2015 summer drought episode provided a good opportunity to further examine the response of spruce forest stands with different climatic conditions to extreme drought stress events.

The eddy covariance (EC) technique provides direct measurements of CO_2 exchange between the atmosphere and the underlying ecosystem [35]. It is a convenient and widely used approach for observation of a forest stand carbon uptake and its dynamic response to environmental variables. Obtained CO_2 fluxes are integrated over hundreds of square meters and resolved to half-hourly intervals [35,36]. Continuous micrometeorological and EC measurements over multiple years allow to detect drought periods and evaluate their impacts on CO_2 exchange.

In this study, we hypothesized that the impact of extreme drought conditions will be more severe on the Norway spruce forest stand located in dry and hot climates, due to higher VPD and the sub-optimal supply of soil moisture in such drier climates. To test this hypothesis, we sought to evaluate the different effects of the 2015 drought during the main growing season in the wet and dry spruce forest ecosystems, compared to normal climatic conditions within the same period of two adjacent years (2014 and 2016). Secondly, we aimed to determine the influence of critical site-specific environmental factors (such as the VPD and soil volumetric water content) on the drought stress response at each of the spruce forest stands located in mountainous (around 900 m a.s.l) and highland (up to 600 m a.s.l) regions within the Czech Republic, using long-term eddy covariance CO_2 flux data.

2. Materials and Methods

2.1. Station Description

The study used multiyear (2014–2016) measurements from Bílý Kříž and Rájec ecosystem stations that are part of the CzeCOS (Czech Carbon Observation System; http://www.czecos.cz/ (accessed on 6 March 2021)) and FLUXNET (Flux Tower Network; https://fluxnet.fluxdata.org/ (accessed on 6 March 2021)) station network. The FLUXNET station IDs for the wet and dry spruce forest are CZ-BK1 and CZ-RAJ, respectively. CZ-BK1 is also a candidate for ICOS (Integrated Carbon Observation System; https://www.icos-cp.eu/ (accessed on 6 March 2021)) network. The main vegetation cover in both stations is the even-aged stand of Norway spruce. Their main characteristics are presented in Table 1. CZ-BK1 is situated in the Moravian-Silesian Beskids Mountains in the Czech Republic and has a cold and humid climate. CZ-RAJ is situated in the Drahany Highland at a lower altitude and has a moderately warmer and drier climate than CZ-BK1. The reference evapotranspiration amount at each forest stand was computed from in situ data and was additionally used for quantifying the atmospheric evaporative demand from 2014 to 2016 (Table 1).

Table 1. Characteristics of the study sites.

Site Name	CZ-BK1	CZ-RAJ
Location	Moravian-Silesian Beskids Mountains	Drahany Highland
Coordinates	49°30′08″ N, 18°32′13″ E	49°26′37″ N, 16°41′48″ E
Elevation (in m a.s.l)	875	625
Topography	Mountainous (13° slope with SSW exposure, located close to a mountain ridge)	Hilly (5° slope with NNE exposure)
Ecosystem Type	Coniferous evergreen forest	Coniferous evergreen forest
Prevailing species	Norway Spruce (*Picea abies* (L.) H.Karst.)	Norway Spruce (*Picea abies* (L.) H.Karst.)
Canopy height (m)	16 (mean, as of 2015)	33 (mean, as of 2015)
Stand age (years)	35 (as of 2016)	113 (as of 2016)
Mean annual air temperature (May–September; °C)	7.2 (2014–2016)	8.3 (2014–2016)
Mean annual precipitation (May–September; mm)	1143 (2014–2016)	610 (2014–2016)
Mean annual reference evapotranspiration (May–September; mm)	569 (2014–2016)	649 (2014–2016)
Soil type	Haplic and Entic Podzol	Modal Cambisol oligotrophic
Max and min fetch (m)	717 (WNW) and 115 (ENE)	697 (SSW) and 96 (NNW)
References	[31,37,38]	[31]

2.2. Eddy Covariance and Ancillary Measurements

At each station, the EC system consisted of an infrared gas analyser (LI-COR, Lincoln, NE, U.S.A.) and an ultrasonic anemometer (Gill, Hampshire, U.K.) measuring at a 20 Hz frequency (models are specified in Table 2). Each system was mounted on a tower several meters above the forest canopy. The EC measurements were complemented by an extensive set of sensors to collect the required auxiliary meteorological data. Measurements included the air temperature (Tair) and humidity profile with EMS33 temperature and humidity sensors (EMS, Brno, Czech Republic) and precipitation using Precipitation Gauge 386C (Met One Instruments, Grants Pass, OR, U.S.A.). Measurements for the incoming photosynthetic active radiation was made using LI-190R Quantum Sensor (LI-COR, Lincoln, NE, U.S.A.) at Bílý Kříž and EMS12 sensor (EMS, Brno, Czech Republic) at Rájec. Additionally, profiles of soil moisture were measured by using the CS616 (Campbell Scientific, Inc., Logan, UT, U.S.A.) sensors. An overall description of the instrumentation at the study sites is given in Table 2.

The *VPD* for each site was computed from Tair and relative humidity (*RH*) as follows [39]:

$$VPD = \frac{100 - RH}{100}(SVP) \quad (1)$$

where *SVP* (Pa) is the saturated vapour pressure given as follows:

$$SVP = 610.7 \times 107.5^{\frac{Tair}{237.3 + Tair}} \quad (2)$$

Table 2. Description of the eddy covariance systems at the investigated sites.

Site Name	CZ-BK1	CZ-RAJ
Ultrasonic Anemometer		
Instrument	Gill HS-50 ultrasonic anemometer, Gill Instruments, Hampshire, U.K.	Gill R3-100 ultrasonic anemometer, Gill Instruments, Hampshire, U.K. but later changed to Gill HS-50 on 5 June 2015
Gas Analyser		
Instrument	LI-7200 enclosed gas analyser, LI-COR, Lincoln, U.S.A.	Initially LI-7000 (IRG-0226) closed-path gas analyser, LI-COR, Lincoln, U.S.A. but later changed to LI-7200 on 5 June 2015
Measurement Height for the Eddy covariance Set-up (m)	Initially 20.5, but later changed to 25 m on 7 June 2016	41
Air Temperature and Humidity Profile		
Instrument	EMS33 temperature and humidity sensor (EMS, Brno, CZ)	EMS33 temperature and humidity sensor (EMS, Brno, CZ)
Measurement Height (m)	2.0, 7.6, 12.6, 13.5, 14.3, 14.8, 15.4, 16.5, 18.7	2.0, 11.0, 23.0, 29.0, 35.0, 42.0
Net Radiation		
Instrument	CNR1 Net Radiometer	CNR1 Net Radiometer
Measurement Height (m)	22	42
Soil Moisture		
Instrument	CS616 (Campbell Scientific, Inc., Logan, UT, U.S.A.)	CS616 (Campbell Scientific, Inc., Logan, UT, U.S.A.)
Measurement Depths (m)	0.05, 0.1, 0.22, 0.34, 0.42	0.05, 0.1, 0.2, 0.5, 0.8

2.3. Data Processing and Analysis

2.3.1. Turbulent Flux Measurements

At both study sites, the EC data from the main growing season (May–September) of 2014–2016 were used for the analysis. Gas analyser measurements of the density of the water vapour in the air and ultrasonic anemometer measurements of three-dimensional wind components and sonic temperature were made [35,36]. The eddy covariance processing was performed, using an open-source software, EddyPro 7.0.6 (Li-COR, Lincoln, NE, U.S.A.). The most recent methods for flux corrections, conversions, and a thorough quality control scheme as proposed by [40] were also applied. This process involves despiking and raw data statistical screening, basic quality checking (QC) of turbulent fluxes (flux stationarity and integral turbulence characteristics tests), coordinate rotation using the planar fit method [41], spectral correction [42–44], detecting and compensating for time lags of signals from the ultrasonic anemometer and the gas analyser, footprint estimations and calculating half-hourly fluxes. In EC data post-processing, determining periods with low turbulent mixing is a critical step [35]. Flux measurements over periods with insufficiently developed turbulence, i.e., low friction velocity (u^*) need to be detected and filtered out. This filtering procedure assured the exclusion of CO_2 measurements not representative of the biotic flux, i.e., net ecosystem exchange (NEE) [31,45,46].

The R software [47] package 'REddyProc' [48] was used to gap-fill the EC data at both sites, using marginal distribution sampling (MDS) [46]. This way, all missing values were replaced by the average value estimated under similar meteorological conditions within a specific time window. If no similar conditions were available within the starting time window of about 7 days, the window was increased. NEE was partitioned into GPP and ecosystem respiration (R_{eco}). The flux partitioning approach by [49] using daytime data was applied to estimate half-hourly GPP (µmol m^{-2} s^{-1}) values. The half-hourly GPP values were aggregated to obtain daily and monthly GPP sums.

2.3.2. Soil Water Content Simulations

Since the soil volumetric water content (SVWC) was not measured throughout the entire period of our analyses (the measurements started at the beginning of 2016), we used a simulated daily SVWC instead. We applied the soil water balance model R-4ET, an R package for Empirical Estimate of Ecosystem EvapoTranspiration as used in [50]. The calibration of the soil water balance model was carried out using Bayesian statistics implemented in the R package BayesianTools [51] with a Differential-Evolution Markov chain Monte Carlo sampler. In Bayesian calibration, the model input parameters are iteratively updated to provide a probability distribution of the calibrated parameters that represents the uncertainty in the measured data and in the model structure. The simulations were repeated 4.8×10^6 times, where the first 1.8×10^6 runs were based on the prior distribution, while the remaining 3×10^6 were constrained by the posterior distribution resulting from the first set of runs. A number of iterations treated as burn-in were set to 0, and the thinning parameter determining the interval in which values are recorded was set to 1. The high number of iterations was necessary to attain a good input parameter convergence with a narrow distribution. For inspecting the convergence, we used Gelman diagnostics with a criterion for the potential scale reduction factor being less than 1.2 for all parameters [52,53]. The final selection of parameters from their probability distributions was conducted by using a maximum a posteriori probability estimate, i.e., the value representing the mode of the posterior distribution.

The main model input variables included meteorological data and the leaf area index (see [50], for more details). The soil was stratified into 12 layers, which increased in thickness with increasing depth down to 3 m. Note that the soil layers were selected in the way that the layers used for optimization matched the depths of all sensors with an extra ±2.5 cm, considering the volume measured by the sensors. At both sites, the available SVWC measurements were available throughout the main part of the root zone. At CZ-BK1, the sensors CS616 (Campbell Scientific, Inc., Logan, UT, U.S.A.) were placed at 0.05, 0.1, 0.22, 0.34, and 0.42 m soil depths and measured since 2016. At CZ-RAJ, the same type of sensors were also placed at 0.05, 0.1, 0.2, 0.5, and 0.8 m depths and measured since 2016. The soil parameters, including SVWC at saturation, field capacity, wilting point, and saturated hydraulic conductivity, were optimized at all 12 depths [50]. Additional single parameters relevant for the SVWC simulations that were optimized included the following: rooting depth with the Beta parameter describing the root profile shape [54], surface resistance and the degree of isohydricity [55], the water interception capacity of leaf and bark area, and curve number representing a runoff parameter [50]. The uniform distribution of priors was used, where the lower and upper limits were set to be within ±50% of the values based on field measurements of the wilting point, field capacity, and saturated water content and ±100% for the remaining parameters that were estimated from the literature or from previous anecdotal analyses. To provide the model with sufficient spin up time to stabilize and ensure reliable and robust parametrization, the simulation was initiated at the beginning of 2010 with initial conditions of SVWC set to the field capacity estimated from soil texture (note that 2010 was one of the wettest years at both sites, according to a computed standardized precipitation-evapotranspiration index over the region). The overall simulation was done for the period 2010–2019 where the observed data for Bayesian calibration spanned the period 2016–2019.

The simulated SVWC averaged over all depths yielded a root mean square error of 0.037 m^3 m^{-3} and 0.017 m^3 m^{-3} for CZ-BK1 and CZ-RAJ, respectively, suggesting realistic SVWC estimates.

2.3.3. Drought Stress Determination

According to [56], drought refers to conditions characterized by low available soil moisture and high atmospheric VPD. As such, in quantifying the dry periods in our study, we used the Standardised Precipitation–Evapotranspiration Index (SPEI), which takes into account both precipitation and potential evapotranspiration for the determination of drought over the main growing season. The SPEI was calculated for various lags (1, 3, 6, 12, and 24 months) from monthly records. We used the period 1981–2010 as the baseline period for the SPEI computation. The computation of SPEI was performed according to [57,58], using the R package SPEI [59]. The SPEI represents an anomaly in the climatological water balance, which is given by the difference between precipitation and reference evapotranspiration. The input data for the SPEI derivation were obtained from climate reanalyses ERA5 monthly averages available at 0.25° spatial resolution, where the reference evapotranspiration was computed from air temperature at 2 m, air dew point temperature in 2 m, wind speed in 2 m converted from the original 10 m height using a logarithmic profile law, and shortwave incoming radiation, following the methodology by [60]. In determining the occurrence of dry and wet events in our study, SPEI classification based on [58] was used (Tables 3 and 4).

Table 3. The Standardised Precipitation-Evapotranspiration Index (SPEI) categories based on the classification of SPEI values by [58].

Drought/ Wet Severity	SPEI Value
Extremely Wet	≥2.00
Severely Wet	1.50–1.99
Moderately Wet	1.00–1.49
Near Normal	−0.99–(0.99)
Moderate Drought	−1.00–(−1.49)
Severe Drought	−1.50–(−1.99)
Extreme Drought	≤−2.00

Table 4. Categorization of dryness/wetness using Standardised Precipitation-Evapotranspiration Index (SPEI) indices for CZ-BK1 and CZ-RAJ stations in years (2014–2016).

YEARS	CZ-BK1		CZ-RAJ	
	SPEI VALUE	CLASS	SPEI VALUE	CLASS
2014	0.94	Near Normal	0.56	Near Normal
2015	−1.75	Severe Drought	−1.55	Severe Drought
2016	−0.20	Near Normal	−0.87	Near Normal

2.3.4. Light Response Curve Fitting

Light response curves (LRC) of daytime GPP were fitted at half-hourly time resolution using the logistic sigmoid approach by [61]:

$$GPP = 2 \times GPP_{max}(0.5 - \frac{1}{1 + exp(\frac{-2\alpha PAR}{GPP_{max}})}) \quad (3)$$

where PAR is the photosynthetically active radiation, α is the apparent quantum yield in mol (CO_2) mol^{-1} (phot.) that describes the effectivity of photosynthesis at low light conditions, and the GPP_{max} is the asymptotic maximum assimilation rate at the light saturation point in μmol m^2 s^{-1}. The fitting was done separately for half-hourly values from years with normal conditions (2014 and 2016) and the year with drought stress conditions (2015). To eliminate night-time measurements, we used a PAR threshold 10 μ, i.e., light compensation irradiance of 10 μmol m^2 s^{-1}, representing light compensation point of NEE.

Since VPD and soil moisture affect GPP [49,62], the effects of VPD and SVWC on the LRC residuals (GPP values after the removal of the strongest PAR dependence) were analysed for both spruce forest sites within the period under study. A piecewise regression was performed on the LRC residuals versus VPD and SVWC to determine the site-specific environmental stress thresholds at which GPP declined during the growing season periods of normal and drought affected years.

2.3.5. Piecewise Regression Analyses for the Assessment of Drought Effect

Since the LRC model does not account for changes in both VPD and SVWC, piecewise regression of residuals against these environmental variables were analysed. This allowed to determine the response of the studied spruce forest ecosystems with contrasting climates to severe drought conditions over the growing season for normal and drought-affected years. A piecewise regression analysis as part of the R package in 'segmented' [63] and the Davies test [64] were used to detect the breakpoints in the regression and also to test for the significant differences in slope parameters of a plot of residuals from the LRC model to drought conditions (high VPD and low SVWC).

3. Results

3.1. *Differences in Meteorological Conditions at the Experimental Sites*

During May–September from 2014 to 2016, the year 2015 was exceptionally dry compared to 2014 and 2016 (Table 4). Additionally, 2015 was characterized by significantly higher mean VPD values ($p < 0.001$; Table 5) during the main growing season as compared to the same period in 2014 and 2016 across both sites. Hence, the mean VPD and SVWC values were the best meteorological indicators of drought, while Tair was comparable for years 2015 and 2016. Consequently, the high atmospheric evaporative demand (high VPD values) and low soil moisture (low SVWC values) recorded in 2015 across both sites during the study period further indicated the dryness stress experienced in the main growing season of 2015.

Table 5. Mean values of vapour pressure deficit (VPD), air temperature (Tair) and soil volumetric water content (SVWC) during the main growing season in years for both spruce forest stands.

YEARS	CZ-BK1			CZ-RAJ		
	VPD (hPa)	Tair (°C)	SVWC (m^3 $^{-3}$)	VPD (hPa)	Tair (°C)	SVWC (m^3 $^{-3}$)
2014	7.6	13.4	0.28	8.7	14.9	0.21
2015	10.2	14.6	0.19	11.2	15.8	0.19
2016	7.7	14.4	0.25	9.3	16.0	0.22

There were observed statistically significant differences ($p < 0.05$; Mann–Whitney–Wilcoxon ranksum test) in the VPD and SVWC values between the normal and drought regimes across both sites (Figure 1). In addition, VPD was higher and SVWC lower at CZ-RAJ than at CZ-BK1, especially during the drought-affected year (DY).

3.2. *Light Response Curves of GPP at Different Climates*

The parameters of LRC for normal years (NY) 2014 and 2016 were compared with those for DY for both experimental forest sites (Figure 2; Table 6). Generally, the apparent quantum yield (α) at CZ-BK1 was found to be higher than that in CZ-RAJ. During the years with normal conditions, α in CZ-BK1 was observed to be 12% higher than in CZ-RAJ when PAR was less than 500 µmol m^{-2} s^{-1}. Moreover, during the DY of 2015, α at CZ-RAJ further declined by 25% as compared to that in CZ-BK1.

The maximum gross primary production at light saturation (GPP_{max}) for CZ-BK1 was found to be higher than that for CZ-RAJ during the entire study period. During the NY period, GPP_{max} in CZ-BK1 was 34% higher than in CZ-RAJ. However, during the dry year period, there were significant reductions in the GPP_{max} values recorded at both spruce forest stations with (18% decline in CZ-BK1 and 17% decline in CZ-RAJ) as compared to those recorded for the years with normal conditions. Additionally, it was further observed that the GPP_{max} value recorded at CZ-RAJ during the DY period was 33%

lower than that in CZ-BK1. Interestingly, the GPP_{max} value recorded in CZ-BK1 during the DY period was still higher than the GPP_{max} value recorded at CZ-RAJ during the years with normal conditions.

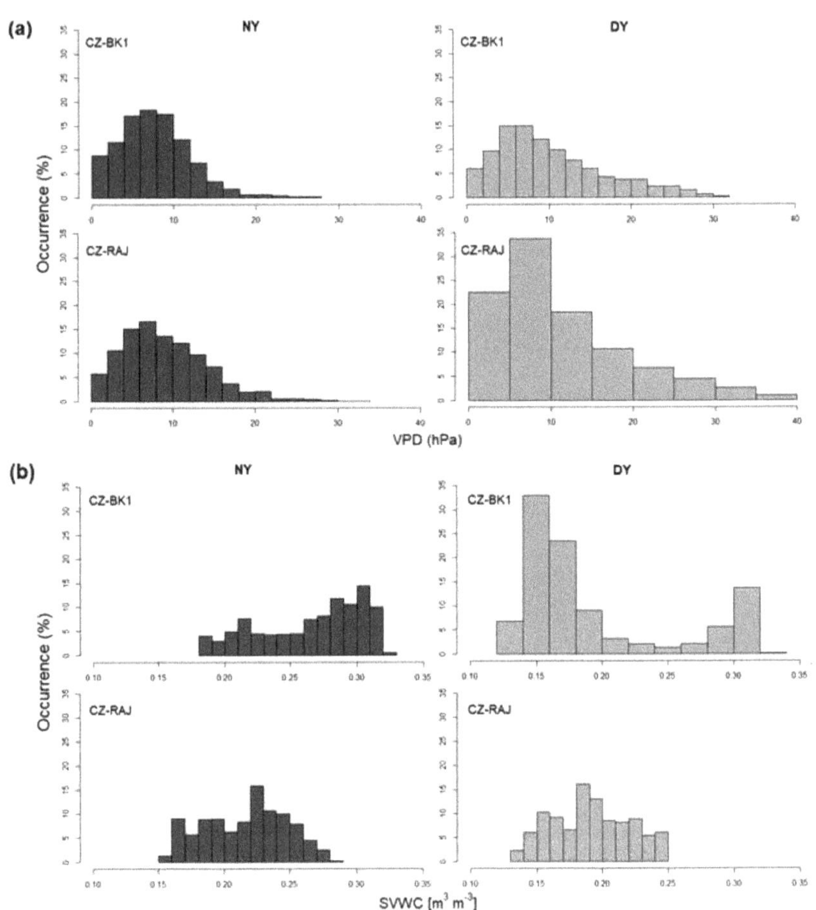

Figure 1. Histogram showing the frequency of occurrence of (**a**) vapour pressure deficit (VPD) and (**b**) soil volumetric water content (SVWC) conditions during May–September between years with normal conditions (dark colour) and drought stress (grey colour) in CZ-BK1 and CZ-RAJ.

Table 6. Light response curve parameters for the normal (2014 and 2016) and dry (2015) years for the wet and dry climates (CZ-BK1 and CZ-RAJ respectively) within the main growing season period of May–September. The apparent quantum yield (α), the maximum gross primary production at light saturation (GPP_{max}) and the coefficient of determination (R^2) are also shown.

Variants	CZ-BK1		CZ-RAJ	
	Years with Normal Conditions (2014 & 2016)	Dry Year (2015)	Years with Normal Conditions (2014 & 2016)	Dry Year (2015)
α [mol (CO_2) mol^{-1} (phot.)]	0.0383 ± 0.0003	0.0446 ± 0.0006	0.0338 ± 0.0003	0.0312 ± 0.0006
GPP_{max}[μmol m^{-2} s^{-1}]	26.91 ± 0.14	22.03 ± 0.13	17.77 ± 0.08	14.75 ± 0.12
R^2	0.88	0.78	0.78	0.73

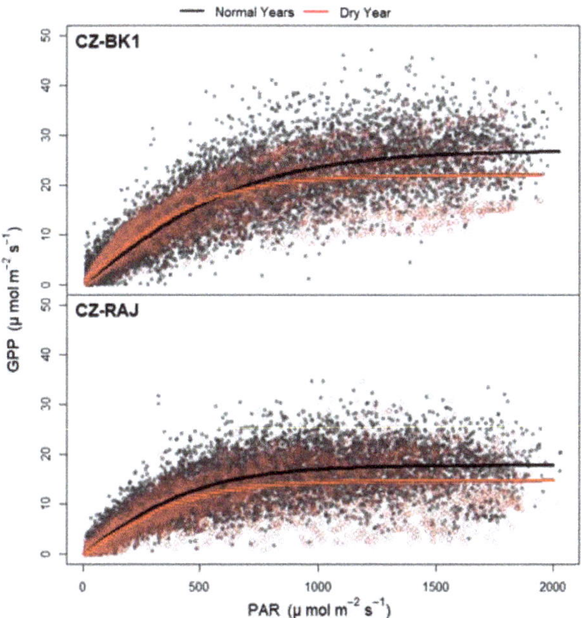

Figure 2. Response of gross primary productions to photosynthetically active radiation during the years with normal conditions (black) and affected by drought stress (red) in CZ-BK1 and CZ-RAJ. The half-hourly GPP values (points) were fitted using the logistic sigmoid light response curves (lines).

3.3. *Response of Light Response Curve Residuals to VPD and SVWC at Different Climates*

The applied LRC model reflects only the light response of GPP and thus, its residuals can be used to assess the influence of additional meteorological parameters (i.e., VPD and SVWC) that are known to affect GPP (Figures 3 and 4). The residual analysis confirmed that the residuals consistently tended to be more negative with increasing VPD and decreasing SVWC (dry conditions). VPD had a significant and stronger effect on GPP than SVWC across both sites, as seen from the piecewise regression analysis, using the residuals from the LRC for the years under study. However, during the DY period, both VPD and SVWC had significant effects on GPP at CZ-RAJ. For CZ-BK1, there was a minimal effect of SVWC on GPP during the years with normal conditions.

Generally, the relationship between the residuals and changes in VPD and SVWC revealed a biphasic response to drought, except for the DY period in CZ-RAJ in Figure 3. All the breakpoints from the piecewise regression were found to be statistically significant (Table 7). However, steeper slopes after the VPD breakpoint values from the initial slope were mostly observed for both DY and NY periods in CZ-RAJ and only for DY period in CZ-BK1. This shows that during all the years under study, GPP decreased at a faster rate after the breakpoint in CZ-RAJ than in CZ-BK1, except for the DY period when there was a significant decline in GPP at a faster rate after the breakpoint across both sites.

3.4. *Impact of Drought on Ecosystem Carbon Fluxes at Monthly Timescale*

The monthly averages of daily sums of GPP and R_{eco} were analysed separately to assess whether the observed net ecosystem productivity (NEP) reduction during drought was mainly a consequence of increased R_{eco}, reduced GPP, or contribution of both across the forest ecosystems (Figures 5–7).

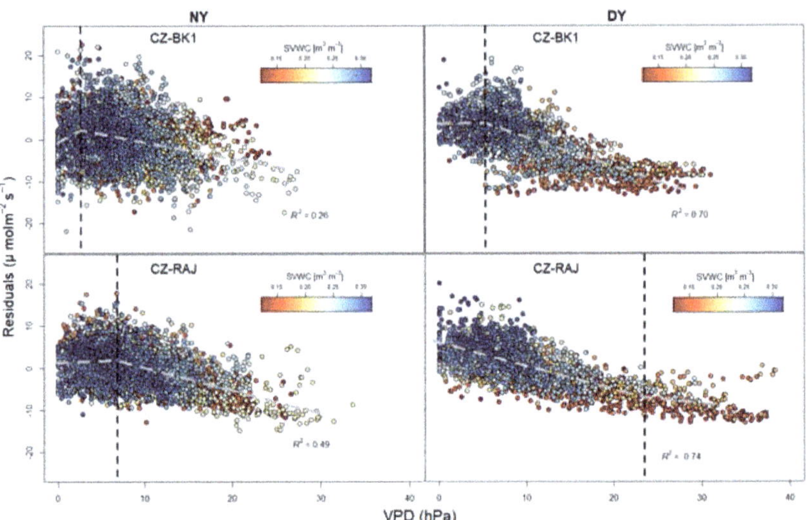

Figure 3. Relationship between the light response curve (LRC) residuals of the light response curve of gross primary production and the vapour pressure deficit (VPD) at the spruce forest sites in CZ-BK1 and CZ-RAJ for the normal (NY) and dry years (DY). Red represents low SVWC conditions and blue shows high SVWC conditions. The grey dashed lines represent the piecewise regression model slope, whereas the black dashed lines show the breakpoint values.

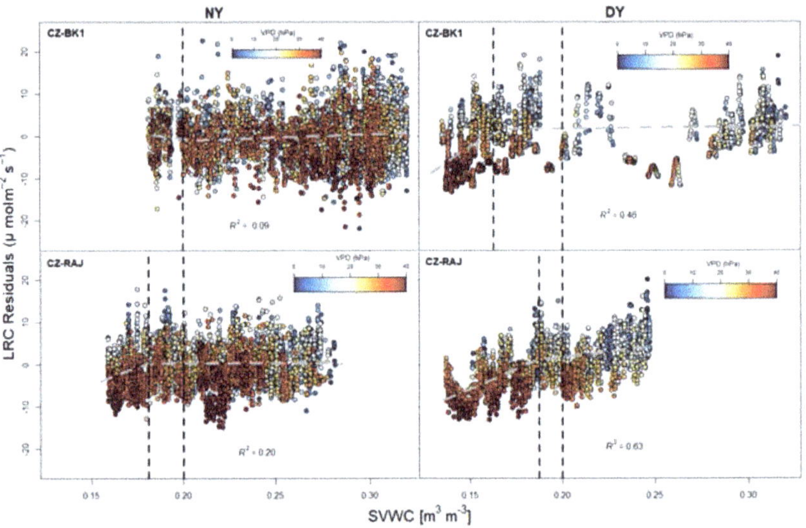

Figure 4. Relationship between the light response curve (LRC) residuals of the light response curve of gross primary production and the soil volumetric water content (SVWC) at the spruce forest sites in CZ-BK1 and CZ-RAJ for the normal (NY) and dry years (DY). Red represents high VPD conditions and blue shows low VPD conditions. The grey dashed lines represent the piecewise regression model slope, whereas the black dashed lines show the breakpoint values.

Table 7. Breakpoints and Slopes from the piecewise regression of the light response curve residuals to vapour pressure deficit (VPD) and soil volumetric water content (SVWC) for the period of May–September.

Variants	CZ-BK1		CZ-RAJ	
	Years with Normal Conditions (2014 & 2016)	Dry Year (2015)	Years with Normal Conditions (2014 & 2016)	Dry Year (2015)
VPD [hPa]	2.6 ***	5.3 ***	6.8 ***	23.5 ***
Slope before breakpoint in VPD	1.12 ± 0.30	0.02 ± 0.12	0.08 ± 0.05	−0.55 ± 0.01
Slope after breakpoint in VPD	−0.39 ± 0.02	−0.70 ± 0.02	−0.54 ± 0.02	−0.22 ± 0.06
SVWC [m^3 m^{-3}]	0.20 ***	0.16 ***	0.18 ***	0.19 ***
Slope before breakpoint in SVWC	−85.49 ± 68.37	305.61 ± 23.43	164.61 ± 25.14	187.22 ± 8.78
Slope after breakpoint in SVWC	14.75 ± 2.60	5.03 ± 2.51	−1.27 ± 2.84	52.16 ± 6.24

Signif. code for the breakpoint values: $p < 0.001$ '***'.

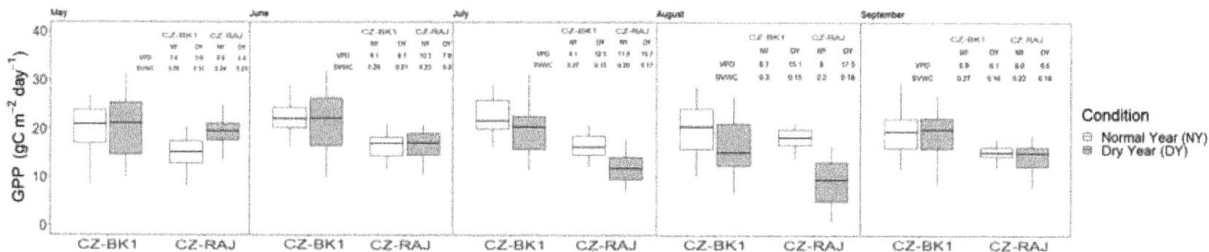

Figure 5. Monthly averages of daily sums of gross primary productivity (GPP) for May–September of the normal years (NY) and dry year (DY) in CZ-BK1 and CZ-RAJ. The tables within the figure represent the mean monthly vapour pressure deficit (VPD) and soil volumetric water content (SVWC) values from May to September for each forest station.

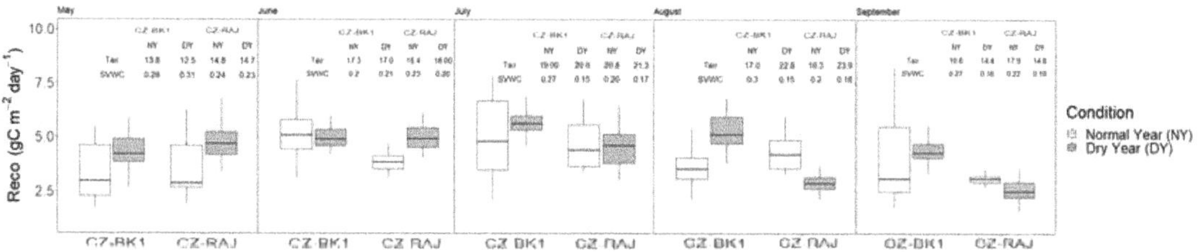

Figure 6. Monthly averages of daily sums of ecosystem respiration (R_{eco}) for May–September of the normal years (NY) and dry year (DY) in CZ-BK1 and CZ-RAJ. The tables within the figure represent the mean monthly air temperature (Tair) and soil volumetric water content (SVWC) values from May to September for each forest station.

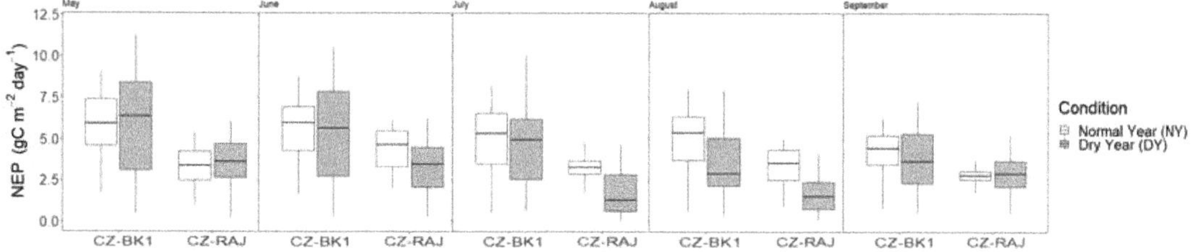

Figure 7. Monthly averages of daily sums of NEP for May–September of the normal years (NY) and dry year (DY) in CZ-BK1 and CZ-RAJ.

It was observed that there was a significant decline in the total GPP by 14% and 6% during the main growing season period of the dry year for CZ-RAJ and CZ-BK1, respectively. There were also observed significant statistical differences ($p < 0.001$) in the mean monthly GPP values recorded for July and August during the dry year as compared to the years characterized by normal conditions at both forest sites, especially for CZ-RAJ (Figure 5). Moreover, this significant decline in the mean monthly GPP values at both spruce forest stands during the months of July and August coincided with high daily mean VPD values (>12 hPa) and low SVWC values (<0.16 m^3 m^{-3}) as compared to the same period for the adjacent years with normal conditions.

However, an increase in the monthly mean R_{eco} values was observed for July to September of the dry year as compared to the adjacent normal years within CZ-BK1 but with no significant statistical differences, except for the month of August (Figure 6). These three months within 2015 were also characterized by high mean Tair (>20 °C) and low mean SVWC (<0.19 m^3 m^{-3}) as compared to the two other adjacent years. In addition, during the DY period within CZ-RAJ, the monthly mean R_{eco} values were found to have declined significantly, as compared to the adjacent NY periods. We also observed a 38% significant decline in the mean monthly NEP values during the dry year period for the dry spruce forest in Rájec, compared to a 12% decrease in the mean monthly NEP values for the humid spruce forest in Bílý Kříž (Figure 7). There were also statistically significant differences ($p < 0.001$) in the mean monthly NEP for the months of July and August across both spruce forest stations. During these months, the observed mean monthly NEP values largely declined during the DY period, as compared to the same period during the NY, especially in CZ-RAJ.

To summarize, the large decline in GPP, R_{eco} and NEP during the dry year period (especially from July to August) for CZ-RAJ showed that the impact of the drought was more severe in CZ-RAJ than in CZ-BK1. However, during July–August of the DY period, R_{eco} at CZ-BK1 significantly increased as compared to that in NY.

4. Discussion

The study sought to assess the different effects of the 2015 summer drought on wet and dry spruce forest ecosystems. Our findings corroborate the earlier published results by [27–29] that the year 2015 was characterized by severe drought conditions across Europe (Figure 1 and Table 4). Furthermore, it was found that the months of July–August of 2015 were the most affected by high mean Tair (>20 °C), high mean VPD (>10 hPa) and low mean SVWC values (<0.19 m^3 m^{-3}) across both spruce forest stands. This shows that the drought effect in 2015 was severe in the months of July–August. Moreover, the results of this study also show that the rate of the forest ecosystem photosynthesis was significantly reduced (likely through the immediate closure of the stomata to protect the tree from desiccation) by the high VPD, Tair and soil water deficit during these two months in 2015. As such, there was an observed strong decline in both forest GPP and NEP across both spruce forest stands during the DY period of 2015, especially in CZ-RAJ compared to CZ-BK1 (M). This strong decline in forest GPP and NEP at CZ-RAJ was mainly due to the high atmospheric evaporative demand coupled with the low SVWC conditions experienced at this forest ecosystem, as compared to that in CZ-BK1 [30]. However, due to the humid climatic conditions at CZ-BK1, an increase in Tair during the months of July–August in 2015 only aided the increase in the kinetics of the enzymes participating in microbial decomposition and root respiration under warm conditions, thereby increasing the overall forest ecosystem respiration [65–71].

Additionally, results from the piecewise regression analysis using the residuals from the LRC model highlight the effects of both VPD and SVWC on forest ecosystem GPP. This is because both photosynthesis and transpiration are mediated by stomatal conductance, which are affected by these environmental variables. However, a steeper decline in forest ecosystem GPP (Figure 3 and Table 7) was observed with high VPD values across both forest stands, even under non-drought years (when SVWC was non-limiting). This is consistent with recent studies, which show that high VPD values aggravate drought effects in forests, due to the abrupt changes over very short timescales within a day, even without dry soil conditions [7,72,73]. This further explains the strong suppression of GPP by high VPD on even non-drought years (2014 and 2016), as the SPEI (for determining dryness) only captures changes in the contributing environmental drivers (temperature and soil moisture) for longer time scales of weeks or months [74,75]. Thus, we would recommend to include the influence of these abrupt changes in VPD on the rate of photosynthesis with SVWC limitations in future LRC models when analysing the impact of drought on GPP, especially for different forest ecosystems that are exposed to regular strong edaphic droughts [49,62,76,77].

5. Conclusions

The study shows that atmospheric constraints increase the vulnerability of the Norway spruce forest to the severity of drought, especially at sites with a moderately dry climate that are characterized by precipitation that is typically equal or smaller than the atmospheric evaporative demand. It also further highlighted the strong influence of VPD on carbon uptake, which was further worsened by the decline in soil moisture. The effect of SVWC on GPP was especially noticeable during severe drought conditions within the DY period. Consequently, with regards to climate change, our results suggest that elevated temperatures will further exacerbate the drought impacts on forest (Norway spruce) ecosystems at sites with precipitation levels equal or smaller than the atmospheric evaporative demand, such as CZ-RAJ. The decline in GPP and NEP in 2015 found in our study thus questions not only the sustainable productivity, but also the existence of Norway spruce per se in such areas, considering the prolonged period of drought in future climatic conditions. The results of this study may help decision makers to quantitatively assess the performance of Norway spruce in future climatic conditions.

Author Contributions: Conceptualisation, C.M. and L.Š.; methodology, C.M., L.Š., M.F., M.P. and M.V.M.; software, C.M., L.K. and L.Š.; validation, M.F., M.A., G.J., J.D.M. and N.K., E.K.N.; formal analysis, C.M. and L.Š.; investigation, C.M. and L.Š.; resources, M.P. and L.Š.; data curation, L.Š. and G.J.; writing—original draft preparation, C.M.; writing—review and editing, L.Š., M.F., M.A., G.J., L.F., J.D.M., N.K., E.K.N. and M.V.M.; visualisation, L.K.; supervision, L.Š. and M.V.M.; project administration, M.P. and L.Š.; funding acquisition, M.P. and M.V.M. All authors have read and agreed to the published version of the manuscript.

Funding: This research and the APC was funded by the Ministry of Education, Youth and Sports of the Czech Republic (CR) within the CzeCOS program, grant number LM2018123.

Institutional Review Board Statement: Not applicable.

Informed Consent Statement: Not applicable.

Data Availability Statement: The data presented in this study are available on FLUXNET and could also be made available on request from the corresponding author.

Acknowledgments: The authors wish to thank Radek Czerný who assisted in the eddy covariance data processing. This work was supported by the Ministry of Education, Youth and Sports of the Czech Republic (CR) within the CzeCOS program, grant number LM2018123. L.Š. was supported by the Ministry of Education, Youth and Sports of CR within Mobility CzechGlobe2 (CZ.02.2.69/0.0/0.0/18 053/0016924). M.F., L.Š. and L.F. acknowledges the support by the Ministry of Education, Youth and Sports of the Czech Republic for SustES—Adaptation strategies for sustainable ecosystem services and food security under adverse environmental conditions (CZ.02.1.01/0.0/0.0/16 019/0000797). M.V.M. acknowledges the support by VEGA project 2/0013/17 (Slovak Agency for Scientific Support): The role of ecosystem services in support of landscape conservation under the global change.

Conflicts of Interest: The authors declare no conflict of interest. The funders had no role in the design of the study; in the collection, analyses, or interpretation of data; in the writing of the manuscript, or in the decision to publish the results.

References

1. Rossi, G.; Cancelliere, A.; Pereira, L.S. (Eds.) *Tools for Drought Mitigation in Mediterranean Regions*; Springer Science and Business Media: Berlin, Germany, 2003; Volume 44.
2. Huntington, T.G. Evidence for intensification of the global water cycle: Review and synthesis. *J. Hydrol.* **2003**, *319*, 83–95. [CrossRef]
3. Bonan, G.B. Forests and climate change: Forcings, feedbacks, and the climate benefits of forests. *Science* **2008**, *320*, 1444–1449. [CrossRef]
4. Allen, C.D.; Macalady, A.K.; Chenchouni, H.; Bachelet, D.; McDowell, N.; Vennetier, M.; Gonzalez, P. A global overview of drought and heat-induced tree mortality reveals emerging climate change risks for forests. *For. Ecol. Manag.* **2010**, *259*, 660–684. [CrossRef]
5. Marek, M.V.; Janouš, D.; Taufarová, K.; Havránková, K.; Pavelka, M.; Kaplan, V.; Marková, I. Carbon exchange between ecosystems and atmosphere in the Czech Republic is affected by climate factors. *Environ. Pollut.* **2011**, *159*, 1035–1039. [CrossRef]
6. Lal, R.; Smith, P.; Jungkunst, H.F.; Mitsch, W.J.; Lehmann, J.; Nair, P.R.; Skorupa, A.L. The carbon sequestration potential of terrestrial ecosystems. *J. Soil Water Conserv.* **2018**, *73*, 145A–152A. [CrossRef]
7. Novick, K.A.; Ficklin, D.L.; Stoy, P.C.; Williams, C.A.; Bohrer, G.; Oishi, A.C.; Papuga, S.A.; Blanken, P.D.; Noormets, A.; Sulman, B.N.; et al. The increasing importance of atmospheric demand for ecosystem water and carbon fluxes. *Nat. Clim. Chang.* **2016**, *6*, 1023–1027. [CrossRef]
8. Kowalska, N.; Šigut, L.; Stojanović, M.; Fischer, M.; Kyselova, I.; Pavelka, M. Analysis of floodplain forest sensitivity to drought. *Philos. Trans. R. Soc. B* **2020**, *375*, 20190518. [CrossRef]
9. Beer, C.; Reichstein, M.; Tomelleri, E.; Ciais, P.; Jung, M.; Carvalhais, N.; Rödenbeck, C.; Arain, M.A.; Baldocchi, D.; Bonan, G.B.; et al. Terrestrial gross carbon dioxide uptake: Global distribution and covariation with climate. *Science* **2010**, *329*, 834–838. [CrossRef]
10. Pan, Y.; Birdsey, R.A.; Fang, J.; Houghton, R.; Kauppi, P.E.; Kurz, W.A.; Phillips, O.L.; Shvidenko, A.; Lewis, S.L.; Canadell, J.G.; et al. A large and persistent carbon sink in the world's forests. *Science* **2011**, *333*, 988–993. [CrossRef]
11. Cox, P.M.; Pearson, D.; Booth, B.B.; Friedlingstein, P.; Huntingford, C.; Jones, C.D.; Luke, C.M. Sensitivity of tropical carbon to climate change constrained by carbon dioxide variability. *Nature* **2013**, *494*, 341. [CrossRef]
12. Le Quéré, C.; Andres, R.J.; Boden, T.; Conway, T.; Houghton, R.A.; House, J.I.; Marland, G.; Peters, G.P.; Van der Werf, G.; Ahlström, A.; et al. The global carbon budget 1959–2011. *Earth Syst. Sci. Data* **2013**, *5*, 165–185. [CrossRef]
13. Krejza, J.; Cienciala, E.; Světlík, J.; Bellan, M.; Noyer, E.; Horáček, P.; Štěpánek, P.; Marek, M.V. Evidence of climate-induced stress of Norway spruce along elevation gradient preceding the current dieback in Central Europe. *Trees* **2020**, *35*, 103–119. [CrossRef]
14. Zeng, N.; Qian, H.; Roedenbeck, C.; Heimann, M. Impact of 1998–2002 midlatitude drought and warming on terrestrial ecosystem and the global carbon cycle. *Geophys. Res. Lett.* **2005**, *32*. [CrossRef]
15. Dai, A. Increasing drought under global warming in observations and models. *Nat. Clim. Chang.* **2013**, *3*, 52. [CrossRef]
16. Reichstein, M.; Bahn, M.; Ciais, P.; Frank, D.; Mahecha, M.D.; Seneviratne, S.I.; Zscheischler, J.; Beer, C.; Buchmann, N.; Frank, D.C.; et al. Climate extremes and the carbon cycle. *Nature* **2013**, *500*, 287. [CrossRef] [PubMed]
17. Wang, L.; Chen, W.; Zhou, W. Assessment of future drought in Southwest China based on CMIP5 multimodel projections. *Adv. Atmos. Sci.* **2014**, *31*, 1035–1050. [CrossRef]
18. Xie, Z.; Wang, L.; Jia, B.; Yuan, X. Measuring and modeling the impact of a severe drought on terrestrial ecosystem CO_2 and water fluxes in a subtropical forest. *J. Geophys. Res. Biogeosci.* **2016**, *121*, 2576–2587. [CrossRef]
19. Zang, C.; Hartl-Meier, C.; Dittmar, C.; Rothe, A.; Menzel, A. Patterns of drought tolerance in major European temperate forest trees: Climatic drivers and levels of variability. *Glob. Chang. Biol.* **2014**, *20*, 3767–3779. [CrossRef]
20. De Boeck, H.J.; Verbeeck, H. Drought-associated changes in climate and their relevance for ecosystem experiments and models. *Biogeosciences* **2011**, *8*, 1121–1130. [CrossRef]
21. Hui, D.; Deng, Q.; Tian, H.; Luo, Y. Climate change and carbon sequestration in forest ecosystems. In *Handbook of Climate Change Mitigation and Adaptation*; Chen, W.Y., Suzuki, T., Lackner, M., Eds.; Springer International Publishing: New York, NY, USA, 2017 ; pp. 555–594.
22. Qie, L.; Lewis, S.L.; Sullivan, M.J.; Lopez-Gonzalez, G.; Pickavance, G.C.; Sunderland, T.; Ashton, P.; Hubau, W.; Salim, K.A.; Aiba, S.I.; et al. Long-term carbon sink in Borneo's forests halted by drought and vulnerable to edge effects. *Nat. Commun.* **2017**, *8*, 1966. [CrossRef]
23. Ciais, P.; Reichstein, M.; Viovy, N.; Granier, A.; Ogée, J.; Allard, V.; Aubinet, M.; Buchmann, N.; Bernhofer, C.; Carrara, A.; et al. Europe-wide reduction in primary productivity caused by the heat and drought in 2003. *Nature* **2005**, *437*, 529. [CrossRef]
24. Granier, A.; Reichstein, M.; Breda, N.; Janssens, I.A.; Falge, E.; Ciais, P.; Grünwald, T.; Aubinet, M.; Berbigier, P.; Bernhofer, C.; et al. Evidence for soil water control on carbon and water dynamics in European forests during the extremely dry year: 2003. *Agric. For. Meteorol.* **2007**, *143*, 123–145. [CrossRef]
25. Trnka, M.; Hlavinka, P.; Možný, M.; Semerádová, D.; Štěpánek, P.; Balek, J.; Bartošová, L.; Zahradníček, P.; Bláhová, M.; Skalak, P.; et al. Czech Drought Monitor System for Monitoring and Forecasting Agricultural Drought and Drought Impacts. *Int. J. Climatol.* **2020**, *40*, 5941–5958. [CrossRef]
26. Orth, R.; Seneviratne, S.I. Introduction of a simple-model-based land surface dataset for Europe. *Environ. Res. Lett.* **2015**, *10*, 0442012.

27. Orth, R.; Zscheischler, J.; Seneviratne, S.I. Record dry summer in 2015 challenges precipitation projections in Central Europe. *Sci. Rep.* **2016**, *6*, 28334. [CrossRef] [PubMed]
28. Van Lanen, H.A.; Laaha, G.; Kingston, D.G.; Gauster, T.; Ionita, M.; Vidal, J.P.; Vlnas, R.; Tallaksen, L.M.; Stahl, K.; Hannaford, J.; et al. Hydrology needed to manage droughts: The 2015 European case. *Hydrol. Process.* **2016**, *30*, 3097–3104. [CrossRef]
29. Ionita, M.; Tallaksen, L.M.; Kingston, D.G.; Stagge, J.H.; Laaha, G.; Van Lanen, H.A.; Scholz, P.; Chelcea, S.M.; Haslinger, K. The European 2015 drought from a climatological perspective. *Hydrol. Earth Syst. Sci.* **2017**, *21*, 1397–1419. [CrossRef]
30. Krupková, L.; Havránková, K.; Krejza, J.; Sedlák, P.; Marek, M.V. Impact of water scarcity on spruce and beech forests. *J. For. Res.* **2018**, *30*, 899–909. [CrossRef]
31. McGloin, R.; Šigut, L.; Havránková, K.; Dušek, J.; Pavelka, M.; Sedlák, P. Energy balance closure at a variety of ecosystems in Central Europe with contrasting topographies. *Agric. For. Meteorol.* **2018**, *248*, 418–431. [CrossRef]
32. Kenderes, K.; Mihók, B.; Standovár, T. Thirty years of gap dynamics in a Central European beech forest reserve. *Forestry* **2008**, *81*, 111–123. [CrossRef]
33. Nadezhdina, N.; Urban, J.; Čermák, J.; Nadezhdin, V.; Kantor, P. Comparative study of long-term water uptake of Norway spruce and Douglas-fir in Moravian upland. *J. Hydrol. Hydromech.* **2014**, *62*, 1–6. [CrossRef]
34. Kmet, J.; Ditmarová, L.; Kurjak, D.; Priwitzer, T. Physiological response of Norway spruce foliage in the drought vegetation period 2009. *Beskydy* **2011**, *4*, 109–118.
35. Aubinet, M.; Vesala, T.; Papale, D. (Eds.) *Eddy Covariance: A Practical Huide to Measurement and Data Analysis*; Springer Science and Business Media: Berlin, Germany, 2012.
36. Baldocchi, D.D. Assessing the eddy covariance technique for evaluating carbon dioxide exchange rates of ecosystems: Past, present and future. *Glob. Chang. Biol.* **2003**, *9*, 479–492. [CrossRef]
37. Sedlák, P.; Aubinet, M.; Heinesch, B.; Janouš, D.; Pavelka, M.; Potužníková, K.; Yernaux, M. Night-time airflow in a forest canopy near a mountain crest. *Agric. For. Meteorol.* **2010**, *150*, 736–744. [CrossRef]
38. Urban, O.; Klem, K.; Holišová, P.; Šigut, L.; Šprtová, M.; Teslová-Navrátilová, P.; Zitová, M.; Špunda, V.; Marek, M.V.; Grace, J. Impact of elevated CO_2 concentration on dynamics of leaf photosynthesis in Fagus sylvatica is modulated by sky conditions. *Environ. Pollut.* **2014**, *185*, 271–280. [CrossRef] [PubMed]
39. Monteith, J.; Unsworth, M. *Principles of Environmental Physics: Plants, Animals, and the Atmosphere*; Academic Press: Cambridge, MA, USA, 2013.
40. Foken, T.; Leuning, R.; Oncley, S.R.; Mauder, M.; Aubinet, M. Corrections and data quality control. In *Eddy Covariance*; Springer: Dordrecht, The Netherlands, 2012; pp. 85–131.
41. Wilczak, J.M.; Oncley, S.P.; Stage, S.A. Sonic anemometer tilt correction algorithms. *Bound. Layer Meteorol.* **2001**, *99*, 127–150. [CrossRef]
42. Ibrom, A.; Dellwik, E.; Larsen, S.E.; Pilegaard, K.I.M. On the use of the Webb-Pearman-Leuning theory for closed-path eddy correlation measurements. *Tellus B Chem. Phys. Meteorol.* **2007**, *59*, 937–946. [CrossRef]
43. Horst, T.W.; Lenschow, D.H. Attenuation of scalar fluxes measured with spatially-displaced sensors. *Bound. Layer Meteorol.* **2009**, *130*, 275–300. [CrossRef]
44. Moncrieff, J.B.; Jarvis, P.G.; Valentini, R. Canopy fluxes. In *Methods in Ecosystem Science*; Springer: New York, NY, USA, 2000; pp. 161–180.
45. Papale, D.; Reichstein, M.; Aubinet, M.; Canfora, E.; Bernhofer, C.; Kutsch, W.; Longdoz, B.; Rambal, S.; Valentini, R.; Vesala, T.; et al. Towards a standardized processing of Net Ecosystem Exchange measured with eddy covariance technique: Algorithms and uncertainty estimation. *Biogeosciences* **2006**, *3*, 571–583. [CrossRef]
46. Reichstein, M.; Falge, E.; Baldocchi, D.; Papale, D.; Aubinet, M.; Berbigier, P.; Bernhofer, C.; Buchmann, N.; Gilmanov, T.; Granier, A.; et al. On the separation of net ecosystem exchange into assimilation and ecosystem respiration: Review and improved algorithm. *Glob. Chang. Biol.* **2005**, *11*, 1424–1439. [CrossRef]
47. R Core Team. *R: A Language and Environment for Statistical Computing*; R Foundation for Statistical Computing: Vienna, Austria, 2018.
48. Wutzler, T.; Lucas-Moffat, A.; Migliavacca, M.; Knauer, J.; Sickel, K.; Šigut, L.; Menzer, O.; Reichstein, M. Basic and extensible post-processing of eddy covariance flux data with REddyProc. *Biogeosciences* **2018**, *15*, 5015–5030. [CrossRef]
49. Lasslop, G.; Reichstein, M.; Papale, D.; Richardson, A.D.; Arneth, A.; Barr, A.; Stoy, P.; Wohlfahrt, G. Separation of net ecosystem exchange into assimilation and respiration using a light response curve approach: Critical issues and global evaluation. *Glob. Chang. Biol.* **2010**, *16*, 187–208. [CrossRef]
50. Fischer, M.; Zenone, T.; Trnka, M.; Orság, M.; Montagnani, L.; Ward, E.J.; Tripathi, A.M.; Hlavinka, P.; Seufert, G.; Žalud, Z.; et al. Water requirements of short rotation poplar coppice: Experimental and modelling analyses across Europe. *Agric. For. Meteorol.* **2018**, *250*, 343–360. [CrossRef]
51. Hartig, F.; Minunno, F.; Paul, S. BayesianTools: General-Purpose MCMC and SMC Samplers and Tools for Bayesian Statistics. R Package Version 0.1.6. 2019. Available online: https://CRAN.R-project.org/package=BayesianTools (accessed on 20 March 2021)
52. Gelman, A.; Rubin, D.B. Inference from iterative simulation using multiple sequences. *Stat. Sci.* **1992**, *7*, 457–472. [CrossRef]

53. Brooks, S.P.; Gelman, A. General methods for monitoring convergence of iterative simulations. *J. Comput. Graph. Stat.* **1998**, *7*, 434–455.
54. Jackson, R.B.; Canadell, J.; Ehleringer, J.R.; Mooney, H.A.; Sala, O.E.; Schulze, E.D. A global analysis of root distributions for terrestrial biomes. *Oecologia* **1996**, *108*, 389–411. [CrossRef] [PubMed]
55. Oren, R.; Sperry, J.S.; Katul, G.G.; Pataki, D.E.; Ewers, B.E.; Phillips, N.; Schäfer, K.V.R. Survey and synthesis of intra-and interspecific variation in stomatal sensitivity to vapour pressure deficit. *Plant Cell Environ.* **1999**, *22*, 1515–1526. [CrossRef]
56. Mishra, A.K.; Singh, V.P. A review of drought concepts. *J. Hydrol.* **2010**, *391*, 202–216. [CrossRef]
57. Beguería, S.; Vicente-Serrano, S.M.; Reig, F.; Latorre, B. Standardized precipitation evapotranspiration index (SPEI) revisited: Parameter fitting, evapotranspiration models, tools, datasets and drought monitoring. *Int. J. Climatol.* **2014**, *34*, 3001–3023. [CrossRef]
58. Vicente-Serrano, S.M.; Beguería, S.; López-Moreno, J.I. A Multiscalar Drought Index Sensitive to Global Warming: The Standardized Precipitation Evapotranspiration Index. *J. Clim.* **2010**, *23*, 1696–1718. [CrossRef]
59. Beguería, S.; Vicente-Serrano, S.M. SPEI: Calculation of the Standardised Precipitation-Evapotranspiration Index. R Package Version 1.7. 2017. Available online: https://CRAN.R-project.org/package=SPEI (accessed on 13 January 2021).
60. Allen, R.G.; Pereira, L.S.; Raes, D.; Smith, M. Crop evapotranspiration—Guidelines for computing crop water requirements. *FAO Irrig. Drain. Pap.* **1998**, *56*, 290.
61. Moffat, A.M. A New Methodology to Interpret High Resolution Measurements of Net Carbon Fluxes between Terrestrial Ecosystems and the Atmosphere. Ph.D. Thesis, Friedrich-Schiller-Universität Jena, Fakultät für Mathematik und Informatik, Jena, Germany, 2012.
62. Migliavacca, M.; Reichstein, M.; Richardson, A.D.; Colombo, R.; Sutton, M.A.; Lasslop, G.; Tomelleri, E.; Wohlfahrt, G.; Carvalhais, N.; Cescatti, A.; et al. Semiempirical modeling of abiotic and biotic factors controlling ecosystem respiration across eddy covariance sites. *Glob. Chang. Biol.* **2011**, *17*, 390–409. [CrossRef]
63. Muggeo, V.M. Estimating regression models with unknown break-points. *Stat. Med.* **2003**, *22*, 3055–3071. [CrossRef]
64. Davies, R.B. Hypothesis testing when a nuisance parameter is present only under the alternative: Linear model case. *Biometrika* **2002**, *89*, 484–489. [CrossRef]
65. Reichstein, M.; Ciais, P.; Papale, D.; Valentini, R.; Running, S.; Viovy, N.; Cramer, W.; Granier, A.; Ogee, J.; Allard, V.; et al. Reduction of ecosystem productivity and respiration during the European summer 2003 climate anomaly: A joint flux tower, remote sensing and modelling analysis. *Glob. Chang. Biol.* **2007**, *13*, 634–651. [CrossRef]
66. Jassal, R.S.; Black, T.A.; Chen, B.; Roy, R.; Nesic, Z.; Spittlehouse, D.L.; Trofymow, J.A. N_2O emissions and carbon sequestration in a nitrogen-fertilized Douglas fir stand. *J. Geophys. Res. Biogeosci.* **2008**, *113*. [CrossRef]
67. Mahecha, M.D.; Reichstein, M.; Carvalhais, N.; Lasslop, G.; Lange, H.; Seneviratne, S.I.; Vargas, R.; Ammann, C.; Arain, M.A.; Cescatti, A.; et al. Global convergence in the temperature sensitivity of respiration at ecosystem level. *Science* **2010**, *329*, 838–840. [CrossRef]
68. Wu, S.H.; Jansson, P.E.; Kolari, P. Modeling seasonal course of carbon fluxes and evapotranspiration in response to low temperature and moisture in a boreal Scots pine ecosystem. *Ecol. Model.* **2011**, *222*, 3103–3119. [CrossRef]
69. Frank, D.; Reichstein, M.; Bahn, M.; Thonicke, K.; Frank, D.; Mahecha, M.D.; Smith, P.; Van der Velde, M.; Vicca, S.; Babst, F.; et al. Effects of climate extremes on the terrestrial carbon cycle: Concepts, processes and potential future impacts. *Glob. Chang. Biol.* **2015**, *21*, 2861–2880. [CrossRef] [PubMed]
70. Van Gorsel, E.; Wolf, S.; Cleverly, J.; Isaac, P.; Haverd, V.; Ewenz, C.; Arndt, S.; Beringer, J.; Resco de Dios, V.; Evans, B.J.; et al. Carbon uptake and water use in woodlands and forests in southern Australia during an extreme heat wave event in the "angry Summer" of 2012/2013. *Biogeosciences* **2016**, *13*, 5947–5964. [CrossRef]
71. von Buttlar, J.; Zscheischler, J.; Rammig, A.; Sippel, S.; Reichstein, M.; Knohl, A.; Jung, M.; Menzer, O.; Arain, M.A.; Buchmann, N.; et al. Impacts of droughts and extreme-temperature events on gross primary production and ecosystem respiration: A systematic assessment across ecosystems and climate zones. *Biogeosciences* **2018**, *15*, 1293–1318. [CrossRef]
72. Ruehr, N.K.; Law, B.E.; Quandt, D.; Williams, M. Effects of heat and drought on carbon and water dynamics in a regenerating semi-arid pine forest: A combined experimental and modeling approach. *Biogeosciences* **2014**, *11*, 4139–4156. [CrossRef]
73. Sulman, B.N.; Roman, D.T.; Yi, K.; Wang, L.; Phillips, R.P.; Novick, K.A. High atmospheric demand for water can limit forest carbon uptake and transpiration as severely as dry soil. *Geophys. Res. Lett.* **2016**, *43*, 9686–9695. [CrossRef]
74. Sheffield, J.; Wood, E.F.; Roderick, M.L. Little change in global drought over the past 60 years. *Nature* **2012**, *491*, 435–438. [CrossRef] [PubMed]
75. Potopová, V.; Boroneant, C.; Možný, M.; Soukup, J. Driving role of snow cover on soil moisture and drought development during the growing season in the Czech Republic. *Int. J. Climatol.* **2016**, *36*, 3741–3758. [CrossRef]
76. Laaha, G.; Gauster, T.; Tallaksen, L.M.; Vidal, J.P.; Stahl, K.; Prudhomme, C.; Heudorfer, B.; Vlnas, R.; Ionita, M.; Van Lanen, H.A.; et al. The European 2015 drought from a hydrological perspective. *Hydrol. Earth Syst. Sci.* **2017**, *21*, 3001–3024. [CrossRef]
77. Pretzsch, H.; Schütze, G.; Biber, P. Drought can favour the growth of small in relation to tall trees in mature stands of Norway spruce and European beech. *For. Ecosyst.* **2018**, *5*, 20. [CrossRef]

 MDPI

Article

A Comparison of the Characteristics of Drought during the Late 20th and Early 21st Centuries over Eastern Europe, Western Russia and Central North America

Anthony R. Lupo [1,*], Nina K. Kononova [2], Inna G. Semenova [3] and Maria G. Lebedeva [4]

1 Atmospheric Science Program and Missouri Climate Center, School of Natural Resources, University of Missouri, Columbia, MO 65211, USA
2 Institute of Geography, Russian Academy of Sciences, 119017 Moscow, Russia; ninakononova@yandex.ru
3 Department of Military Training, Odessa State Environmental University, 65016 Odessa, Ukraine; in_home@ukr.net
4 Department of Geography and Geoecology, Belgorod State University, 308015 Belgorod, Russia; lebedeva_m@bsu.edu.ru
* Correspondence: lupoa@missouri.edu; Tel.: +1-573-489-8457

Abstract: The character of the atmospheric general circulation during summer-season droughts over Eastern Europe/Western Russia and North America during the late twentieth and early twenty first century is examined here. A criterion to examine atmospheric drought events that encompassed the summer season (an important part of the growing season) was used to determine which years were driest, using precipitation, evaporation, and areal coverage. The relationship between drought and the character of the atmosphere, using the Dzerzeevsky weather and climatic classification scheme, atmospheric blocking, teleconnections, and information entropy, was used to study the atmospheric dynamics. The National Centers for Environmental Prediction (NCEP) re-analyses dataset archived at the National Center for Atmospheric Research (NCAR) in Boulder, CO, USA, is used to examine the synoptic character and calculate the dynamic quantities for these dry events. The results demonstrate that extreme droughts over North America are associated with a long warm and dry period of weather and the development of a moderate ridge over the Central USA driven by surface processes. These were more common in the late 20th century. Extreme droughts over Eastern Europe and Western Russia are driven by the occurrence of prolonged blocking episodes, as well as surface processes, and have become more common during the 21st century.

Keywords: meteorological drought; agricultural drought; atmospheric circulation; elementary circulation mechanism (ECM); information entropy; atmospheric blocking

Citation: Lupo, A.R.; Kononova, N.K.; Semenova, I.G.; Lebedeva, M.G. A Comparison of the Characteristics of Drought during the Late 20th and Early 21st Centuries over Eastern Europe, Western Russia and Central North America. *Atmosphere* **2021**, *12*, 1033. https://doi.org/10.3390/atmos12081033

Academic Editors: Andrzej Walega and Agnieszka Ziernicka-Wojtaszek

Received: 6 July 2021
Accepted: 10 August 2021
Published: 12 August 2021

Publisher's Note: MDPI stays neutral with regard to jurisdictional claims in published maps and institutional affiliations.

1. Introduction

Drought is a complicated and interdisciplinary problem that has been the subject of many studies in recent years, especially in connection with climate change (e.g., References [1,2] and the references therein). These events impact more people globally than any other natural hazard [2]. Drought can also be classified as meteorological, agricultural, hydrological, and socioeconomic drought, and these definitions can be found in meteorological climate textbooks (e.g., References [3,4]). In general, meteorological drought will onset soonest, followed by agricultural and then hydrologic (e.g., Reference [5]). Socioeconomic drought defines the impact of the other drought types on the supply and demand of economic goods [4].

Formally, drought-type recognition requires the analysis of atmospheric dynamics and hydrological indicators characterizing the atmospheric and soil moisture (e.g., Palmer Drought Indices [6]). However, the use of standardized drought indices (e.g., Standardized Precipitation Index—SPI [7]; Standardized Precipitation Evapotranspiration Index—SPEI [8]) is recommended—the former by WMO for operational use [9]. This allows the drought classification by considering the temporal scales of observed precipitation deficits and the impact on usable water sources. Thus, precipitation deficits for one-to-two months generally defines the onset of atmospheric or meteorological drought. Continued deficits spanning three-to-six months leads to a deficit of soil moisture content, which corresponds to agricultural drought.

In the USA, drought is monitored collaboratively by the National Drought Mitigation Center at the University of Nebraska–Lincoln and the United States Department of Agriculture [10]. They produce a map showing the severity of drought which is used by stakeholders. In the Russian Federation, for example, drought is monitored through the All-Russian Research Institute of Agricultural Meteorology within Roshydromet [11]. Their automated system assesses drought weekly from May through September, using a blend surface data and satellite imagery (Normalized Difference Vegetation Index—NDVI).

In Ukraine, information about droughts is represented by the Ukrainian Hydrometeorological Center [12], namely the Department of Agrometeorology, which provides information to agricultural decision and policymakers about the occurrence, development, and intensity of drought by regions. Satellite monitoring of droughts is being carried out experimentally by the Ukrainian National Space Facilities Control and Test Center [13], which provides public information about spatiotemporal distribution of the vegetation indices (e.g., NDVI) through their geoinformation portal.

In Europe, drought is monitored by the European Drought Observatory (EDO; see Reference [14]), which is maintained by the Joint Research Centre—EU Science Hub [15]. The EDO provides public drought-relevant information, such as maps of indicators derived from different data sources (e.g., precipitation measurements, satellite measurements, and modeled soil moisture content), as well as different tools for displaying and analyzing the information and drought reports.

Studying drought is difficult, since it occurs on timescales associated with the general circulation (sub-seasonal timescales or greater), but the spatial scale can vary from as small as the lower part of the meso-scale up to the planetary-scale. Even temporally, the drought period may not overlap with calendar months, but drought can have an impact on important phenological times in agriculture [16,17]. During different stages of development, the plant water demand depends on prevailing weather conditions, defined by using meteorological parameters (e.g., temperature, precipitation, and wind). For prolonged meteorological drought, anomalies in characteristics, such as potential evapotranspiration, soil water, or groundwater and reservoir levels, are observed [18]. For example, Reference [16] demonstrated that the important part of the growing season (reproductive stage) for corn and soybean in the Central USA is July and August, respectively, even though the entire growing season can impact growth and yield. Then Reference [17] shows the impact of dry periods on winter and spring wheat in the Missouri River Basin. Additionally, the large-scale economic, agricultural, and societal impact of drought typically determines what years tend to be remembered as "drought years" in the mind of public opinion.

Drought has such a large impact on agriculture and the economy [16–20] and even human mortality [20]. All populations are vulnerable to a certain extent, and different societies are more resilient to the stress of drought. Many studies (e.g., References [21,22]) have developed methodologies for assessing the relative vulnerability of populations to drought. For example, Reference [21] measures exposure multidimensionally in the USA by including drought frequency, population, and freshwater ecosystems affected. Others, such as Reference [22], used socioeconomic indicators, such as available natural resources, economic capacity, human resources, and infrastructure and technology, to compare drought vulnerability across Africa.

Many studies examine the meteorological factors contributing to drought. Over North America, studies such as References [23,24] (and the references therein) attributed the drought of 1980 to an accentuated and persistent upper air (500 hPa) ridge over the center of the USA, and strong troughs off the Pacific and Atlantic coasts, respectively. This drought was preceded by cooler sea surface temperature (SST) anomalies over the Central Tropical

Pacific for about two seasons previous to the onset of drought, and the 500 hPa flow pattern had already been established by spring. This drought is memorable for the impact on agriculture and the economy of the USA (e.g., References [23,24]).

Later, Reference [25] supported the view of References [23,24] regarding the spring (precursor) 500 hPa pattern and identified an unusual Pacific North American (PNA) teleconnection pattern associated with a shorter wavelength while studying a late spring blocking event over North America. Studies such as References [26,27] (and the references therein) associated certain Pacific Ocean Basin SST anomalies with upper air patterns over North America, and these could be correlated with temperature and precipitation anomalies for much of the middle of the USA and in particular, the State of Missouri. The work of References [28–31] extended that of Reference [26] and found that prominent SST patterns varied in association with interannual (e.g., El Niño and Southern Oscillation—ENSO) and interdecadal variability (e.g., Pacific Decadal Oscillation—PDO).

Additionally, many others have examined drought and the variability across the USA or North America [32–38]. Certain ENSO-related SST anomalies correlated with warm and dry temperature anomalies for the Upper Midwest, Central USA, and the Gulf Coast [34–36]. Winter dryness is associated with weak El Niño events, while La Niña summers are associated with summer dryness [28,29,38] in Missouri, but Reference [30] demonstrated that the ENSO impact on precipitation is different for the Upper Midwest versus the Lower Mississippi Valley. In the west, Reference [38] found long-lived drought is linked to La Niña, and these droughts can extend into the Central USA. Moreover, Reference [29] associated cooler and wetter summers during 1971–2002, with more blocking and blocking days versus drier and warmer summers.

However, References [32,33] demonstrated that streamflow in the Upper Midwest and Gulf of Mexico regions revealed wet (dry) conditions during El Niño (La Niña) during the mid- to late-20th century. Then Reference [35] found than in general, the Southeast USA is dry during El Niño years throughout the 20th century. Initially, these results seem contradictory, but Reference [37] found that streamflow, as an indicator, has a long memory. Lastly, Henson et al. [31] (and the references therein) studied the relationship between interannual and interdecadal variability of Pacific Region SSTs, growing season conditions, and corn and soybean yields in Missouri. They found that yields were generally less during the transition toward La Niña years, as these were associated with warmer and drier summers.

Within Europe and Asia, a series of dry years has been observed within the last 15 years (e.g., Reference [39]), and several of these have influenced Eastern Europe and Russia (e.g., Reference [40]). Ionita et al. [39] demonstrated that past "megadrought epochs" over Central Europe were decadal in nature. They also determined these could be linked to solar activity, the Atlantic Multidecadal Oscillation, and atmospheric circulation epochs, including those that are associated with atmospheric blocking. Additionally, they demonstrated that the current spate of Eurasian drought at the start of the 21st century is similar to multidecadal droughts in the past and within the range of past variability. The work of Reference [41] found that the trend in drought and drought severity over the globe is steady since the mid-20th century, and Reference [42] found a similar result for Central and Eastern Europe. The latter study did find some local trends within the broader region.

In the southern part of the East European plain, Cherenkova et al. [43] demonstrated the average yield for winter wheat, spring wheat, and spring barley in the westerly phase of the Quasi-Biennial Oscillation (QBO) exceeded the same yield for the eastward QBO. They attributed the difference to different rainfall patterns and drought frequency over this region for opposing phases of the QBO. In a follow-up article, Reference [44] found an increase in the frequency of dangerous atmospheric droughts during the 1995–2014 period when compared to 1963–1994. They found this was due to an increase in the frequency and intensity of blocking anticyclones in the Atlantic–European region during spring and summer for the period 1963–2016, especially around 30° E longitude.

The 2010 drought in Eastern Europe and Russia has been attributed to summer season blocking episodes (e.g., References [45,46] and the references therein), and these studies examined the dynamic interaction between the synoptic-scale and large-scale in supporting these blocking events [45]. The latter [46] associated five well-known dry and wet years with ENSO variability across the region. They linked the transition of ENSO with dry summers. Then Reference [47] (which Reference [44] supports) linked a warmer drier climate of Southwestern Russia with the increase in certain weather types, including blocking.

The association of Northern Hemisphere (NH) weather types with dry and wet periods was championed by the work of N.K. Kononova (e.g., References [48–52]). These NH weather types, termed elemental circulation mechanisms (ECMs), were first proposed by Reference [53]. This work identified 13 different NH flow types and 41 subtypes based on the amplitude, location, and number of waves on the polar front jet stream [54]. These could be grouped into four general NH flow types, namely two zonal and two meridional types. The studies of References [48–50] used these four NH flow types in their studies of NH flow regime climatologies and dynamics.

Further, References [48–52] would associate an epoch of relatively zonal flow during the early to mid-20th century with the dry years over North America and parts of Russia during the 1930s. Then References [54,55] relate global and regional temperature changes, respectively, to changes in the dominant type of NH ECM. The relationship between the ECM and NH teleconnection activity, as well as the theoretical work of Lorenz and others, was discussed in more detail in Reference [55]. Briefly, the authors [55] argued that these ECM may be subjectively classified quasi-steady states within NH flow regimes. These are similar to teleconnections which are generally local phenomena except for the Arctic Oscillation (AO). They also demonstrate that long-term epochs of relatively zonal or meridional ECM occurrences correspond to epochs of lower or higher AO. Moreover, teleconnections such as the PNA are associated with long period Rossby Wave trains (e.g., Reference [56]). Additionally, Reference [54] demonstrated that the concept of information entropy could be applied to the relative occurrence of zonal versus meridional flow types.

This classification is also convenient for the definition of synoptic processes influencing the occurrence of regional weather hazards often related to drought, such as a dry hot wind, called "*Sukhovey*" in Ukraine and Russia [57]. This dry hot wind is widespread throughout steppe areas of Eastern Europe during the warm season of the year. They may accompany drought seasons or may occur during other years, but they are always associated with the periphery of anticyclones. Droughts in this region usually occur under large-scale anticyclones or ridges [58].

The goal of this study is to examine the occurrence of summer season meteorological and agricultural drought and the atmospheric circulation and dynamics from 1970 to 2020 over agriculturally sensitive regions of North America (NA) and Eastern Europe and Western Russia (EE/WR). The work of Reference [46] selected only five years from the latter region based on the precipitation anomalies only. Moreover, drought has been more frequent during the previous decade in both regions providing motivation for study. Here, we develop an objective criterion for summer-season drought based on precipitation and potential evaporation anomalies, as well as the amount of area impacted by using composite re-analyses that are comparable to time-series indexes in identifying drought. The occurrence of summer-season drought is here related to longer-term atmospheric variability; we also relate these droughts to the frequency and strength of atmospheric blocking and the relative occurrence of zonal versus meridional ECMs during summer. We also examine whether drought in each region is commonly associated with precursors and differentiate between the atmospheric circulation character of extreme drought summers versus moderate drought summers.

2. Data and Methods

2.1. Data

The NH 500 hPa height (m), precipitation rate (mm day^{-1}) (P), and potential evaporation (W m^{-2}) (E) data were retrieved from the National Centers for Environmental Prediction (NCEP)/National Center for Atmospheric Research (NCAR) re-analyses [59], available through the NOAA Earth System Research Laboratory (ESRL) website. These data are available at time intervals from 6 h to monthly and on a 2.5° latitude by 2.5° longitude grid from 1948 to 2020. The potential evaporation data were converted to mm day^{-1} by dividing E by the latent heat of vaporization (L = 2.5 × 10^6 J kg^{-1}) and the density of water (1000 kg m^{-3}). The study period was 1970–2020.

The observed atmospheric blocking information was obtained from the blocking archive housed in the University of Missouri Weather Analysis and Visualization (WAV) laboratory by the Global Climate Change Group (GCC) [60]. Briefly, a generic definition for atmospheric blocking is that these events are persistent quasi-stationary anticyclones or ridges in the mid-latitude jet stream [61]. The blocking information used was duration (days) and intensity (BI) and is available from 1968 to 2021. Blocking events within the EE/WR were examined consistent with References [46,47] (20°–60° E) and during the spring and summer seasons. Over the NA region, blocking is relatively rare over the continental region (e.g., References [25,62]), and extreme weather over NA, especially the central region, has traditionally been associated with blocking over the Eastern North Pacific Region (e.g., References [25,62,63]). Here we define the Eastern Pacific as 180°–100° W, consistent with Reference [62] and other studies. The teleconnection indexes, specifically the AO, the North Atlantic Oscillation (NAO), and PNA, were downloaded from the Climate Prediction Center website [64].

These teleconnections were chosen, since these are commonly associated with weather and climate in the study regions, which are defined below. Moreover, the AO does show correspondence with zonal versus meridional ECM groups (e.g., Reference [55]). The daily classifications of the ECM, as well as their monthly and annual statistics since 1899, are available for download via the Russian Academy of Sciences, Institute of Geography website, "Fluctuations in the Atmospheric Circulation of the Northern Hemisphere in the 20th and Early 21st Century" [65].

The definition for ENSO used in this study is described in Reference [62] and the references therein, and a brief description is given here. The Japanese Meteorological Agency (JMA) ENSO index is available via the Center for Ocean and Atmospheric Prediction Studies (COAPS) from 1868 to present [66]. The JMA classifies ENSO phases by using SST within the bounded region of 4° S to 4° N, 150° W to 90° W and defines the start of an ENSO year as 1 October, and its conclusion on 30 September of the following year. This index is used in many other published works (see Reference [62] and the references therein), and a list of years is provided below (Table 1). This index is useful since it acknowledges the longevity of ENSO events, but it may produce different classification for years versus other definitions. For example, Reference [67] found that, while the JMA index is more sensitive to La Niña events than other definitions, it is less sensitive than other indices to El Niño events.

2.2. Methods

This study examines meteorological drought and the accompanying atmospheric circulation within agriculturally sensitive regions of North America (NA) and Eurasia, specifically Eastern Europe and Western Russia (EE/WR). However, given the definition of agricultural drought and the variables used here [18], the results can be extended to this type of drought. These regions were bounded by the boxes outlined below (Figure 1) and following the flowchart in Figure 2. Over NA, this box is bounded by 30° N and 50° N, and 100° W and 80° W, including much of the corn, wheat, soybean belts in NA (e.g., References [17,31]). Over EE/WR, this box is bounded by 40° N and 60° N, and 20° E and 50° E, including the major wheat growing regions of Eastern Europe, Ukraine, and Western Russia. Soybeans, corn, sugar beet, and sunflower are also primary crops of Southwestern Russia. In Ukraine, for example, winter wheat, spring barley, and corn are the main grain crops. Sunflowers and sugar beets are also major crops. Winter wheat and spring barley accounts for about 90–95% percent of the total area for relevant crops and grows throughout Ukraine. Corn is the third most important feed grain, planted in areas located predominantly in Eastern and Southern Ukraine [68]. In the neighboring countries of Eastern Europe, the same grain crops (wheat and barley) are grown, taking into account latitudinal differences in agroclimatic conditions comparing to Ukraine.

Table 1. List of ENSO years used here. The years below are taken from References [62,66].

El Niño (EN)	Neutral (NEU)	La Niña (LN)
1969	1968	1967
1972	1977–1981	1970–1971
1976	1983–1985	1973–1975
1982	1989–1990	1988
1986–1987	1992–1996	1998–1999
1991	2000–2001	2007
1997	2003–2005	2010
2002	2008	2017
2006	2011–2013	2020
2009	2016	
2014–2015	2019	
2018		

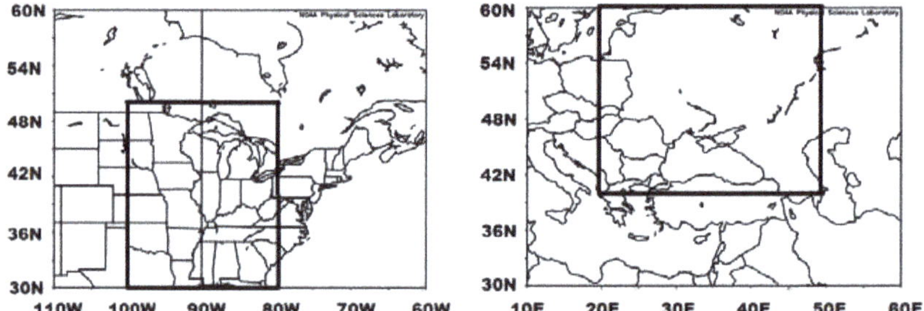

Figure 1. The study regions used here, NA (**left**) and EE/WR (**right**).

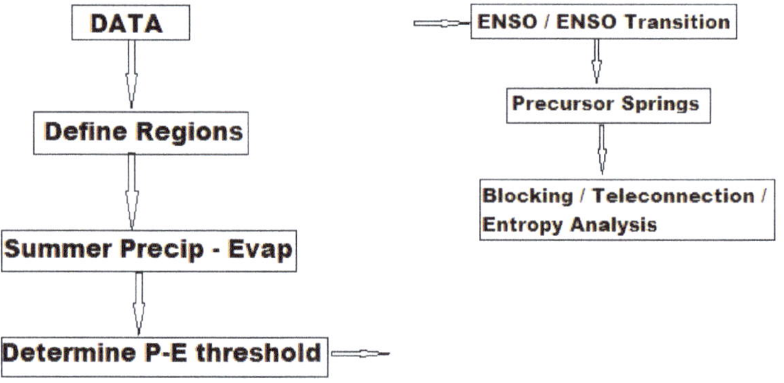

Figure 2. Flowchart presentation of the study performed following the format of Reference [32].

Additionally, these areas are large enough to identify a regional circulation signal as it relates to atmospheric teleconnection activity. However, as demonstrated by

References [28–30] and others, the regions may be large enough to produce different ENSO signals in the temperature and precipitation regimes for different parts of the region. The total area of the NA study region is 3.786×10^6 km^2 and the EE/WR study region is 4.765×10^6 km^2, which means the latter region is approximately 25% larger.

A description and the history of ECM developed by Reference [53] can be found in References [48–54]. The original work of Reference [53] included four circulation groups which are displayed in Table 2 (adapted from Reference [54]—their Table 1). As described above, the ECM comprises 13 main circulation types separated into 41 subtypes identified subjectively based on surface maps originally. Then Reference [69] used 500 hPa maps when these became routinely available by the late 1940s. Many of these references (e.g., References [54,55]) show maps of the NH as examples of these flow regime types. The subtypes are identified based on the amplitude of the NH mid-latitude flow daily and then categorized these based on the timing (warm or cold season), the geographical location, and number of ridge–trough couplets in the jet stream. In relation to the onset of extreme hydrometeorological phenomena, such as droughts or floods, the type of ECM identified make it possible to examine the geographical distribution of key synoptic and large-scale atmospheric features including blocking (e.g., Reference [70]).

Table 2. Adapted from Reference [54] Table 1, global atmospheric circulation groups of Reference [53].

Circulation Group	ECMs Included	Atmospheric Pressure at the North/South Pole	Number of Amplified Waves
Zonal (Type 1)	1–2	High	0
Zonal Breaking (Type 2)	3–7	High	1
Amplified Ridging (Type 3)	8–12	High	2–4
Equatorward Troughs (Type 4)	13	Low	3 or 4

The study of meteorological drought in EE/WR by Reference [46] defined drought based on the precipitation amounts in the Moscow and Belgorod region alone during the summer season and based on five years which were known to be dry and impact regional agriculture. In providing an objective criterion accounting for precipitation and evaporation, the following definition is used: (1) the largest difference between the maximum precipitation anomaly (mm day^{-1}) and the potential evaporation anomaly (mm day^{-1}—near the location of the former) within the study region; and (2) the areal coverage of the negative precipitation anomaly. Using precipitation- and evaporation-based indexes (e.g., the Palmer Drought Severity Index—PDSI) for meteorological drought is common (e.g., References [34–38]). These indexes are generally derived as long-term time series, whereas here we developed seasonal composite spatial maps of precipitation and potential evaporation.

The study regions defined above were large enough to have some portion of the region be drier than normal during approximately 80% of the entire study period. In providing for a large enough sample of summer seasons, but avoid making most summer seasons drought summers, the criterion for drought was refined (Tables 3 and 4). If criterion one was greater than -9 mm day^{-1}, the year was labelled as an extreme drought season. If criterion one was less than -9 mm day^{-1} and the areal extent was larger than 1.893×10^6 km^2 (50% of the NA region and 40% of the EE/WR), that year was considered a moderate drought year. Using this criterion (-9 mm day^{-1} and the area) includes nearly all years recognized by the respective societies and studies cited in section one as impactful drought years. The extreme drought sample size was ten and eight years for the NA and EE/WR study regions, respectively. The respective moderate drought sample size was 12 and eight years. Additionally, using a threshold to define drought has been used by other studies (e.g., References [35,36,71]).

Table 3. The occurrence of extreme and moderate North American (NA) drought. Columns one two and three show the maximum precipitation minus evaporation anomaly (mm day^{-1}), the percent of the study region covered by -1 mm day^{-1} or greater P–E, and the product of the column two and three/rank, respectively. Columns four, five, and six show the ENSO phase, as well as the type of ENSO transition taking place during the summer, and whether the preceding spring showed drought conditions, respectively.

Year	Pre-Evaporation Anomaly	% Area Covered	Col 2 × 3/Rank	ENSO Phase	Transition	Precursor
Extreme						
1972	−14.5	60	−8.7/2	LN	LN to EN	Yes
1976	−9.0	90	−8.1/3	LN	LN to EN	Yes
1978	−9.0	70	−6.3/6	NEU	NEU to NEU	Yes
1980	−14.0	75	−10.5/1	NEU	NEU to NEU	Yes
1982	−10.0	25	−2.5/21	NEU	NEU to EN	No
1990	−11.0	40	−4.4/9	NEU	NEU to NEU	Yes
1998	−9.0	40	−3.6/14	EN	EN to LN	Yes
2000	−10.0	75	−7.5/4	LN	LN to NEU	Yes
2011	−13.5	30	−4.1/12	LN	LN to NEU	Yes
2012	−9.0	30	−2.7/20	NEU	NEU to NEU	Yes
Moderate						
1970	−7.0	75	−5.3/7	EN	EN to LN	Yes
1973	−7.0	60	−4.2/10	EN	EN to LN	No
1974	−8.0	85	−6.8/5	LN	LN to LN	Yes
1975	−5.0	50	−2.5/22	LN	LN to LN	No
1977	−7.0	55	−3.9/13	EN	EN to NEU	Yes
1985	−5.0	60	−3.0/18	NEU	NEU to NEU	Yes
1988	−7.0	50	−3.5/15	EN	EN to LN	Yes
1999	−6.0	70	−4.2/11	LN	LN to LN	Yes
2006	−5.0	60	−3.0/19	NEU	NEU to EN	Yes
2015	−7.0	50	−3.5/16	EN	EN to EN	No
2017	−5.5	60	−3.3/17	NEU	NEU to LN	No
2020	−8.0	60	−4.8/8	NEU	NEU to LN	No

Table 4. As in Table 3, except for Eastern Europe/Western Russia (EE/WR). Column 4 was multiplied by 0.8 to normalize the study area to the area from Table 3.

Year	Pre-Evaporation Anomaly	% Area Covered	Col 2 × 3/Rank	ENSO Phase	Transition	Precursor
Extreme						
1972	−9.0	30	−2.2/12	LN	LN to EN	No
1992	−10.0	70	−5.6/4	EN	EN to NEU	Yes
1994	−10.0	80	−6.4/3	NEU	NEU to NEU	Yes
2002	−10.0	70	−5.6/5	NEU	NEU to EN	Yes
2008	−9.0	35	−2.0/14	LN	LN to NEU	No
2010	−11.5	50	−4.6/6	EN	EN to LN	No
2015	−10.5	80	−6.7/1	EN	EN to EN	No
2017	−10.0	80	−6.4/2	NEU	NEU to LN	Yes
Moderate						
1971	−5.5	40	−1.8/16	LN	LN to LN	Yes
1975	−5.5	40	−1.8/15	LN	LN to LN	Yes
1979	−5.0	50	−2.2/13	NEU	NEU to NEU	Yes
1993	−7.0	40	−2.2/11	NEU	NEU to NEU	Yes
1996	−5.0	75	−3.0/8	NEU	NEU to NEU	Yes
2000	−8.0	40	−2.6/9	LN	LN to NEU	No
2009	−7.0	40	−2.2/14	NEU	NEU to EN	Yes
2020	−8.0	65	−4.2/7	NEU	NEU to LN	Yes

The work of Reference [54] used the concept of Shannon or information entropy to discuss the relative frequency in the occurrence of zonal versus meridional flows. In Table 2, Type 1 and Type 2 flows were considered zonal NH flow regimes, while Type 3 and 4 are meridional NH flow regimes. The formula used for information entropy was as follows:

$$H(x) = -\sum_{i=1}^{n} p(x_i) \log_b p(x_i) \tag{1}$$

where b is the base logarithm and $p(x)$ is the probability of a certain outcome. Entropy is described as a measure of predictability, structure, or organization within a system [72] (and references therein). Moreover, Reference [54] discusses information entropy in more detail. Briefly, an unbiased coin should result in $H(x) = 1.00$. As described in Reference [54] and using the NH flow types of Reference [53], the information entropy for the NH observed flow would be 0.99 if all 41 of the NH flow types in Reference [53] were equally likely (Type 1 and Type 2 account for 18 of the 41 subtypes or 44%). The observed information entropy from 1899 to the present was 0.90, and the years since the late 20th century and early 21st century were 0.76 and 0.57, respectively. These were significant departures from the overall sample [54]. All of these values for information entropy are for calendar years from 1899 to the present. Thus, here we examine spring and summer season values of information entropy, work which was not done in Reference [54].

3. Results

In this section, the occurrence of summer-season drought over the study regions shown in Figure 1 and Tables 3–5 are examined. A comparison to Reference [46] and previous work is performed where relevant. Many of the summer seasons over the period 1970–2020 that were regarded as very dry and impactful by other studies (see Introduction) to agriculture are identified by the criterion used here. Six summers were drought summers in both regions simultaneously (Tables 3 and 4).

Table 5. As in Table 3, except for NA and EE/WR wet years.

Year	ENSO Phase	Transition	Precursor
NA			
1979	NEU	NEU to NEU	No
1992	EN	EN to NEU	No
1996	NEU	NEU to NEU	No
1997	NEU	NEU to EN	No
2003	NEU	NEU to NEU	No
2004	NEU	NEU to NEU	Yes
2008	LN	LN to NEU	Yes
2010	EN	EN to LN	Yes
EE/WR			
1973	EN	EN to LN	No
1974	LN	LN to LN	No
1977	EN	EN to NEU	No
1978	NEU	NEU to NEU	No
1985	NEU	NEU to NEU	No
1988	EN	EN to LN	No
1989	LN	LN to NEU	No
1995	NEU	NEU to NEU	Yes

3.1. Interannual and Interdecadal Variability

The forecasting of drought one year in advance across the Missouri River Basin and the USA in general based in decadal scale variability was calculated to be $80M and $1.1B by References [73,74], respectively. Thus, interannual and interdecadal variability in the large-scale flow patterns and their relationship to local temperature and precipitation is important for generating long-range forecasts. A cursory examination of the NA study region summer droughts demonstrated that 13 of the 22 years identified occurred during the decade of the 1970 and 2010s (Table 3). A similar result can be seen for the EE/WR region as 8 of 16 summer-season droughts occurred during these same decades. The extreme summer droughts occurred most often during the 1970s for NA and the 2010s for EE/WR. In the latter region, these can be related teleconnections, such as the AMO (e.g., Reference [39]), PDO (e.g., Reference [37]), or interdecadal variability of blocking (e.g., References [62,75]). In the NA region, the earlier decade overlaps with the changeover of the PDO from negative (1949–1976) to the positive (1977–1998) phase. For the later decade it is not clear that the PDO has changed from the current negative phase yet. This suggests a PDO signal which is consistent with References [2,31], who found the negative PDO years were drier in this region. However, Reference [75] finds a 20-year cycle in NA continental drought. Moreover, References [37,76] found a PDO signal, but Reference [76] related this to the positive PDO for drought over the USA. Additionally, a cursory examination of the decadal epochs of the AO and NAO (see Reference [64]) demonstrates that dry summer season are associated generally with certain values of these indexes.

When examining the ENSO related variability over NA (Table 3) from 1970 to 2020, a complex pattern emerges. Four of the ten extreme summer droughts were associated with LN years. If moderate drought years are included, seven of 11 LN (22% of all study years) occurred during the study period. On the other hand, five of the moderate summer drought years were EN years (14 total years or 27%). Thus, it is apparent that the distribution of extreme summer drought years is skewed toward LN years, while moderate summer droughts are skewed toward EN years.

However, when testing these distributions by using a simple Chi-square goodness of fit test (e.g., Reference [77]) only the result for extreme summer drought distribution is different from that of the total sample, but not at standard levels of statistical significance. This may be due to the small sample size and low number of bins. Across the study region, Reference [29] shows that EN years are dry, while for LN years this depends on the phase of the PDO (drier during the negative PDO). These results found here would be consistent as the extreme drought summers occur during LN and NEU years, as in Reference [38], but also during the negative PDO. The moderate dry summers occurring over more than 50% of the region is consistent with Reference [29], as well. The connection of dry years to LN years is supported by References [38,76].

Further, examining the ENSO transition summers indicates that five of the extreme summer-season droughts involved a transition in the positive direction (Table 3), defined as toward EN since EN are associated with positive SST anomalies. Six of the moderate summer drought years as associated with negative transitions. There was a total of 15 positive and 14 negative transitions during the 51-year period of study.

Within the EE/WR region (Table 4), there is only a slight tilt toward EN years for extreme summer-season drought, while the opposite was true for moderate summer drought years. Spatially, for the extreme summer droughts, the precipitation deficit was larger in the northern part of the study region, while the opposite was generally true for moderate summer droughts (not shown; e.g., Reference [46]). There were no EN years associated with moderate drought summers. The Chi-square test was applied here and for this region, the moderate drought summer distribution was not the same as the total distribution but not at standard levels of significance. There was no preference found for positive versus negative summer ENSO transitions overall across the study region. These results are similar to those of Reference [46], which found no real preference for transitions in the positive or negative direction in the Moscow region. They also found a preference for the transition toward EN years for the Belgorod Region.

Interestingly, when considering the bioclimatic potential by using indexes (e.g., hydrothermal coefficient or bioclimatic potential) that combine both surface temperature and precipitation during the growing season for the time period 1988–2014, Reference [78] found that in the Belgorod and Missouri USA region, these indexes were lower (associated with drought) for both EN and LN years and higher for NEU years. Their results were significant at $p = 0.01$ for LH growing seasons in Belgorod and EN years for Missouri in the USA.

Table 5 shows the wet years for both study regions. There are no statistical preferences for EN versus LN years or ENSO phase transitions. In Tables 3–5, a majority of the summer drought years in both regions were preceded by a dry spring. However, for the extreme summer droughts over EE/WR, only half of these were preceded by a dry spring. Thus, the results of Reference [46], which concluded that summer droughts were not necessarily preceded by a dry spring in the EE/WR region, may have been biased by the fact that their small sample of years were mainly in the extreme category here. The majority of the wet years were not preceded by wet springs in either region.

3.2. *Synoptic–Dynamic Analysis*

In References [52,54], the annual frequency of zonal versus meridional ECM or 500 hPa NH flow regimes were examined to define long-term circulation epochs. Both of these studies related the occurrence of these epochs with global annual temperature, blocking, and hazardous weather. Neither of these references examined the seasonal frequency of zonal and meridional flow epochs and this can be examined here as it relates to summer-season drought in our study regions. As argued in Reference [54], if each of the ECM defined originally by Reference [53] were equally likely, then the frequency of occurrence for zonal to meridional ECM would be 0.46 to 0.54. However, in Reference [54] the frequency of occurrences from 1899 to 2019 was 0.31 to 0.69, yielding a value of 0.90 for the information entropy.

From 1970 to 2020, the frequency of occurrence for zonal to meridional flows was 0.20 to 0.80 and the value of the information entropy was 0.73 which is similar to the 1899–2019 distribution [54] at $p = 0.1$ when using the Kolmogorov–Smirnov or Chi-Square goodness of fit test. However, the 1970–2020 values are similar to those of 1957–2019 (see Reference [54]). If the frequency of zonal and meridional flow types is examined over the course of a year [65], it is apparent that zonal flows occur more often during the summer season and less so during the cold season. The frequency of occurrence for zonal flows to meridional flows during spring and summer were 0.18 to 0.82 and 0.25 to 0.75, respectively. The information entropy was 0.68 and 0.82, respectively, but the difference is not statistically significant.

In the previous section, it was established that the majority of summer-season droughts (about 70%) in both regions followed after drier spring seasons. An examination of the relative occurrence of blocking, teleconnections, and zonal versus meridional NH flow regimes for spring and summer is shown in Figures 3–5.

3.2.1. Preceding Spring Seasons

During the preceding spring seasons (Figure 5A,B), the occurrence of zonal versus meridional NH flows or ECM for extreme and moderate drought years, as well as wet years, was similar to those of spring seasons, overall, from 1970 to 2020. However, regionally, there are some strong differences in the flow field (Figure 6) between springs preceding summer drought or wet summers. In Figure 6A,B, the Pacific Region flow was clearly more zonal with an anomalously strong Aleutian Low present. This explains fewer Pacific Region blocking events (Figure 3A) and the negative PNA Index within NA region for drought springs (Figure 4A). However, for the springs preceding wet summers, the PNA Index is positive (Figure 4A) and there was significantly more blocking (Figure 3A). This is reflected in the strong positive 500 hPa height anomaly over the East Pacific (Figure 6C).

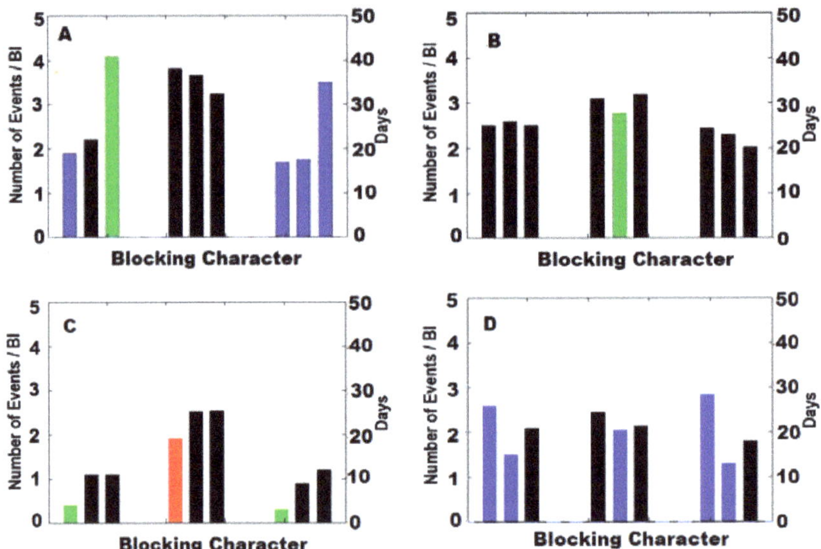

Figure 3. The character of blocking events for (**A**) NA spring, (**B**) EE/WR spring, (**C**) NA summer, and (**D**) EE/WR summer. The left-hand ordinate is mean block occurrence (number—leftmost bars) and mean block intensity (BI—middle bars), and the right-hand ordinate is mean blocking days (rightmost bars). Blue, green, and red bars stand for statistical significance at $p = 0.1$, 0.05, and 0.01, respectively. For each set of bars, the display represents extreme drought, moderate drought, and wet years from left to right.

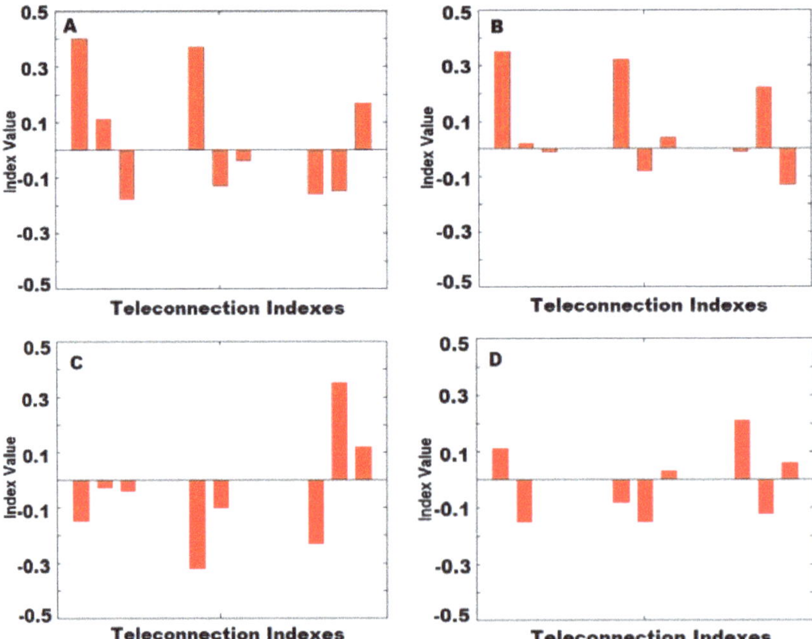

Figure 4. The major teleconnection indexes for (**A**) NA spring, (**B**) EE/WR Spring, (**C**) NA Summer, and (**D**) EE/WR summer, where the left, middle, and right group of bars are the AO, NAO, and PNA indexes, respectively. The ordinate is the value of the index.

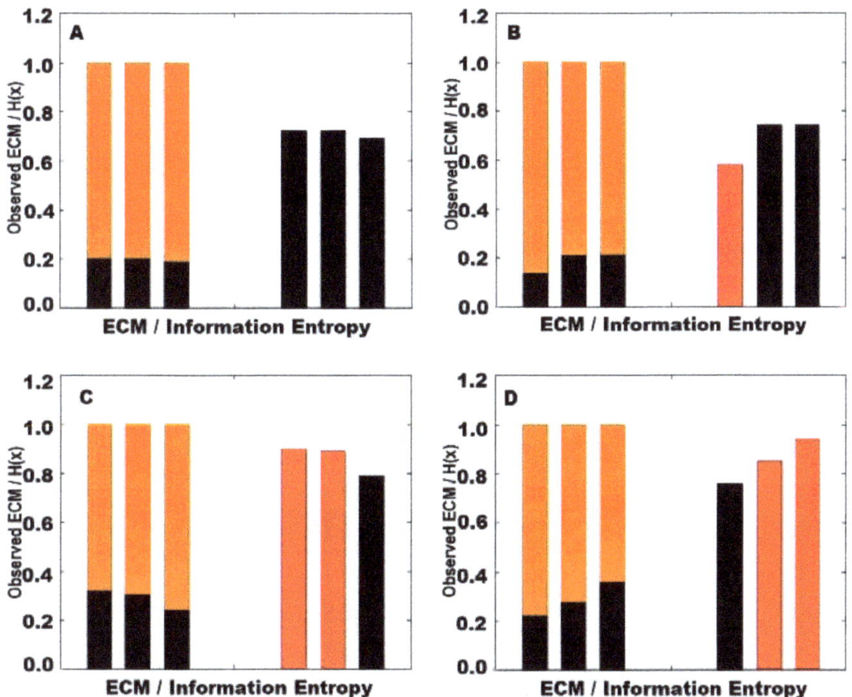

Figure 5. The fraction of the observed ECM (ordinate) that are zonal and meridional (left bars) and the information entropy ($H(x)$—ordinate) (right bars) for (**A**) NA spring, (**B**) EE/WR Spring, (**C**) NA Summer, and (**D**) EE/WR summer. For the left-hand bars, the black and orange colors represent the fraction of observed zonal versus meridional ECM.

Only during the spring preceding the extreme summer drought for the EE/WR study region was the NH flow more meridional relative to other spring seasons during the study period at a statistically significant level ($p = 0.05$) (Figure 5B). This is reflected in a strongly positive NAO (and AO) in Figure 4B, but also in Figure 7A. Lebedeva et al. [55] demonstrated the long-term correspondence between the NAO and AO, as these teleconnections were highly correlated. The correlations between the AO and NAO are similarly high here for spring and summer seasons and for each subsample in Figure 4 ($p = 0.01$). There is also a negative correlation between the AO and PNA at $p = 0.05$, but only in extreme and moderate dry springs and summers. For wet seasons, the PNA and AO do not correlate.

Within the EE/WR, the NAO is weakly negative for springs preceding moderate drought summers (Figures 4B and 7B). However, a negative (EU1—see Reference [56]) Eurasian (EU) teleconnection pattern emerges (e.g., References [79,80]) which is somewhat clear in Figure 7A also. A negative EU is characterized by a positive 500 hPa height anomaly near 20° E and 145° E and a trough near 75° E. Thus, the reversal of the NAO in spring may be an indicator of an extreme dry summer versus a moderately dry summer in the presence of a weakly negative EU. However, for spring seasons preceding wet summers, there is little signal in the NAO but a very distinctly positive EU (EU2—see Reference [56]) pattern (Figure 7C). Finally, there was no preference toward more or fewer blocking events in any of the spring seasons (Figure 3B); however, weaker blocking in the study region during the spring seasons preceding moderate drought summers and negative NAO phase is consistent with Reference [62].

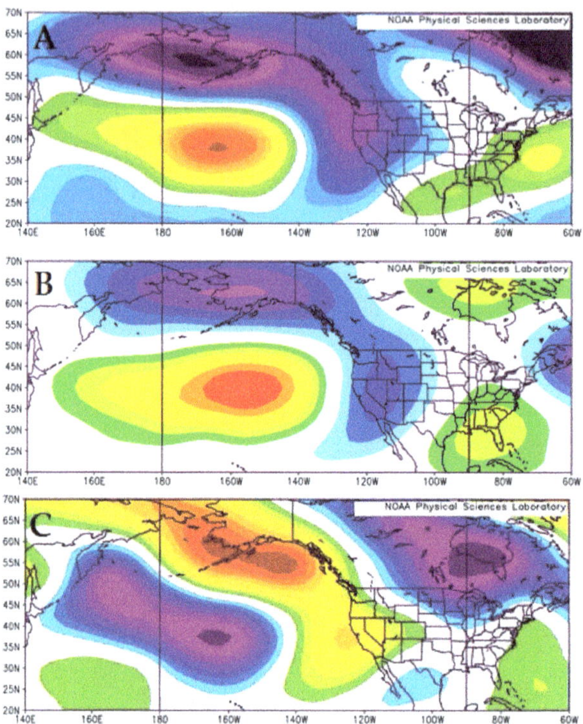

Figure 6. The 500 hPa height anomalies (m) in the NA region during the spring season (March–May) preceding (**A**) extreme drought (Table 3), (**B**) moderate drought (Table 3), and (**C**) wet summer seasons (Table 5). The warm (cold) colors are positive (negative) height anomalies. The contour interval is (**A**,**C**) 2.5 m and (**B**) 5 m.

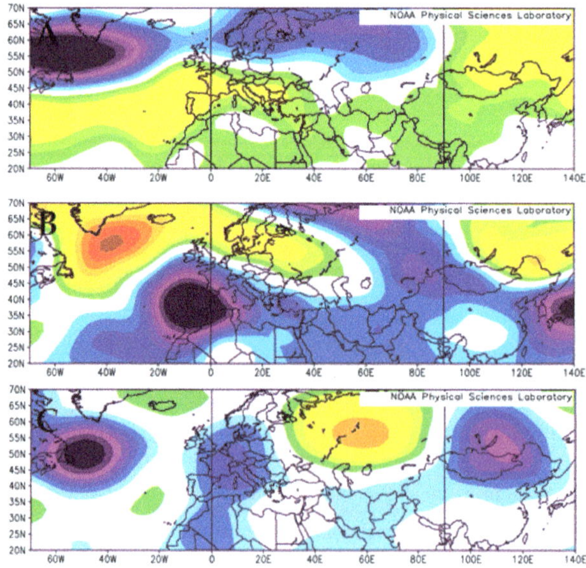

Figure 7. As in Figure 6, except for the EE/WR spring seasons in Tables 4 and 5. The contour interval in (**A**,**B**) is 2.5 m, and in (**C**), it is 5 m.

3.2.2. Dry Summer Seasons

During the summer season (Figure 5C,D), the occurrence of more zonal NH flow regimes during both extreme and moderate drought and wet seasons was consistent with summer seasons overall. During all drought summers for the NA region, the flow was clearly even more zonal, and this result was significant at $p = 0.01$ (Figure 5C). While extreme drought summers were accompanied by larger maximum dry precipitation anomalies than moderate drought summers (-3.9 mm day^{-1} versus -2.8 mm day^{-1}) as anticipated, the absolute value of the maximum 500 hPa height anomalies were of similar strength (18.2 m versus 22.5 m). As the flow was more zonal the maximum 500 hPa height anomalies were not always positive height anomalies within the NA study region. This can be seen in Figure 8A,B.

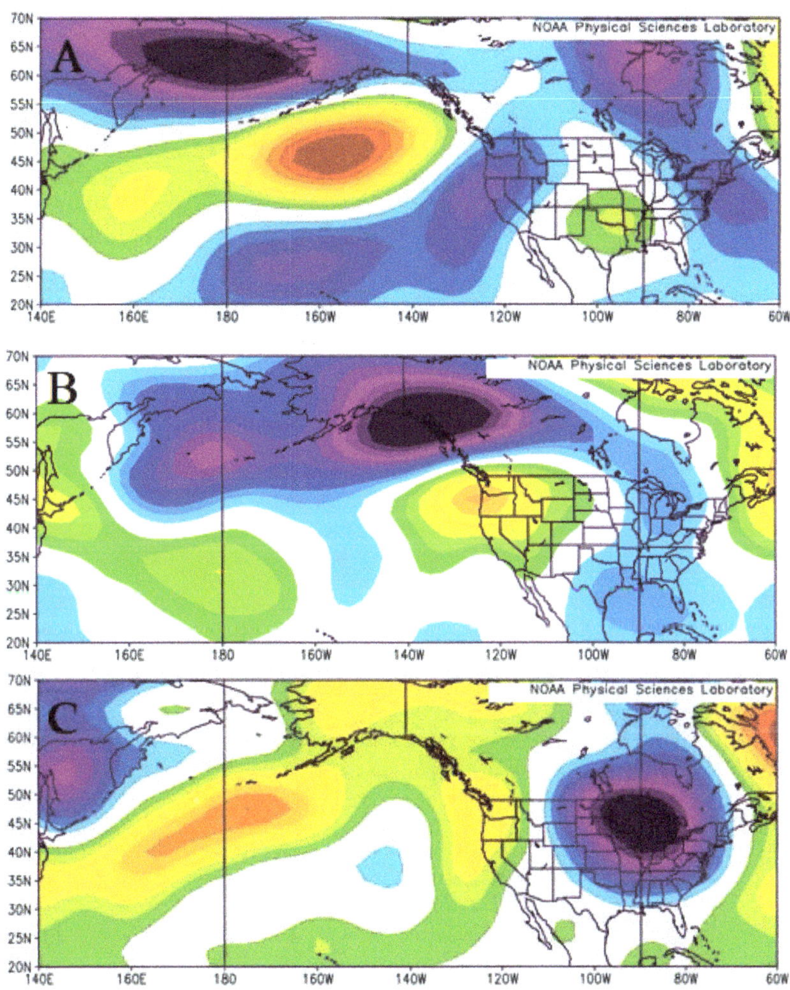

Figure 8. As in Figure 6, except for the NA summer seasons. The contour interval is (**A**) 2.5 m and (**B**,**C**) 1.5 m.

In Figure 8A, the extreme dry years featured a positive 500 hPa height anomaly within the study region and the continuation of negative PNA values from the spring season (Figure 4A). This is accompanied by significantly fewer ($p = 0.05$) and weaker ($p = 0.01$)

blocking events in association with an anomalously strong Aleutian Low (Figure 3A). Additionally, note the troughs off each coast of North America which validates the results of References [23–25]. For moderate drought years (Figure 8B), the pattern changes to a positive PNA pattern (Figure 4B). The positive height anomaly is now over the Western NA region which strengthens the climatological ridge located there (not shown). A negative height anomaly is located over the Eastern USA (and study region). This configuration places the study region in the convergent region between a 500 hPa trough and ridge with high pressure at the surface. In fact, for extreme drought summers eight of ten were positive height anomalies while only six of 12 moderate drought years were similar. In both cases (Figure 8A,B), the dominant PNA pattern has a shorter wavelength for dry summers as posited by Reference [25]. Finally, there is a strong trough over the eastern 2/3 of USA for wet summers (Figure 8C). These results here support more strongly the results of Reference [29], which differentiated between wet and dry summers over the Midwest USA.

In the EE/WR study region (Figure 5D), the extreme dry years were more meridional over the entire NH, but more zonal during moderate drought and wet years ($p = 0.01$). Like the NA region, the maximum dry precipitation anomalies are larger during the extreme drought summers than during moderate drought summers (-3.1 versus -2.4 mm day^{-1}). However, the maximum 500 hPa height anomalies for both drought groups (53.3 versus 26.0 m) were larger overall than over the NA region. This is consistent with the higher relative occurrence of meridional flow types (Figure 5C,D). Almost all of the dry years (13 of 16) were associated with positive 500 hPa height anomalies within the study region.

Several studies (e.g., References [47,49,55,58,78]) demonstrated that the EE/WR region has been associated with an increase in the occurrence of more meridional ECM or NH flow regimes, especially during the summer (Figure 5D). The study region has also been associated with an increase in atmospheric blocking (e.g., References [44–47,55,57,58]) and studies of drought associated with the extreme summer-season drought of 2010 over this region associated extreme dry summers with atmospheric blocking.

However, during the summer season (Figure 3D), extreme droughts were associated with more blocking events and days, and this is significant at $p = 0.10$, while moderate droughts were associated with fewer blocking events, blocking days, and weaker events, all significant at $p = 0.10$. These results are consistent with Section 3.1 and the occurrence of blocking versus ENSO in the Atlantic Region [62].

Furthermore, during extreme drought summers over the EE/WR region, the negative EU pattern of the spring season continued (Figure 9A versus Figure 7A) but is stronger, such that the largest positive 500 hPa height anomaly is located over the study region. In fact, the strongly negative EU pattern during extreme dry summer seasons is reminiscent of the quasi-stationary sub-seasonal and seasonal Rossby Wave trains that accompany the NH flow in the PNA region and other parts of the globe (e.g., References [56,81–85]). During wet summer seasons (Figure 9C), the EU pattern is positive. Moderate drought years are associated with a negative NAO, similar to that of the spring season, but the EU pattern becomes less organized.

Lastly, the discussion above implies that atmospheric dynamics are the primary drivers of summer-season drought in both study regions. In order to determine strength of the contributions of surface processes to these droughts, we examine the potential evaporation (E). For the NA region, the average P–E (Table 3) was -10.9 mm day^{-1} for extreme dry summers and the average maximum precipitation anomaly was -3.9 mm day^{-1}. Thus, the maximum potential evaporation anomaly was 7 mm day^{-1}. For, moderate drought years the P–E anomaly was -6.5 mm day^{-1}, and the maximum potential evaporation was 3.7 mm day^{-1}. Thus, the extreme dry summer potential evaporation was larger in both an absolute sense and relative to the total P–E anomaly. Given the analysis here, this suggests that the extreme drought years were driven strongly by surface processes, as well. The same analysis for the EE/WR region produces mean P–E values of -9.9 and -6.4 mm day^{-1} for extreme and moderate dry summers, respectively. Using the precipitation anomalies cited in Section 3.2.2 suggests a similar result (maximum potential evaporation of 6.4 and 4 mm day^{-1}, respectively), that extreme dry years are driven to a greater extent by surface process. This is in spite of the fact that 50% of these summers were not preceded by a dry spring.

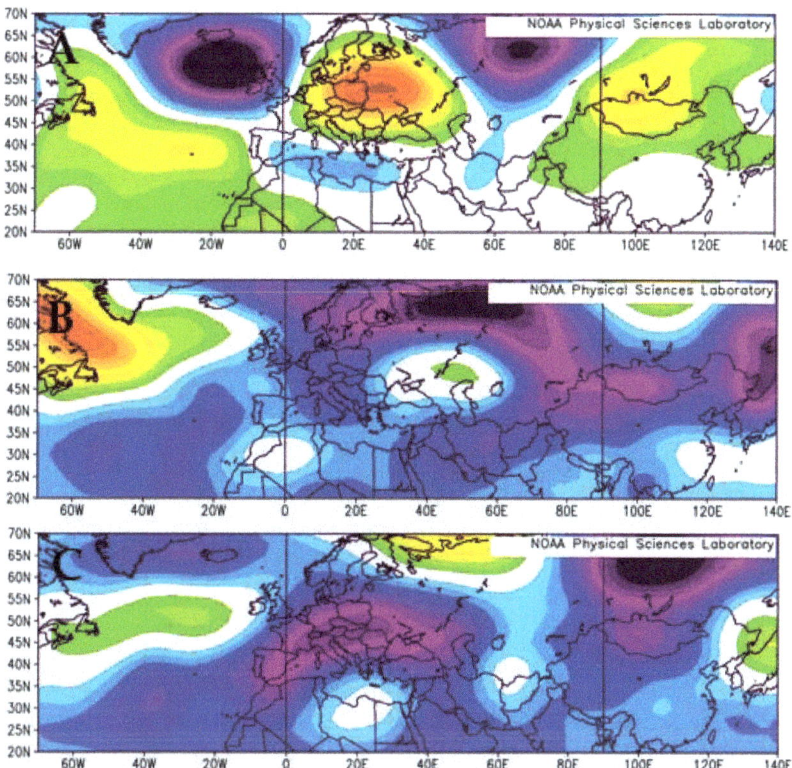

Figure 9. As in Figure 7, except for the EE/WR summer seasons. The contour interval is (**A**,**C**) 2.5 m and (**B**) 1.5 m.

4. Summary and Conclusions

Drought is a difficult topic to study, as it is typically challenging to define precisely the timescales and space scales for this phenomenon. It is also an important topic because of the impact on agricultural and economic activity. Here summer-season drought was examined for the agriculturally important regions of the Central USA and Eastern Europe and Western Russia during the late 20th and early 21st centuries. Here summer-season drought was defined by the seasonal mean composite precipitation minus potential evaporation anomalies. This criterion was developed to separate extreme dry summers from those of moderately dry summers and then compared to wet summers. The criterion was successful in identifying major drought summers that time-series indexes have identified as impactful droughts. This study produced the following results.

Summer-season drought within the agriculturally sensitive regions of NA and EE/WR occurred more often during the 1970s and the 2010s, with minima during the 1990s and 2000s (NA) and 1980s (EE/WR).

Summer seasons defined as extreme drought were accompanied by larger maximum precipitation deficits and greater maximum potential evaporation. The extreme-drought years were accompanied also by potential evaporation values that were a greater percentage of the maximum precipitation minus evaporation total.

In the NA region, extreme dry summers occurred more often during LN years, while moderate drought occurred more often during EN years. The opposite was found within the EE/WR region. The only result supported by statistical testing was the distribution of extreme summer drought in the NA region, but this result was weak.

Examining the synoptic–dynamic character of the NA drought summers demonstrated that these occurred in association with the more frequent occurrence of zonal NH flow regimes or ECMs (significant at $p = 0.01$). Extreme dry summers were separated from moderate dry summers by significantly fewer and weaker blocking events ($p = 0.05$ and stronger) and a negative PNA regime. For extreme dry summers, the positive 500 hPa anomaly was located above the study region rather than to the west as for moderate drought summers.

The synoptic–dynamic character of extreme drought within the EE/WR region showed a very strong negative EU teleconnection pattern similar to quasi-stationary long period Rossby Wave Trains found in other regions of the world. These summers were accompanied by significantly more blocking, although not necessarily stronger blocking events, as well as relatively more meridional NH flow regimes or ECMs. For moderate drought years, there was significantly weaker and less blocking, as well as a significantly more zonal NH flow.

In both regions, extreme dry summers were a continuation of the atmospheric flow regimes from the spring season, whereas, for moderate dry summers, the spring season flow regime was different from that of the summer season. In addition, for both regions, wet summer seasons displayed synoptic–dynamic characteristics that were either opposite in terms of the teleconnections, blocking character, or associated with more meridional (NA) or zonal (EE/WR) ECMs.

In summary, this study demonstrated the utility of the criterion used and the results provide recognizable and distinct atmospheric circulation patterns that could be used to identify possible drought summers. Thus, these results would have use for seasonal and sub-seasonal forecasting application.

Author Contributions: Conceptualization, N.K.K. and A.R.L.; methodology, all co-authors; software, A.R.L.; validation, all co-authors; formal analysis, all co-authors; investigation, all co-authors; resources, all co-authors; data curation, N.K.K. and A.R.L.; writing—original draft preparation, A.R.L.; writing—review and editing, A.R.L., I.G.S. and M.G.L.; visualization, all co-authors; supervision, not applicable; project administration, not applicable; funding acquisition, not applicable. All authors have read and agreed to the published version of the manuscript.

Funding: This research received no external funding.

Institutional Review Board Statement: Not applicable.

Informed Consent Statement: Not applicable.

Data Availability Statement: The analyzed data can be found on the computers within the Global Climate Change Laboratory at the University of Missouri. Other information, such as the archive for the ECM or blocking, can be found in the reference section.

Acknowledgments: The authors acknowledge, with deep gratitude, the contribution of the two anonymous reviewers whose time and effort in providing comments has improved this paper substantially.

Conflicts of Interest: The authors declare no conflict of interest.

References

1. Intergovernmental Panel on Climate Change (IPCC). Climate Change 2013: The Physical Scientific Basis. 2013. Available online: http://www.ipcc.ch (accessed on 8 June 2021).
2. Wilhite, D.A. *Drought: A Global Assessment*; Routledge: London, UK, 2000; 752p.
3. Ahrens, C.D.; Henson, R. *Meteorology Today: An Introduction to Weather, Climate, and the Environment*, 11th ed.; Cengage Learning: Boston, MA, USA, 2000.

4. American Meteorological Society (AMS). Drought. 2013. Available online: https://www2.ametsoc.org/ams/index.cfm/about-ams/ams-statements/statements-of-the-ams-in-force/drought/ (accessed on 28 July 2021).
5. Cherenkova, Y.A. Quantitative evaluation of atmospheric drought in federal districts of the European Russia. *Proc. Russ. Acad. Sciences. Geogr. Ser.* **2013**, *6*, 76–85. (In Russian)
6. Heim, R.R., Jr. A review of Twentieth-Century drought indices used in the United States. *Bull. Am. Meteorol. Soc.* **2002**, *83*, 1149–1165. [CrossRef]
7. McKee, T.B.; Doesken, N.J.; Kleist, J. The Relationship of Drought Frequency and Duration to Time Scales. In Proceedings of the 8th Conference on Applied Climatology, Anaheim, CA, USA, 17–23 January 1993; American Meteorological Society: Boston, MA, USA. Available online: https://www.droughtmanagement.info/literature/AMS_Relationship_Drought_Frequency_Duration_Time_Scales_1993.pdf (accessed on 11 August 2021).
8. Vicente-Serrano, S.M.; Begueria, S.; Lopez-Moreno, J.I. A multi-scalar drought index sensitive to global warming: The Standardized Precipitation Evapotranspiration Index. *J. Clim.* **2010**, *23*, 1696–1718. [CrossRef]
9. World Meteorological Organization. *Standardized Precipitation Index User Guide (WMO-No.1090)*; World Meteorological Organization: Geneva, Switzerland, 2012; Available online: http://www.droughtmanagement.info/literature/WMO_standardized_precipitation_index_user_guide_en_2012.pdf (accessed on 20 June 2021).
10. National Drought Mitigation Center. 2021. Available online: https://www.drought.unl.edu/ (accessed on 3 June 2021).
11. Kleshchenko, A.D.; Dolgii-Trach, V.A. The development strategy of agrometeorological monitoring. *Russ. Meteorol. Hydrol.* **2011**, *36*, 492. [CrossRef]
12. Ukrainian Hydrometeorological Center. 2021. Available online: https://meteo.gov.ua (accessed on 5 July 2021).
13. Ukrainian National Space Facilities Control and Test Center. 2021. Available online: https://spacecenter.gov.ua (accessed on 5 July 2021).
14. European Drought Observatory (EDO). 2021. Available online: https://edo.jrc.ec.europa.eu/edov2/php/index.php?id=1000. (accessed on 5 July 2021).
15. The Joint Research Centre—EU Science Hub. 2021. Available online: https://ec.europa.eu/jrc/en (accessed on 5 July 2021).
16. Hu, Q.; Buyanovsky, G. Climate effects on corn yield in Missouri. *J. Appl. Meteorol.* **2003**, *42*, 1626–1635. [CrossRef]
17. Mehta, V.M.; Mendoza, K.; Daggupati, P.; Srinivasan, R.; Rosenberg, N.J.; Deb, D. High resolution simulations of decadal climate variability impacts on water yield in the Missouri River Basin with the Soil and Water Assessment Tool (SWAT). *J. Hydrometeorol.* **2016**, *17*, 2455–2476. [CrossRef]
18. Mannocchi, F.; Todisco, F.; Vergini, L. Agricultural drought: Indices, definition and analysis. In *The Basis of Civilization—Water Science? Proceedings of the UNESCO/IAHS/IWIIA Symposium, Rome, Italy, December 2003*; IAHS Publication 286; IAHS: Wallingford, UK, 2004; pp. 246–254. Available online: http://hydrologie.org/redbooks/a286/iahs_286_0246.pdf (accessed on 30 July 2021).
19. Ding, Y.; Hayes, M.J.; Widhalm, M. Measuring economic impacts of drought: A review and discussion. *Disaster Prev. Manag. Int. J.* **2011**, *20*, 434–446. [CrossRef]
20. Atlas of Mortality and Economic Losses from Weather and Climate Extremes 1970–2012. Available online: https://public.wmo.int/en/resources/library/atlas-mortality-and-economic-losses-weather-and-climate-extremes-1970-2012 (accessed on 28 July 2021).
21. Engström, J.; Jafarzadegan, K.; Moradkhani, H. Drought vulnerability in the United States: An integrated assessment. *Water* **2020**, *12*, 2033. [CrossRef]
22. Naumann, G.; Barbosa, P.; Garrote, L.; Iglesias, A.; Vogt, J. Exploring drought vulnerability in Africa: An indicator based analysis to be used in early warning systems. *Hydrol. Earth Syst. Sci.* **2014**, *18*, 1591. [CrossRef]
23. Namias, J. Anatomy of great plains protracted heat waves (especially the 1980 US. summer drought). *Mon. Weather Rev.* **1982**, *110*, 824–838. [CrossRef]
24. Namias, J. Some causes of the United States drought. *J. Clim. Appl. Meteorol.* **1983**, *22*, 30–39. [CrossRef]
25. Lupo, A.R.; Bosart, L.F. An analysis of a relatively rare case of continental blocking. *Q. J. R. Meteorol. Soc.* **1999**, *125*, 107–138. [CrossRef]
26. Kung, E.C.; Chern, J.-G. Prevailing anomaly patterns of the Global Sea Surface temperatures and tropospheric responses. *Atmósfera* **1995**, *8*, 99–114.
27. Gershanov, A.; Barnett, T.P. Interdecadal modulation of ENSO teleconnections. *Bull. Am. Meteorol. Soc.* **1998**, *79*, 2715–2725. [CrossRef]
28. Lupo, A.R.; Kelsey, E.P.; Weitlich, D.K.; Mokhov, I.I.; Akyuz, F.A.; Guinan, P.E.; Woolard, J.E. Interannual and interdecadal variability in the predominant Pacific Region SST anomaly patterns and their impact on a local climate. *Atmosfera* **2007**, *20*, 171–196.
29. Lupo, A.R.; Kelsey, E.P.; Weitlich, D.K.; Davis, N.A.; Market, P.S. Using the monthly classification of global SSTs and 500 hPa height anomalies to predict temperature and precipitation regimes one to two seasons in advance for the mid-Mississippi region. *Natl. Weather Dig.* **2008**, *32*, 11–33.
30. Birk, K.; Lupo, A.R.; Guinan, P.E.; Barbieri, C.E. The interannual variability of midwestern temperatures and precipitation as related to the ENSO and PDO. *Atmofera* **2010**, *23*, 95–128.
31. Henson, C.B.; Lupo, A.R.; Market, P.S.; Guinan, P.E. ENSO and PDO-related climate climate variability impacts on Midwestern United States crop yields. *Int. J. Biometeorol.* **2017**, *61*, 857–867. [CrossRef]

32. Kahya, E.; Dracup, J.A. US streamflow patterns in relation to the El Niño/Southern Oscillation. *Water Resour. Res.* **1993**, *29*, 2491–2503. [CrossRef]
33. Dracup, J.A.; Kahya, E. The relationships between US streamflow and La Niña events. *Water Resour. Res.* **1994**, *30*, 2133–2141. [CrossRef]
34. Ting, M.; Wang, H. Summertime U.S. Precipitation Variability and Its Relation to Pacific Sea Surface Temperature. *J. Clim.* **1997**, *10*, 1853–1873. [CrossRef]
35. Mo, K.C.; Schemm, J.E. Relationships between ENSO and drought over the southeastern United States. *Geophys. Res. Lett.* **2008**, *35*, L15701. [CrossRef]
36. Mo, K.C.; Schemm, J.E. Drought and persistent wet spells over the United States and Mexico. *J. Clim.* **2008**, *21*, 980–994. [CrossRef]
37. Rajagopalan, B.; Cook, E.; Lall, U.; Ray, B.K. Spatiotemporal Variability of ENSO and SST Teleconnections to Summer Drought over the United States during the Twentieth Century. *J. Clim.* **2000**, *13*, 4244–4255. [CrossRef]
38. Cook, E.R.; Seager, R.; Cane, M.A.; Stahle, D.W. North American drought: Reconstructions, causes, and consequences. *Earth Sci. Rev.* **2007**, *81*, 93–134. [CrossRef]
39. Ionita, M.; Dima, M.; Nagavciuc, V.; Scholz, P.; Lohmann, G. Past megadroughts in central Europe were longer, more severe, and less warm that modern droughts. *Nat. Commun. Earth Environ.* **2021**, *2*, 61. [CrossRef]
40. Zolotokrylin, A.N. Droughts and desertification in subboreal landscapes of Russia. *Izv. Russ. Acad. Sci. Geogr.* **2013**, *5c*, 64–73.
41. Spinoni, J.; Barbosa, P.; Bucchignani, E.; Cassano, J.; Cavazos, T.; Christensen, J.H.; Christensen, O.B.; Coppola, E.; Evans, J.; Geyer, B.; et al. Future global meteorological drought hot spots: A study based on CORDEX data. *J. Clim.* **2020**, *33*, 3635–3661. [CrossRef]
42. Jaagus, J.; Aasa, A.; Aniskevich, S.; Boincean, B.; Bojariu, R.; Briede, A.; Danilovich, I.; Castro, F.D.; Dumitrescu, A.; Labuda, M.; et al. Long-term changes in drought indices in eastern and central Europe. *J. Clim.* **2021**, *34*. [CrossRef]
43. Cherenkova, E.; Semenova, I.; Bardin, M.; Zolotokrylin, A.N. Drought and grain crop yields over the East European Plain under influence of quasibiennial oscillation of global atmospheric processes. *Int. J. Atmos. Sci.* **2015**, *2015*, 932474. [CrossRef]
44. Cherenkova, E.A. Dangerous atmospheric droughts in the European part of Russia under the conditions of summer warming. *Fundam. Appl. Climatol.* **2017**, *2*, 130–143. (In Russian) [CrossRef]
45. Lupo, A.R.; Mokhov, I.I.; Akperov, A.G.; Chernokulsky, A.V.; Hussain, A. A dynamic analysis of the role of the planetary and synoptic scale in the summer of 2010 blocking episodes over the European part of Russia. *Adv. Meteorol.* **2012**, *2012*, 584257. [CrossRef]
46. Lupo, A.R.; Mokhov, I.I.; Chendev, Y.G.; Lebedeva, M.G.; Akperov, M.; Hubbart, J.A. Studying summer season drought in western Russia. *Adv. Meteorol.* **2014**, *2014*, 942027. [CrossRef]
47. Lebedeva, M.G.; Krymckaya, O.V.; Lupo, A.R.; Chendev, Y.G.; Petin, A.N.; Solovyov, A. Trends in summer season climate for Eastern Europe and Southern Russia in the early 21st century. *Adv. Meteorol.* **2016**, *2016*, 5035086. [CrossRef]
48. Kononova, N.K. *The Classification of Northern Hemisphere Circulation Mechanisms According to B.L. Dzerdzeevskii*; Izdatel'stvo "Voyentekhinizdat": Moscow, Russia, 2009.
49. Kononova, N.K. Circulation of the atmosphere in the European sector of the northern hemisphere in the 21st century and fluctuations in air temperature in the Crimea. *Geopolit. Ecogeodyn. Reg.* **2014**, *10*, 633–639.
50. Kononova, N.K. Circulating epochs in the sectors of the northern hemisphere from 1899–2014. *Geopolit. Ecogeodyn. Reg.* **2015**, *11*, 56–66.
51. Kononova, N.K. Fluctuations in the global atmospheric circulation in the 20th–21st centuries. *Complex Syst.* **2016**, *4*, 22–37.
52. Kononova, N.K.; Lupo, A.R. 2019: Investigation of the variability of circulation regimes and dangerous weather phenomena in Russia in the 21st century. *IOP Conf. Ser. Earth Environ. Sci.* **2020**, *606*, 012023. [CrossRef]
53. Dzerdzeevskii, B.L.; Kurganskaya, V.M.; Vitviskaya, Z.M. The Classification of Circulation Mechanisms in the Northern Hemisphere and the Characteristics of Synoptic Seasons. In *Synoptic Meteorology*; Gidrometizdat: Leningrad, Russia, 1946.
54. Kononova, N.K.; Lupo, A.R. Changes in the Dynamics of the Northern Hemisphere Atmospheric Circulation and the Relationship to Surface Temperature in the 20th and 21st Centuries. *Atmosphere* **2020**, *11*, 255. [CrossRef]
55. Lebedeva, M.G.; Lupo, A.R.; Chendev, Y.G.; Krymskaya, O.V.; Solovyev, A.B. Changes in the atmospheric circulation conditions and regional climatic characteristics in two remote regions since the mid-20th century. *Atmosphere* **2019**, *10*, 11. [CrossRef]
56. Barnston, A.G.; Livesey, R.E. Classification, seasonality and persistence of low-frequency atmospheric circulation patterns. *Mon. Weather Rev.* **1987**, *115*, 1083–1126. [CrossRef]
57. Slizhe, M.; Semenova, I.; Pianova, I.; El Hadri, Y. Dynamics of macrocirculation processes accompanying by the dry winds in Ukraine in the present climatic period. *Hrvat. Meteorološki Časopis* **2018**, *53*, 17–29. Available online: https://hrcak.srce.hr/231265 (accessed on 30 July 2021).
58. Semenova, I.; Slizhe, M. Synoptic conditions of droughts and dry winds in the Black Sea Steppe Province under recent decades. *Front. Earth Sci.* **2020**, *8*, 69. [CrossRef]
59. Kalnay, E.; Kanamitsu, M.; Kistler, R.; Collins, W.; Deaven, D.; Gandin, L.; Iredell, M.; Saha, S.; White, G.; Woollen, J.; et al. The NCEP/NCAR 40-year reanalysis project. *Bull. Am. Meteorol. Soc.* **1996**, *77*, 437–471. [CrossRef]
60. University of Missouri Blocking Archive. 2021. Available online: http://weather.missouri.edu/gcc/ (accessed on 17 June 2021).
61. Lupo, A.R. Atmospheric Blocking Events: A Review. *Ann. N. Y. Acad. Sci.* **2020**. [CrossRef]
62. Lupo, A.R.; Jensen, A.D.; Mokhov, I.I.; Timazhev, A.V.; Eichler, T.; Efe, B. Changes in global blocking character during the most recent decades. *Atmosphere* **2019**, *10*, 92. [CrossRef]

63. Nunes, M.J.; Lupo, A.R.; Lebedeva, M.G.; Chendev, Y.G.; Solovyov, A.B. The occurrence of extreme monthly temperatures and precipitation in two global regions. *Pap. Appl. Geogr.* **2017**, *3*, 143–156. [CrossRef]
64. National Oceanic and Atmospheric Administration (NOAA) Climate Prediction Center (CPC) Teleconnections. Available online: https://www.cpc.ncep.noaa.gov/products/precip/CWlink/daily_ao_index/month_ao_index.shtml (accessed on 9 June 2021).
65. Kononova, N.K. Fluctuations in the Atmospheric Circulation of the Northern Hemisphere in the 20th and Early 21st Century. Available online: https://atmospheric-circulation.ru/ (accessed on 17 June 2021).
66. Center for Ocean and Atmosphere Prediction Studies. Available online: http://www.coaps.fsu.edu (accessed on 10 June 2021).
67. Hanley, D.E.; Bourassa, M.A.; O'Brien, J.J.; Smith, S.R.; Spade, E.R. A Quantitative Evaluation of ENSO Indices. *J. Clim.* **2003**, *16*, 1249–1258. [CrossRef]
68. Ukraine: Agricultural Overview. World Data Center for Geoinformatics and Sustainable Development—WDC Ukraine. 2021. Available online: http://wdc.org.ua/en/node/29 (accessed on 2 July 2021).
69. Dzerdzeevskii, B.L. Circulation Mechanisms in the Atmosphere of the Northern Hemisphere in the Twentieth Century. In *Materials of Meteorological Research*; Institute of Geography of the USSR Academy of Sciences and the Interagency Geophysical, Committee under the Presidium of the USSR Academy of Sciences: Moscow, Russia, 1968; p. 240.
70. Kononova, N.K. Meteorological extremums in Siberia in 2019 and their connection with circulation of the atmosphere. *Environ. Dyn. Glob. Clim. Chang.* **2019**, *10*, 110–119. [CrossRef]
71. Svoboda, M.D.; LeComte, D.; Hayes, M.; Heim, R.; Gleason, K.; Angel, J.; Rippey, B.; Tinker, R.; Palecki, M.; Stooksbury, D.; et al. The drought monitor. *Bull. Am. Meteorol. Soc.* **2002**, *83*, 1181–1190. [CrossRef]
72. Jensen, A.D.; Lupo, A.R.; Mokhov, I.I.; Akperov, M.G.; Reynolds, D.D. Integrated regional enstrophy and block intensity as a measure of Kolmogorov Entropy. *Atmosphere* **2017**, *8*, 237. [CrossRef]
73. Fernandez, M.A.; Huang, P.; McCarl, B.; Mehta, V.M. Value of decadal climate variability information for agriculture in the Missouri River Basin. *Clim. Chang.* **2016**, *139*, 517–533. [CrossRef]
74. Rhodes, L.A.; McCarl, B.A. The value of ocean decadal climate variability information to United States agriculture. *Atmosphere* **2020**, *11*, 318. [CrossRef]
75. Stambaugh, M.C.; Guyette, R.P.; McMurry, E.R.; Cook, E.R.; Meko, D.M.; Lupo, A.R. Drought duration and frequency in the U.S. Corn Belt during the last millennium (AD 992–2004). *Agric. For. Meteorol.* **2011**, *151*, 154–162. [CrossRef]
76. Cook, B.I.; Smerdon, J.E.; Seagar, R.; Cook, E.R. Pan-Continental Droughts in North America over the Last Millennium. *J. Clim.* **2014**, *27*, 383–397. [CrossRef]
77. Wilks, D.S. *Statistical Methods in the Atmospheric Sciences*, 2nd ed.; International Geophys Series Number 91; Academic Press: Cambridge, MA, USA, 2006; p. 627.
78. Lebedeva, M.G.; Lupo, A.R.; Henson, C.B.; Solovyov, A.B.; Chendev, Y.G.; Market, P.S. A Comparison of Bioclimatic potential of Two Global Regions during the Late 20th Century and Early 21st Century. *Int. J. Biometeorol.* **2017**, *62*, 609–620. [CrossRef]
79. Wallace, J.M.; Gutzler, D.S. Teleconnections in the geopotential height field during the Northern Hemisphere winter. *Mon. Weather Rev.* **1981**, *109*, 784–812. [CrossRef]
80. Wang, N.; Zhang, Y. Evolution of Eurasian teleconnection pattern and its relationship to climate anomalies in China. *Clim. Dyn.* **2015**, *44*, 1017–1028. [CrossRef]
81. Renwick, J.A.; Revell, M.J. Blocking over the South Pacific and Rossby wave propagation. *Mon. Weather Rev.* **1999**, *127*, 2233–2247. [CrossRef]
82. Jiang, X.; Lau, N.-C. Intraseasonal teleconnection between North American and Western North Pacific monsoons with 20-day time scale. *J. Clim.* **2008**, *21*, 2664–2679. [CrossRef]
83. Wang, Y.; Lupo, A.R. An extratropical air-sea interaction over the North Pacific in association with a preceding El Niño episode in early summer. *Mon. Weather Rev.* **2009**, *137*, 3771–3785. [CrossRef]
84. Zhao, P.; Cao, Z.; Chen, J. A summer teleconnection pattern over the extratropical Northern Hemisphere and associated mechanisms. *Clim. Dyn.* **2010**, *35*, 523–534. [CrossRef]
85. Wang, Y.; Lupo, A.R.; Qin, J. A response in the ENSO cycle to an extratropical forcing mechanism during the El Niño to La Niña transition. *Tellus Ser. A Dyn. Meteorol. Oceanogr.* **2013**, *65*, 22431. [CrossRef]

 *atmosphere*

Article

The Analysis of Long-Term Trends in the Meteorological and Hydrological Drought Occurrences Using Non-Parametric Methods—Case Study of the Catchment of the Upper Noteć River (Central Poland)

Katarzyna Kubiak-Wójcicka [1,2,*], Agnieszka Pilarska [1] and Dariusz Kamiński [3,4]

[1] Faculty of Earth Sciences and Spatial Management, Nicolaus Copernicus University, Lwowska 1, 87-100 Toruń, Poland; apilarska@umk.pl
[2] Centre for Underwater Archaeology, Nicolaus Copernicus University, Lwowska 1, 87-100 Toruń, Poland
[3] Faculty of Biological and Veterinary Sciences, Nicolaus Copernicus University, Lwowska 1, 87-100 Toruń, Poland; daro@umk.pl
[4] Centre for Climate Change Research, Nicolaus Copernicus University, Lwowska 1, 87-100 Toruń, Poland
* Correspondence: kubiak@umk.pl

Abstract: The study aims to identify long-term trends in the changes of drought occurrences using the Mann-Kendall (MK) test and the Theil-Sen estimator. Trend research was carried out on the example of the catchment area of the Upper Noteć River, which covers an agricultural area of Poland with some of the lowest water reserves. The meteorological droughts were identified based on the Standardized Precipitation Index (SPI), while the hydrological droughts were determined on the basis of the Standardized Runoff Index (SRI) in various time scales (1, 3, 6, 9 and 12 months) in the period of 1981–2016. The relationship between SPI and SRI was determined on the basis of the Pearson correlation analysis. The results showed that statistically significant trends (at the significance level of 0.05) were identified at 3 out of 8 meteorological stations (downward trend at Kłodawa station and upward trend for drought at Sompolno and Kołuda Wielka stations). Statistically significant hydrological droughts showed an increase in occurrences at the Łysek station, while a downward trend was noted at the Noć Kalina station. No trend was found at the Pakość station. The analysis of the correlation between meteorological and hydrological droughts showed a strong relationship in dry years. The maximum correlation coefficient was identified in longer accumulation periods i.e., 6 and 9 months. The example of the catchment of the Upper Noteć River points to the necessity of using several indicators in order to assess the actual condition of the water reserves.

Keywords: meteorological drought; hydrological drought; trends; central Poland

Citation: Kubiak-Wójcicka, K.; Pilarska, A.; Kamiński, D. The Analysis of Long-Term Trends in the Meteorological and Hydrological Drought Occurrences Using Non-Parametric Methods—Case Study of the Catchment of the Upper Noteć River (Central Poland). *Atmosphere* **2021**, *12*, 1098. https://doi.org/10.3390/atmos12091098

Academic Editors: Andrzej Walega and Agnieszka Ziernicka-Wojtaszek

Received: 16 July 2021
Accepted: 20 August 2021
Published: 25 August 2021

Publisher's Note: MDPI stays neutral with regard to jurisdictional claims in published maps and institutional affiliations.

1. Introduction

Drought is a natural disaster characterized by long-term water scarcity [1–3]. Drought is one of the most serious natural threats that causes damage to various aspects of the environment, society, and economy [4,5]. No universal definition of drought has been established due to the wide variability in water supply and demand worldwide [6,7]. There are four categories of drought in the literature: meteorological, agricultural, hydrological, and socioeconomic [8–13]. Drought damage is a serious issue in numerous countries around the world. Due to its nature, drought is difficult to monitor, and its effects are often poorly documented. Among the various sectors of the economy, agriculture is one of the most vulnerable to drought, where its effects are also most noticeable [14–16]. Numerous reports and scientific articles indicate that forecasts of future climate conditions suggest an increase in the frequency and intensity of droughts in some regions of the world [17–21]. The increase in the frequency of drought occurrence in recent years has not been limited to arid and semi-arid regions [22–26], but has been gradually becoming more common in regions with a temperate and humid climate [27–32]. Poland, which has one of the most limited water resources in Europe [33], is among those regions experiencing an increase in drought frequency [34]. The distribution of water resources in Poland has been diversified in terms of time and space. In the current climate, many regions of the country often suffer from water scarcity. In the future, this scarcity may become even more serious, and the availability of water resources might become limited. In recent years, only slight changes in annual precipitation totals have been recorded in Poland, however a noticeable shift has been observed in seasonal and monthly precipitation distribution [35,36]. Moreover, there have been quite significant changes in thermal conditions, characterized by a great increase in air temperature over a multi-year period [37]. As a result of these changes, temporary difficulties in water supply have been recorded in some areas of Poland [38]. This problem will be particularly harmful for the agricultural regions of the country. Polish agriculture is largely dependent on precipitation, which is highly variable both in terms of its temporal and spatial characteristics. Plant production is reliant mainly on water obtained from precipitation and available to plants by means of water-retentive soil [39]. In the event of a drought in agricultural areas, crop irrigation is required, especially during the growing season [40].

Drought studies in Poland have been conducted at a regional and local level. Previously published drought analyses mainly refer to the classification of drought types using various drought indicators [41–43], monitoring of drought conditions [44–47], as well as the characteristics of the drought, including its duration, intensity, size and frequency [48].

This study focuses on the research of drought trends in the period of 1981–2016, in a particularly drought-susceptible area of Poland. The analysis of drought trends in the long term might indicate the direction of possible near-future changes. The case study focuses on the catchment of the Upper Noteć River, which is a heavily exploited agricultural area with some of the lowest reserves of water.

The main objectives of this study included:

(1) the identification of meteorological droughts based on SPI indicators, and hydrological droughts based on SRI indicators in various time scales (1, 3, 6, 9 and 12 months)
(2) trend determination using the Mann-Kendall (MK) test and Theil-Sen estimator
(3) the determination of a relationship between SPI and SRI by means of Pearson's correlation analysis.

The obtained results will help the departments of state administration, responsible for water resources management, make informed decisions and establish a long-term local development strategy, regulating the sustainable management of water resources.

2. Materials and Methods

2.1. Study Area and Dataset

The research was carried out in Poland, a region located in a temperate climate zone with a predominance of polar-sea air masses. The amount of precipitation in Poland varies temporally and spatially. The average annual precipitation in Poland recorded in the period of 1981–2010 was 603 mm. The lowest annual precipitation total of 500 mm was observed in the central part of the country, and the highest—970 mm, in the south of Poland [49]. The area covered by the analysis includes the Upper Noteć catchment, closed by the water gauge in Pakość (Figure 1). The catchment area up to the Pakość station is 2301.98 km^2. This catchment is located in the historical region of Kujawy, which is extremely important for agriculture. Arable land within the catchment area accounts for 76.07% of the catchment area, while forest areas account for 11.4%. It is also the region with the lowest annual precipitation in Poland and the area with the highest water shortages in agriculture [50].

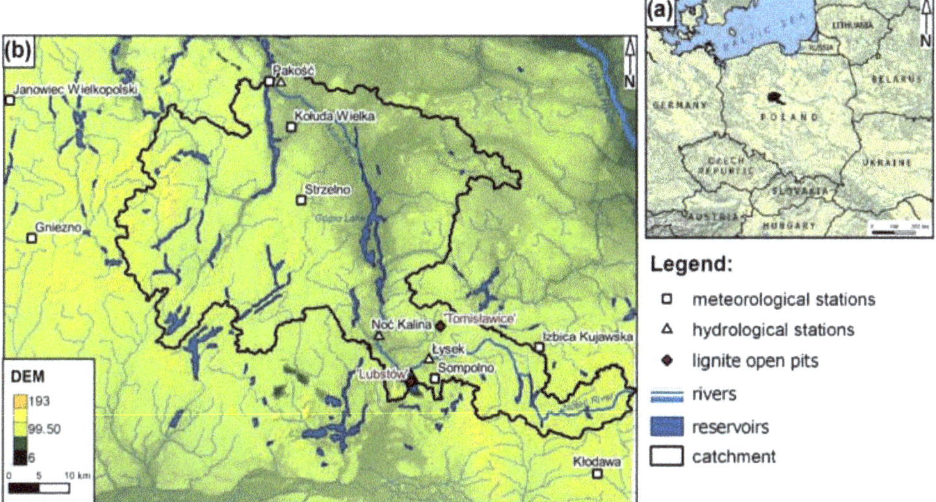

Figure 1. Study area. Source: (**a**) own elaboration made with [51,52]. Map elaborated in the coordinate system: WGS 84/UTM zone 34N; (**b**) own elaboration made with [52–54], location of the stations based on [55,56] and Table 1, location of the lignite open pits based on [56,57]. Meteorological station Izbica Kujawska is out of catchment area according to coordinates from Table 1. Location of the hydrological station Pakość is determined by cartographic issues. Map elaborated in the coordinate system: WGS 84/UTM zone 34N.

Table 1. Annual sum of atmospheric precipitation in the period of 1980–2016.

Meteorological Station	Altitude (m a.s.l.)	Latitude	Longitude	Total Precipitation during the Year (mm)		
				Average	Maximum/Year	Minimum/Year
Izbica Kujawska	120	52°26′ N	18°46′ E	529.3	817.6/2010	309.6/2011
Pakość	75	52°48′ N	18°05′ E	513.5	704.1/2010	291.4/1989
Kołuda Wielka	85	52°44′ N	18°09′ E	500.1	810.7/1980	212.5/1989
Strzelno	105	52°38′ N	18°11′ E	542.5	816.9/1980	246.6/1989
Sompolno	96	52°23′ N	18°31′ E	516.0	847.6/2010	302.7/1989
Gniezno	124	52°33′ N	17°34′ E	506.6	708.5/2010	282.2/1982
Janowiec Wielkopolski	95	52°46′ N	17°29′ E	519.6	760.2/2010	275.5/1982
Kłodawa	120	52°15′ N	18°55′ E	531.2	763.2/2001	306.4/1989

The data used in this work are derived from historical series of daily precipitation totals recorded at 8 meteorological stations (Izbica Kujawska, Kołuda Wielka, Pakość, Sompolno, Strzelno, Gniezno, Janowiec Wielkopolski, Kłodawa). Air temperature measurements were obtained from the Kołuda Wielka station. Discharge data were obtained from 3 hydrological stations: Łysek, Noć Kalina and Pakość. Meteorological and hydrological data for the period of 1980–2016 were obtained from the Institute of Meteorology and Water Management—National Research Institute. Daily values were converted into monthly values for the purpose of extended calculations.

In terms of annual precipitation, the Upper Noteć area is one of the regions with the lowest precipitation in Poland. Total annual precipitation in the period of 1980–2016 ranged from 500.1 mm (Kołuda Wielka) to 542.5 mm (Strzelno) (Table 1). The highest precipitation totals were recorded in 2010 at most meteorological stations. The exceptions were Kłodawa, with the maximum precipitation occurring in 2001, and Kołuda Wielka and Strzelno, with the maximum recorded precipitation in 1980. The lowest precipitation totals were recorded at most stations in 1989, with the exception of the stations in Gniezno and Janowiec, where the lowest annual precipitation were recorded in 1982, and Izbica Kujawska, where the lowest totals occurred in 2011. Air temperature measurements were carried out only at the meteorological station in Kołuda Wielka. The average annual air temperature at the meteorological station in Kołuda Wielka in 1980–2016 was 8.5 °C, and the highest average annual air temperature was recorded in 1989 (9.8 °C) (Figure 2).

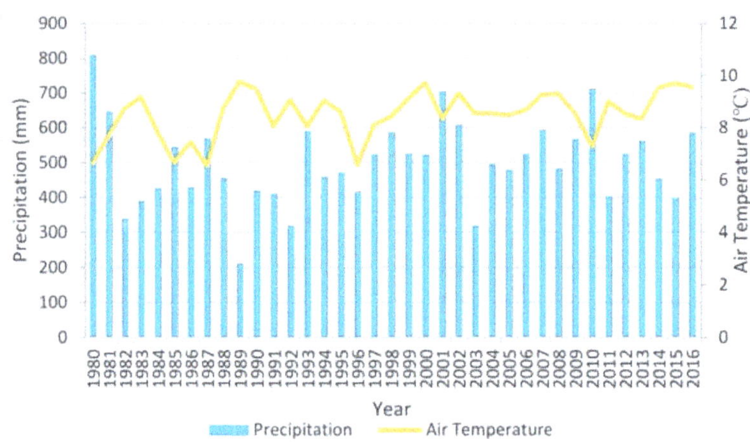

Figure 2. Average annual air temperature and average precipitation sums at the Kołuda Wielka meteorological station in the period of 1980–2016.

In terms of hydrology, the catchment area of the Upper Noteć is categorised as one of the areas with the most limited water resources [58]. According to Tomaszewski and Kubiak-Wójcicka [59], the average long-term unit runoff of the Noteć measured at the Pakość water gauge in the period of 1951–2015 amounted to 2.41 $dm^3 \cdot s^{-1} \cdot km^{-2}$. This is the lowest unit runoff value recorded in Poland, with an average of 5.5 $dm^3 \cdot s^{-1} \cdot km^{-2}$ [34]. Both precipitation and discharge in the catchment area of the Upper Noteć River are among the lowest in Poland. During the analysed period of 1980–2016, the unit runoff of the Noteć River at the Pakość station was 2.21 $dm^3 \cdot s^{-1} \cdot km^{-2}$. The highest values of the maximum discharge were recorded in July 1980 at all hydrological stations. The maximum discharge at the Pakość station was 69.3 $m^3 \cdot s^{-1}$ (unit runoff 30.1 $dm^3 \cdot s^{-1} \cdot km^{-2}$), and the lowest discharge was recorded in October 2003 (0.23 $dm^3 \cdot s^{-1} \cdot km^{-2}$) (Table 2). At the Noć Kalina station, the lowest discharge was recorded in September 1989 and August 1992. However, the lowest discharge at the Łysek station was recorded from August to December 2015 and in January 2016, at which time virtually no discharge occurred in the watercourse bed.

Table 2. Hydrological characteristics of the Noteć River in the period of 1980–2016—annual discharge.

Hydrological Station	The Catchment Area (km^2)	Average Multi-Year Discharge ($m^3 \cdot s^{-1}$)	Maximum Discharge ($m^3 \cdot s^{-1}$)	Minimum Discharge ($m^3 \cdot s^{-1}$)
Łysek	303.32	0.74	10.7	0.001
Noć Kalina	426.11	1.47	16.7	0.05
Pakość	2301.98	5.08	69.3	0.53

2.2. Standardized Precipitation Index (SPI) and Standardized Runoff Index (SRI)

Two indicators were used to determine droughts: the Standardized Precipitation Index (SPI), which defines meteorological droughts, and the Standardized Runoff Index (SRI), which defines hydrological droughts. Data from 8 meteorological stations: 5 located within the catchment area of the Noteć River (Pakość, Strzelno, Sompolno, Izbica Kujawska, Kołuda Wielka) and 3 in its close vicinity (Kłodawa, Gniezno and Janowiec Wielkopolski) were used to determine meteorological droughts.

The Standardized Precipitation Index (SPI) is one of the most frequently used indicators of a meteorological drought, and was developed on the basis of the normalization of precipitation probabilities [60]. This indicator defines a precipitation deficit and allows the monitoring of droughts in different time frames. The SPI is recommended by the World Meteorological Organisation (WMO) for determining the phenomenon of drought [61]. For more information on the formulation of SPI, its advantages and limitations, see papers [62–64].

The SPI was calculated on the basis of monthly precipitation totals for 5 meteorological stations in the upper catchment area of the Noteć River (Izbica Kujawska, Strzelno, Kołuda Wielka, Sompolno and Pakość) and 3 stations in its close vicinity (Kłodawa, Gniezno and Janowiec Wielkopolski). The daily precipitation data were aggregated into monthly time scales, and fitted to a two-parameter gamma distribution function. The SPI was calculated for each month at different timescales. Therefore, 5 different time series were analysed, i.e., 1-, 3-, 6-, 9- and 12-months. SPI values define the deviation from the median expressed in units of standard deviation, which was calculated according to the formula:

$$SPI,\ SRI = \frac{f(x) - u}{\sigma} \quad (1)$$

SPI, SRI—Standardized Precipitation Index, Standardized Runoff Index
$f(x)$—transformed sum of precipitation, discharges
μ—mean value of the normalized index x
σ—standard deviation of index x

In order to calculate the SPI, the compliance of the distribution of the transformed variable $f(P)$ with the normality distribution was tested using the x^2—Pearson normality test [65].

The Standardized Runoff Index (SRI) is calculated according to the same procedure as the SPI, however it is based on the discharge data [66–69]. A 2-parameter logarithmic function was used as a normalizing function when calculating the SRI [70]. The detailed calculation method is presented in the study [9,47]. The probabilities were transformed into standard normal distribution. The application of SPI allows for the differentiation of the intensity of a drought using a set of SPI thresholds: −1, −1.5, −2 and 1.0, 1.5, 2 for moderate, severe and extreme droughts and precipitations, respectively [28].

The proposed approach is based on the assessment of water resources under different hydroclimatic conditions and the determination of different intensity classes. The adoption of standardized indicators allowed for the classification of drought intensity, which is presented in Table 3. Extreme events were identified for the indicator values above 1.0, when rainy periods occur, and for the indicator values below −1.0, when there are droughts.

Table 3. The classification scale for SPI and SRI values.

SPI, SRI Value	Category
SPI/SRI ≥ 2.0	Extremely wet
2.0 > SPI/SRI ≥ 1.5	Severely wet
1.5 > SPI/SRI ≥ 1.0	Moderately wet
1.0 > SPI/SRI > −1.0	Normal
−1.0 ≥ SPI/SRI > −1.5	Moderately dry
−1.5 ≥ SPI/SRI > −2.0	Severely dry
SPI/SRI ≤ −2.0	Extremely dry

2.3. Mann–Kendall Test

The Mann–Kendall test [71,72] is a non-parametric statistical method used to determine whether a time series has a monotonic upward or a downward trend. It is a rank-based procedure that is particularly suitable for data with abnormal distribution that contain outliers and non-linear trends [73].

The Mann–Kendall S statistic is described with the Formula (2):

$$S = \sum_{k=1}^{n-1} \sum_{j=k+1}^{n} sgn(x_j - x_k) \quad (2)$$

$$sgn(x_j - x_k) = \begin{cases} +1\ if\ (x_j - x_k) > 0 \\ 0\ if\ (x_j - x_k) = 0 \\ -1\ if\ (x_j - x_k) < 0 \end{cases} \quad (3)$$

where:

x_j and x_k—values of the variable in individual years j and k, where $j > k$,
n—the series count (number of years).

Positive "S" values represent an upward trend, while negative values indicate a downward one. The calculation of "$sgn\ (x_j - x_k)$" is done via Equation (3).

The S statistic shows a tendency to quickly move towards normality, and for $n > 10$ this statistic has an approximately normal distribution with the mean of 0 and the variance described by the Formula (4):

$$Var(S) = [n(n-1)(2n+5)]/18 \quad (4)$$

The normalized Z test statistic is determined by the Formula (5):

$$Z = \begin{cases} \frac{S-1}{\sqrt{Var(S)}}\ if\ S > 0 \\ 0\ if\ S = 0 \\ \frac{S+1}{\sqrt{Var(S)}}\ if\ S < 0 \end{cases} \quad (5)$$

In the Mann–Kendall test, the null hypothesis is that there is no significant trend in the data series. The trend is significant if the null hypothesis cannot be accepted. The acceptance region at the significance level of $\alpha = 0.05$ is defined by the range of $-1.96 \leq Z \leq 1.96$ (no significant trend), while the rejection region was determined by $Z < -1.96$ (significant downward trend) and $Z > 1.96$ (significant upward trend), where Z is the normalized test statistic [47].

The non-parametric Mann–Kendall test is commonly used to quantify trends in hydrometeorological time series [74,75], despite some limitations [76–79].

2.4. Sen's Slope

The Mann–Kendall test is an effective method of identifying trends in a time series, but does not indicate the magnitude of the trends. The test might be supplemented with a non-parametric Sen's method. In order to estimate the actual slope of the existing trend, the non-parametric Sen's method was used [80]. The main advantage of the Sen's slope estimator is its resistance to the presence of extreme values [81].

The slope β expressed by the Theil–Sen estimator (β) is described by the Formula (6):

$$\beta = Median\left((x_j - x_k)/(j-k)\right) \quad (6)$$

A positive value of β indicates an upward (increasing) trend, and a negative value indicates a downward (decreasing) trend in the time series.

The Mann–Kendall test and Theil–Sen estimator were performed by means of a RStudio [82] with packages: "readxl" [83] and "trend" [84]. Information about what equations are used in the "trend" package for Mann–Kendall test and Theil–Sen estimator is available at [85].

QGIS ver. 3.10.9 and QGIS ver. 3.10.9 with GRASS ver. 7.8.3. were used in the study. Additionally, GIMP ver. 2.10.18 and Inkscape ver. 1.0.1 were used as graphic tools. In preparation of the Figure 3 the Inverse Distance Weighting (IDW) interpolation was used.

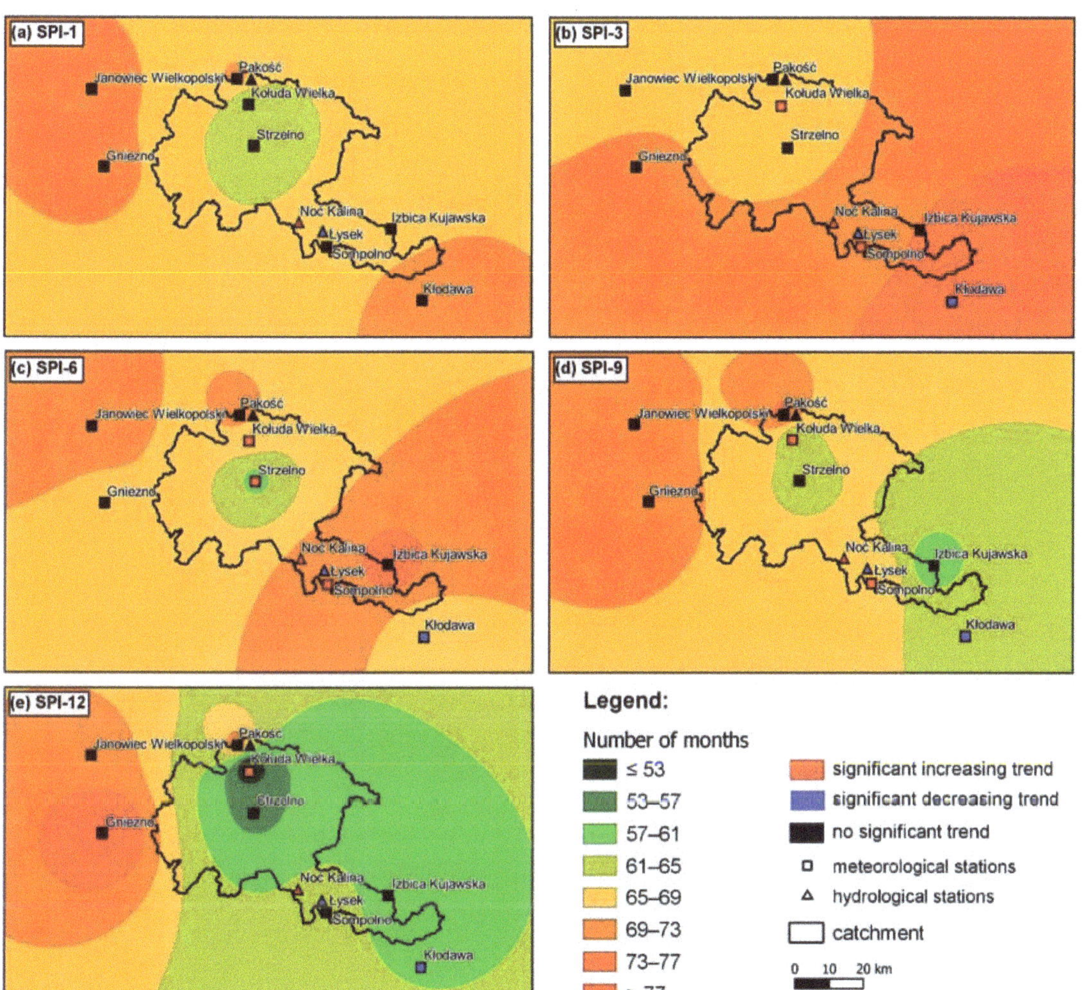

Figure 3. Spatial distribution of the number of months with SPI ≤ -1.0 in the period of 1981–2016. Source: (**a**–**e**) own elaboration made with [52], Table 4, Tables 6 and 7. Location of the stations based on [55,56] and Table 1. Meteorological station Izbica Kujawska is out of catchment area according to coordinates from Table 1. Location of the hydrological station Pakość is determined by cartographic issues. Maps elaborated in the coordinate system: WGS 84/UTM zone 34N.

Table 4. Meteorological drought parameters (SPI) in different time scales in the period of 1981–2016.

Parameters of Droughts	SPI−1	SPI−3	SPI−6	SPI−9	SPI−12
		Izbica Kujawska			
Number of months with SPI ≤ −1.0	67	73	75	59	59
Number of months with SPI ≥ 1.0	58	68	68	63	65
Minimum value of the index	−3.00	−2.81	−2.30	−2.93	−2.53
Maximum value of the index	3.59	2.83	2.53	2.47	2.33
		Sompolno			
Number of months with SPI ≤ −1.0	66	71	72	68	63
Number of months with SPI ≥ 1.0	61	57	66	67	66
Minimum value of the index	−3.50	−2.61	−2.70	−2.55	−2.57
Maximum value of the index	3.25	2.67	2.65	2.93	2.77
		Strzelno			
Number of months with SPI ≤ −1.0	62	68	60	64	56
Number of months with SPI ≥ 1.0	70	67	66	69	68
Minimum value of the index	−3.27	−3.01	−3.09	−3.13	−2.99
Maximum value of the index	3.37	2.94	2.39	2.90	2.28
		Kołuda Wielka			
Number of months with SPI ≤ −1.0	62	66	66	62	49
Number of months with SPI ≥ 1.0	65	68	72	55	55
Minimum value of the index	−3.43	−2.71	−3.32	−3.47	−3.38
Maximum value of the index	3.16	2.90	2.37	2.75	2.62
		Pakość			
Number of months with SPI ≤ −1.0	70	69	71	74	71
Number of months with SPI ≥ 1.0	66	70	69	69	72
Minimum value of the index	−3.54	−2.83	−2.80	−2.67	−2.69
Maximum value of the index	2.80	2.64	2.35	2.90	2.12
		Gniezno			
Number of months with SPI ≤ −1.0	70	70	68	72	79
Number of months with SPI ≥ 1.0	64	65	67	67	61
Minimum value of the index	−3.43	−2.95	−2.73	−2.73	−2.66
Maximum value of the index	2.50	2.37	2.45	2.70	2.10
		Janowiec Wielkopolski			
Number of months with SPI ≤ −1.0	71	67	72	70	71
Number of months with SPI ≥ 1.0	66	62	66	65	68
Minimum value of the index	−3.22	−2.83	−2.70	−2.79	−2.72
Maximum value of the index	2.73	2.26	2.35	2.09	2.07
		Kłodawa			
Number of months with SPI ≤ −1.0	73	81	66	65	61
Number of months with SPI ≥ 1.0	52	72	74	68	69
Minimum value of the index	−3.09	−2.92	−2.60	−2.72	−2.46
Maximum value of the index	3.26	2.26	2.35	2.58	2.35

2.5. Pearson's Correlation Analysis

The Pearson correlation coefficient (r) was used to detect the relationship between meteorological droughts and hydrological droughts. With the use of this coefficient, it was possible to determine the linear relationship between the SPI and SRI variables in different accumulation periods. The value of the correlation coefficient is in a closed range [−1, 1]. The greater its absolute value, the stronger the linear relationship between the variables. 0—means there is no linear relation, 1—means a positive relation, and −1—means a negative relation between the variables.

3. Results and Discussion

3.1. The Characteristics of Droughts in the Period of 1981–2016

SPI values were calculated separately for all eight weather stations for the time scales of 1, 3, 6, 9 and 12 months. Depending on the selected accumulation period, the range of SPI values was different for individual meteorological stations, as follows: SPI−1 from −3.54 to 3.59, SPI−3 from −3.01 to 2.94, SPI−6 from −3.32 to 2.65, SPI−9 from −3.47 to 2.93 and SPI−12 from −3.38 to 2.77 (Table 4). The longer the indicator's accumulation period, the lower the drought intensity. The most intense meteorological droughts were recorded in the following periods: 1982–1985, 1989–1996, 2002–2006, 2008–2009. The number of months with drought occurrences varied, depending on the period of accumulation and distribution of individual meteorological stations. In the 1-month accumulation period, the number of months with drought occurrences (values ≤ -1.0) ranged from 62 to 73 months i.e., from 14.3% to 16.9% of all the months in the analysed multi-year period of 1981–2016. On the other hand, wet months with SPI values ≥ 1.0 lasted from 52 to 71 months i.e., from 12.0% to 16.4% of the analysed time period. In the case of SPI−1, the droughts lasted the longest in Kłodawa, and were the shortest in Kołuda and Strzelno. The highest number of months with drought occurrences (SPI ≤ -1.0) in the 3-month accumulation period was recorded at the Kłodawa station (81 months), and the lowest in Kołuda Wielka (66 months). In the longer accumulation periods i.e., 6, 9 and 12 months, the number of months was similar and amounted to 60 to 75 months for SPI−6, 59 to 74 months for SPI−9 and 59 to 79 months for SPI−12.

Greater differentiation was recorded in the cases of hydrological drought occurrences (SRI) at the station in Łysek, Noć Kalina and Pakość (Table 5). The SRI values in Łysek were characterized by the largest range i.e., from −5.22 to 2.00. The most intense droughts were recorded from April to December of 2016, which was related to relatively minor discharges of the Noteć River during this period. At the Noć Kalina station, the SRI values recorded were from −3.16 to 2.49. The most intense hydrological droughts were recorded from May to September of 1990, depending on the period of accumulation. The SRI values in Pakość were characterized by a lower amplitude and ranged from −2.33 to 3.05. The number of months with SRI values below −1.0 ranged from 75 to 81 months, and in Noć Kalina—from 77 to 101 months.

Table 5. The Parameters of a hydrological drought (SRI) in different time scales in the period of 1981–2016.

Parameters of Droughts	SRI−1	SRI−3	SRI−6	SRI−9	SRI−12
		Pakość			
Number of months with SPI ≤ -1.0	75	77	77	81	79
Number of months with SPI ≥ 1.0	68	67	73	75	76
Minimum value of the index	−2.33	−2.20	−2.20	−1.93	−1.71
Maximum value of the index	2.34	2.30	2.83	3.05	2.71
		Noć Kalina			
Number of months with SPI ≤ -1.0	77	83	86	101	99
Number of months with SPI ≥ 1.0	72	76	67	58	59
Minimum value of the index	−3.16	−3.10	−2.79	−2.23	−2.25
Maximum value of the index	2.49	2.31	1.97	2.05	1.92
		Łysek			
Number of months with SPI ≤ -1.0	29	28	28	30	36
Number of months with SPI ≥ 1.0	26	31	31	32	40
Minimum value of the index	−5.22	−5.18	−5.09	−4.96	−4.77
Maximum value of the index	1.52	1.64	1.40	2.00	1.77

The hydrological droughts (SRI ≥ -1.0) at the station in Łysek were characterized by a short duration, from 6.5% to 8.3% of the analysed period, and high intensity. The number of wet months (SRI ≥ 1.0) ranged from 6.0% to 9.3% of all the months of the analysed multiannual period. The largest number of months with SRI values ≤ -1.0 was recorded at the Noć Kalina station i.e., from 17.9% to 23.4% of the analysed multi-year period, while the wet months constituted between 13.4% and 17.6% of this period. The hydrological station in Pakość closes the catchment area of the Upper Noteć River. The lower intensity of hydrological droughts at the Pakość station and their much shorter duration compared to the Noć Kalina and Łysek stations may have resulted from anthropogenic activities carried out in the area. These activities were mainly related to the lignite open pits, which ran south of the Łysek and Noć Kalina stations (Lubstów), as well as the operation of a new open pit mine located north of the Łysek station (Tomisławice).

The spatial distribution of the number of months with meteorological drought occurrences in the catchment area of the Upper Noteć is shown in Figure 3. The number of months with droughts in the accumulation period of 1 month does not present large spatial differentiation. Larger differences are noticeable in the period of 3- and 12-month accumulation. In the case of SPI−3, the months with droughts in the southern part of the catchment lasted the longest, while in the SPI−12 period they were the longest in the western part of the catchment.

Figure 4 shows the average values of SPI from 8 stations as well as the development of hydrological drought occurrences (SRI) recorded at the station in Pakość. The number of months with averaged meteorological droughts is between 13.4 and 15% of all the analysed months. As presented in Figure 4, the development of hydrological droughts is related to meteorological droughts.

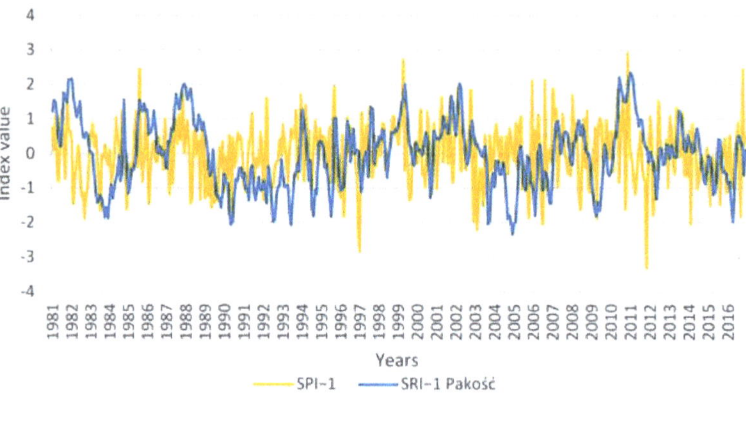

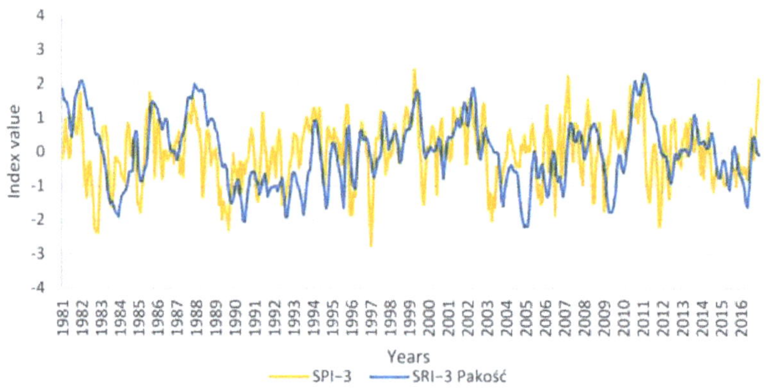

Figure 4. *Cont.*

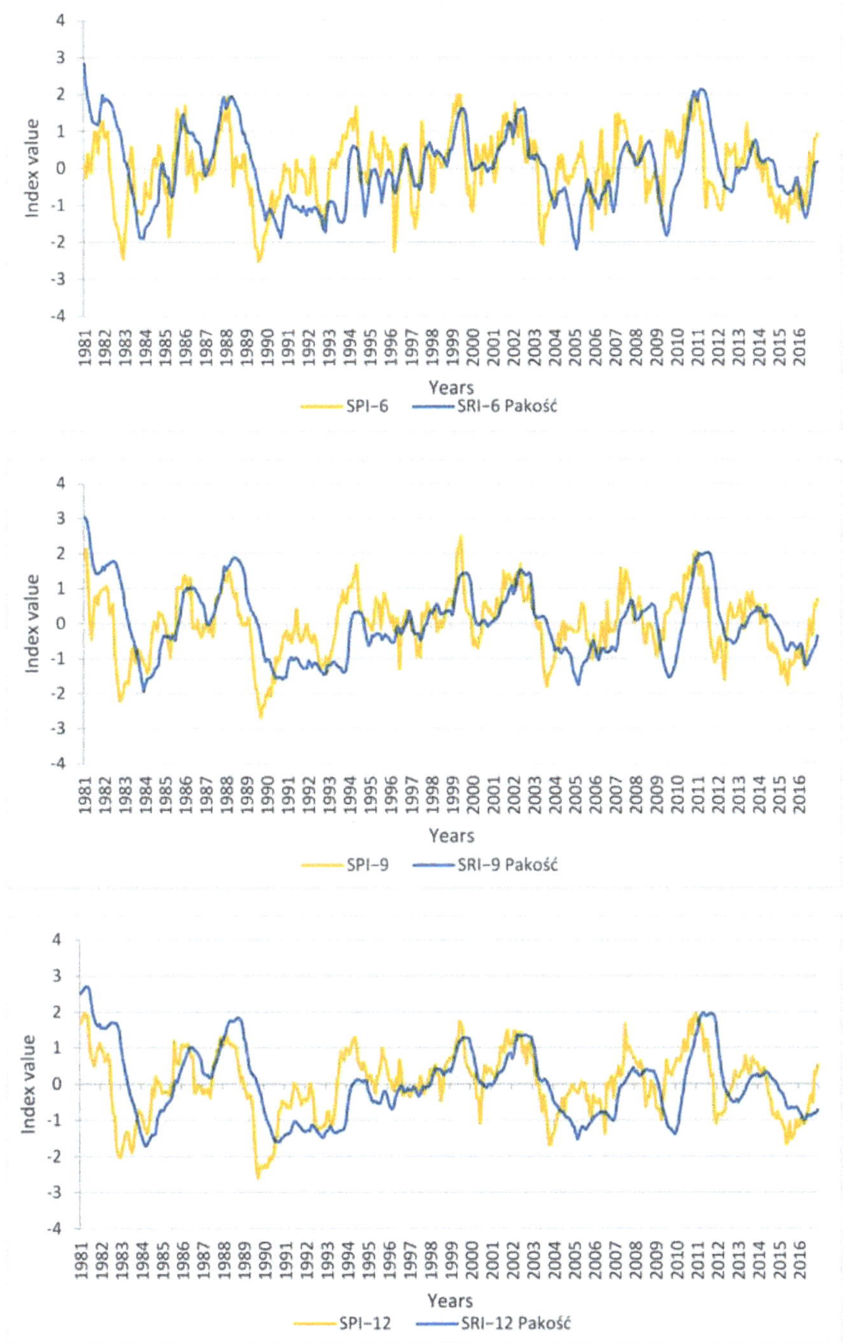

Figure 4. The development of meteorological and hydrological droughts in the period of 1981–2016 for different accumulation periods (n = 1, 3, 6, 9 and 12), SPI—average values from 8 stations, SRI for Pakość.

3.2. Trends in Meteorological and Hydrological Drought Occurrences

The results of the trend analysis for a series of SPI values (time scales of 1, 3, 6, 9 and 12 months) for individual meteorological stations are presented in Table 6. A statistically insignificant trend was noted at most meteorological stations. Out of eight meteorological stations, a statistically significant upward trend was recorded at only two stations, and a downward trend at one station (the drought intensified). The downward trend in the SPI value for the meteorological station in Kłodawa means that the meteorological drought increases in the period of accumulation of 3–12 months. The Z value obtained for the Kłodawa station was relatively high and ranged from −1.817 to −4.317, and the Theil–Sen slope was between −0.0007 and −0.0018. The test results for the Kłodawa station (Tables 6 and 7) show that the expected Z value is negative for the indices, and the Theil–Sen slope is also always negative.

Table 6. Results of trend analysis SPI in different time scales in the period of 1981–2016 at meteorological stations.

Stations	Parameters	SPI				
		SPI−1	SPI−3	SPI−6	SPI−9	SPI−12
Kłodawa	Z	−1.817	−2.429	−3.179	−3.755	−4.318
	S	-5.45×10^3	-7.28×10^3	-9.53×10^3	-1.12×10^4	-1.29×10^4
	var_S	8.99×10^6	8.99×10^6	8.99×10^6	8.99×10^6	8.99×10^6
	p-value	0.0692	0.0151	0.0015	0.0002	1.58×10^{-5}
	Sen's slope	−0.0007	−0.0010	−0.0013	−0.0016	−0.0018
		N	D	D	D	D
Izbica Kujawska	Z	−0.434	0.125	0.251	0.033	−0.451
	S	-1.30×10^3	3.76×10^2	7.54×10^2	9.90×10^1	-1.35×10^3
	var_S	8.99×10^6	8.99×10^6	8.99×10^6	8.99×10^6	8.99×10^6
	p-value	0.6646	0.9005	0.8017	0.9739	0.6523
	Sen's slope	−0.0002	5.4386×10^{-5}	0.0001	1.50×10^{-5}	−0.0002
		N	N	N	N	N
Sompolno	Z	1.253	2.180	2.413	2.090	1.648
	S	3.76×10^3	6.54×10^3	7.24×10^3	6.27×10^3	4.94×10^3
	var_S	8.99×10^6	8.99×10^6	8.99×10^6	8.99×10^6	8.99×10^6
	p-value	0.2102	0.0293	0.0158	0.0366	0.0994
	Sen's slope	0.0005	0.0009	0.0010	0.0009	0.0007
		N	I	I	I	N
Strzelno	Z	1.223	1.728	2.034	1.341	0.908
	S	3.67×10^3	5.18×10^3	6.10×10^3	4.02×10^3	2.72×10^3
	var_S	8.99×10^6	8.99×10^6	8.99×10^6	8.99×10^6	8.99×10^6
	p-value	0.2214	0.0840	0.0412	0.1801	0.3641
	Sen's slope	0.0005	0.0007	0.0008	0.0006	0.0004
		N	N	I	N	N
Gniezno	Z	0.453	1.272	1.916	1.790	1.506
	S	1.36×10^3	3.81×10^3	5.75×10^3	5.37×10^3	4.52×10^3
	var_S	8.99×10^6	8.99×10^6	8.99×10^6	8.99×10^6	8.99×10^6
	p-value	0.6503	0.2035	0.0553	0.0734	0.1322
	Sen's slope	0.0002	0.0005	0.0008	0.0007	0.0006
		N	N	N	N	N
Janowiec Wlkp.	Z	−0.109	−0.100	−0.048	−0.683	−0.931
	S	-3.27×10^2	-3.01×10^2	-1.45×10^2	-2.05×10^3	-2.79×10^3
	var_S	8.99×10^6	8.99×10^6	8.99×10^6	8.99×10^6	8.99×10^6
	p-value	0.9134	0.9203	0.9617	0.4946	0.3517
	Sen's slope	-4.40×10^{-5}	-3.78×10^{-5}	-1.81×10^{-5}	−0.0003	−0.0004
		N	N	N	N	N

Table 6. *Cont.*

Stations	Parameters	SPI				
		SPI−1	SPI−3	SPI−6	SPI−9	SPI−12
Kołuda Wielka	Z	1.700	3.251	4.178	5.084	4.336
	S	5.10×10^3	9.75×10^3	1.25×10^4	1.52×10^4	1.30×10^4
	var_S	8.99×10^6	8.99×10^6	8.99×10^6	8.99×10^6	8.99×10^6
	p-value	0.08918	0.0012	2.95×10^{-5}	3.69×10^{-7}	1.45×10^{-5}
	Sen's slope	0.0007	0.0013	0.0016	0.0019	0.0016
		N	I	I	I	I
Pakość	Z	0.175	0.834	1.207	0.844	0.284
	S	5.27×10^2	2.50×10^3	3.62×10^3	2.53×10^3	8.51×10^2
	var_S	8.99×10^6	8.99×10^6	8.99×10^6	8.99×10^6	8.99×10^6
	p-value	0.8607	0.4044	0.2274	0.3986	0.7768
	Sen's slope	7.48×10^{-5}	0.0003	0.0005	0.0003	0.0001
		N	N	N	N	N

Description: N—no significant trend, D—significant decreasing trend, I—significant increasing trend.

Table 7. Results of trend analysis SRI in different time scales in the period of 1981–2016 at hydrological stations.

Stations	Parameters	SRI				
		SRI−1	SRI−3	SRI−6	SRI−9	SRI−12
Łysek	Z	−5.342	−5.412	5.692	−5.974	−6.2735
	S	-1.60×10^4	-1.62×10^4	-1.71×10^4	-1.79×10^4	-1.88×10^4
	var_S	8.99×10^6	8.99×10^6	8.99×10^6	8.99×10^6	8.99×10^6
	p-value	9.19×10^{-8}	5.97×10^{-8}	1.25×10^{-8}	2.32×10^{-9}	3.53×10^{-10}
	Sen's slope	−0.0013	−0.0014	−0.0015	−0.0017	−0.0019
		D	D	D	D	D
Noć Kalina	Z	4.076	3.798	3.032	2.444	2.214
	S	1.22×10^4	1.14×10^4	9.09×10^3	7.33×10^3	6.64×10^3
	var_S	8.99×10^6	8.99×10^6	8.99×10^6	8.99×10^6	8.99×10^6
	p-value	4.58×10^{-5}	0.00014	0.0024	0.0145	0.0268
	Sen's slope	0.0017	0.0016	0.0013	0.0010	0.0008
		I	I	I	I	I
Pakość	Z	−0.784	−1.032	−1.134	−1.428	−1.488
	S	-2.35×10^3	-3.10×10^3	-3.40×10^3	-4.28×10^3	-4.46×10^3
	var_S	8.99×10^6	8.99×10^6	8.99×10^6	8.99×10^6	8.99×10^6
	p-value	0.433	0.3019	0.2566	0.1533	0.1368
	Sen's slope	−0.0003	−0.0004	−0.0005	−0.0006	−0.0007
		N	N	N	N	N

Description: N—no significant trend, D—significant decreasing trend, I—significant increasing trend.

The upward trend in the SPI value at the stations in Sompolno and Kołuda Wielka indicates that the phenomenon of meteorological drought is decreasing. The highest Z values, between 3.25 and 5.08, were obtained during the 3-, 6-, 9- and 12-month accumulation periods, while the Theil–Sen slope values were between 0.009 and 0.0019.

In the case of hydrological droughts, a statistically significant downward trend was recorded at the Łysek station in all analysed periods of accumulation. This means a noticeable increase in drought occurrences in the analysed multi-year period of 1981–2016. Z values were negative and ranged from −5.34 to −6.27, while Theil–Sen values were between −0.0013 to −0.0019. An upward trend was recorded at the Noć Kalina station. Z values ranged from 2.21 to 4.07, while the Theil–Sen slope ranged from 0.0008 to 0.0017. In the case of Pakość station, the trend was statistically insignificant. The Z value calculated in the 12-month accumulation period in Pakość approaches the region of a trend acceptance, which might indicate that the trend in Pakość in the longer accumulation period is determined by the occurrence of hydrological droughts at the Łysek station.

Spatial distributions of trends (significant and insignificant) for a series of SPI and SRI values for the 1, 3-, 6-, 9- and 12-month time scales are shown in Figure 3.

3.3. Correlations between SPI and SRI Values

In order to establish the relationship between meteorological droughts occurring in the catchment area of the Upper Noteć and hydrological droughts, an analysis of the correlation between the SPI and SRI indices was carried out using the Pearson correlation analysis. The results showed that the strongest correlation between SPI and SRI in the analysed period of 1981–2016 was obtained at the 12-month time scale (r = 0.51) (Figure 5). In the case of individual years, the highest correlation indicators between hydrological and meteorological droughts varied depending on the length of the accumulation period. The strength of the relationship between SPI and SRI in the catchment area of the Upper Noteć River was higher for long accumulation periods (6 and 9 months), and lower for the short ones (1 and 3 months). The highest correlation values for the 1-month accumulation period were recorded in 1998 (r = 0.76), while for the 3-month accumulation period the highest values were recorded in 1982 (r = 0.83) and 1996 (r = 0.81). For the 6-month accumulation period, the maximum correlation index was recorded in 1987 (r = 0.94) and in 1982 (r = 0.91). High values were obtained in 1987 for the 9-month accumulation period (r = 0.94) and the 12-month accumulation period (r = 0.90).

3.4. Discussion

Understanding the changes in the intensity of droughts in the past and being able to predict expected changes over different time scales is incredibly important, as precipitation-driven hydrological processes (e.g., evapotranspiration and surface and groundwater discharge) affect all water reserves [86]. In this study, we found that in the analysed period of 1981–2016, there was a relationship between the occurrence of meteorological and hydrological droughts. The strength of this relationship varied. The analysed multi-year period of 1981–2016 showed a high variability, from dry years (SPI ≤ -1.0) to wet years (SPI ≥ 1.0), which can be concluded from the SPI values in various time scales. The driest years included 1982, 1989, 1992, 2003 and 2015. In these years, meteorological droughts covering not only the region of Poland, but parts of Europe, were recorded. Meteorological droughts which have occurred in Europe since the beginning of the 21st century, and were accompanied by heat waves in 2003, 2006, 2010, 2015, are great examples of such phenomena [87–91].

In the studied area, an increase in the intensity of meteorological droughts (downward trend) was observed at only one out of eight meteorological stations. A statistically significant, clear upward trend in SPI drought was identified at two stations. More distinct trends, but opposite in direction, were observed in the case of hydrological droughts recorded at the stations in Łysek and Noć Kalina. The obtained statistics for the Pakość station, calculated in the 12-month accumulation period, point to the rejection of the null hypothesis on the lack of a statistically significant trend. The same direction of changes in the trend was recorded at the Łysek and Pakość stations. This means an increase in the intensity of hydrological droughts. In the longer accumulation period, the occurrence of hydrological droughts at the Łysek station determines the hydrological droughts at the Pakość station.

The strength of the relationship between meteorological droughts and hydrological droughts shows significant variation. This variation is not only the result of the size of the annual sums of precipitation, but also if an increase in air temperature in the analysed area (Figure 2), which leads to an increase in evapotranspiration [92]. Anthropogenic activities related to the operation of a lignite open pit have a significant impact on the analysed area. Some of the water from the mine drainage was directed to the Noteć River above the Noć Kalina station. The amount of water varied in individual years and depended on the location of the exploitation operations. According to Wachowiak [93], in the period of 1995–2009 the Upper Noteć was flooded with some of the mine water from the drainage of the Lubstów open pit. Since 2009, there have been cases of mine water discharge from the Tomisławice open pit via the Pichna River. The correlations between meteorological and hydrological droughts were variable, in some years the strength of the relationship was high (positive correlation), while in other years the relationship was low (negative correlation).

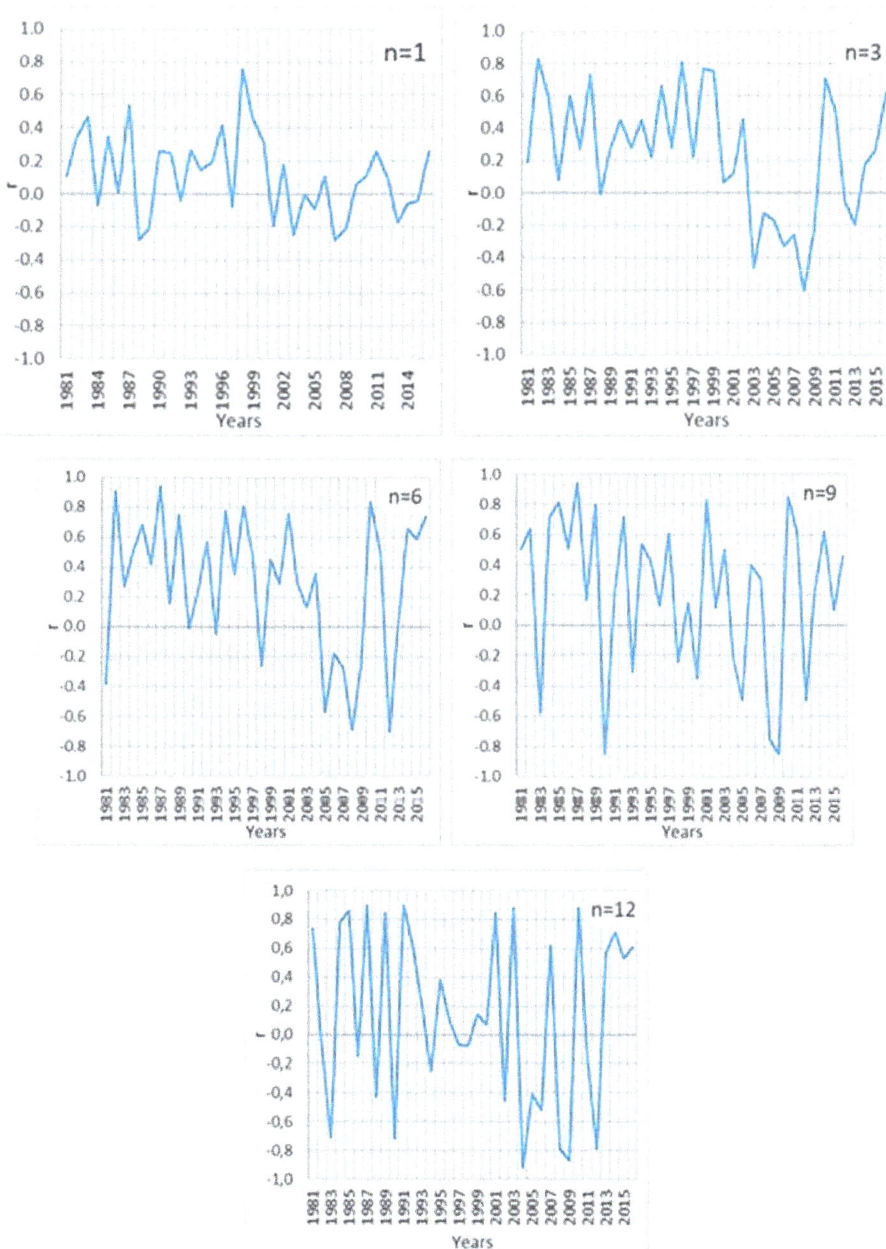

Figure 5. Correlation coefficient r between average SPI and SRI in Pakość for different periods of accumulation in the period of 1981–2016 (*n*—number of accumulated months).

The Pearson correlation analysis shows that there is a relationship between meteorological and hydrological droughts in the study area. However, it should be emphasized that these results should not be directly interpreted. In the correlation analysis, a non-linear relationship can be inadequately described or undetected [94]. Non-linear models (polynomial, exponential and logarithmic) for the relationship between meteorological and hydrological droughts were analysed in the research conducted by Salimi et al. [95].

The research on the correlation between droughts conducted by Tokarczyk and Szalińska [44] for catchments with large areas showed that the largest correlations between SPI and SRI occurred for longer periods of accumulation. Similar relationships between meteorological and hydrological droughts were obtained for other catchments in Poland [43]. In the case of the catchment area of the Upper Noteć River, the relationships between meteorological and hydrological conditions are not natural. The flow regime depends on the amount of water discharged in particular periods, and on the retention capacity of lakes, which is particularly noticeable at the Pakość station. The amount of water accumulated in the Pakość reservoir, through which the Noteć flows, is regulated by a water accumulating weir.

4. Conclusions

The study analysed the trends in meteorological and hydrological drought occurrences in the long-term period of 1981–2016, for the catchment area of the Upper Noteć River. The identification of a meteorological drought was carried out with the use of an SPI, based on monthly precipitation totals from eight meteorological stations. Hydrological drought was determined by means of an SRI for the monthly discharges of the Noteć, which were obtained from three hydrological stations. Non-parametric Mann–Kendall tests and the Sen slope were used to determine trends. The following conclusions might be drawn:

- Statistically significant trends, at the significance level of 0.05, were identified at three out of eight meteorological stations, based on the Mann–Kendall test and the Sen slope.
- An increase in meteorological drought occurrences was recorded at the Kłodawa station (downward trend), while a decrease in droughts was recorded at the Sompolno and Kołuda Wielka stations.
- Hydrological droughts showed an upward trend at the Łysek station, while a decrease in the trend was recorded at the Noć Kalina station, and both were statistically significant. No changes in the trend were found at the Pakość station.
- The analysis of the correlation between meteorological and hydrological droughts in individual years showed a strong relationship in dry years e.g., 1982 and 1989. The maximum correlation index was 0.94 and was identified over longer accumulation periods i.e., 6 and 9 months.
- The anthropogenic effects related to the operation of an open cast lignite mine may have had an impact on the relationship between droughts.

The example of the catchment area of the Upper Noteć River indicates that the management of water resources requires the use of at least several indicators that will allow an assessment of the actual state of water reserves. Using the SPI to detect meteorological droughts can be used as a drought warning system [96]. In some cases, the value of the SRI depends on the way water is managed within the catchment area. The size of the runoff may be disturbed by anthropogenic factors. Effective water resource management strategies require constant monitoring of water reserves, which should be consulted among various stakeholder groups related in particular to industry, and agriculture.

Author Contributions: Conceptualization, K.K.-W., A.P. and D.K.; methodology, K.K.-W. and A.P.; software, K.K.-W. and A.P.; validation, K.K.-W. and A.P.; formal analysis, K.K.-W., A.P. and D.K.; resources, K.K.-W. and A.P.; data curation, K.K.-W. and A.P.; writing—original draft preparation, K.K.-W.; writing—review and editing, K.K.-W., A.P. and D.K.; visualization, A.P.; supervision,

K.K.-W., A.P. and D.K.; project administration, K.K.-W. All authors have read and agreed to the published version of the manuscript.

Funding: This research received no external funding.

Institutional Review Board Statement: Not applicable.

Informed Consent Statement: Not applicable.

Data Availability Statement: The datasets used and/or analyzed during the current study are available from the corresponding author on reasonable request.

Conflicts of Interest: The authors declare no conflict of interest.

References

1. Nalbantis, I.; Tsakiris, G. Assessment of Hydrological Drought Revisited. *Water Resour. Manag.* **2009**, *23*, 881–897. [CrossRef]
2. Chen, S.-T.; Kuo, C.-C.; Yu, P.-S. Historical trends and variability of meteorological droughts in Taiwan. *Hydrol. Sci. J.* **2009**, *54*, 430–441. [CrossRef]
3. Roodari, A.; Hrachowitz, M.; Hassanpour, F.; Yaghoobzadeh, M. Signatures of human intervention—Or not? Downstream intensification of hydrological drought along a large Central Asian river: The individual roles of climate variability and land use change. *Hydrol. Earth Syst. Sci.* **2021**, *25*, 1943–1967. [CrossRef]
4. Bae, H.; Ji, H.; Lim, Y.-J.; Ryu, Y.; Kim, M.-H.; Kim, B.-J. Characteristics of drought propagation in South Korea: Relationship between meteorological, agricultural, and hydrological droughts. *Nat. Hazards* **2019**, *99*, 1–16. [CrossRef]
5. Faridatul, M.; Ahmed, B. Assessing Agricultural Vulnerability to Drought in a Heterogeneous Environment: A Remote Sensing-Based Approach. *Remote Sens.* **2020**, *12*, 3363. [CrossRef]
6. Dai, A. Drought under global warming: A review. *Wiley Interdiscip. Rev. Clim. Chang.* **2011**, *2*, 45–65. [CrossRef]
7. Tian, L.; Yuan, S.; Quiring, S. Evaluation of six indices for monitoring agricultural drought in the south-central United States. *Agric. For. Meteorol.* **2018**, *249*, 107–119. [CrossRef]
8. Wilhite, D.A.; Glantz, M.H. Understanding: The Drought Phenomenon: The Role of Definitions. *Water Int.* **1985**, *10*, 111–120. [CrossRef]
9. Bąk, B.; Kubiak-Wójcicka, K.K. Assessment of meteorological and hydrological drought in Torun (central Poland town) in 1971–2010 based on standardized indicators. In Proceedings of the Water Resources and Wetlands Conference, Tulcea, Romania, 8–10 September 2016; Gastescu, P., Bretcan, P., Eds.; pp. 164–170. Available online: http://www.limnology.ro/wrw2016/proceedings/22_Bak_Kubiak.pdf (accessed on 15 July 2021).
10. Hasan, H.H.; Razali, S.F.M.; Muhammad, N.S.; Hamzah, F.M. Assessment of probability distributions and analysis of the minimum storage draft rate in the equatorial region. *Nat. Hazards Earth Syst. Sci.* **2021**, *21*, 1–19. [CrossRef]
11. Zhao, F.; Wu, Y.; Sivakumar, B.; Long, A.; Qiu, L.; Chen, J.; Wang, L.; Liu, S.; Hu, H. Climatic and hydrologic controls on net primary production in a semiarid loess watershed. *J. Hydrol.* **2018**, *568*, 803–815. [CrossRef]
12. Wu, J.; Chen, X.; Love, C.A.; Yao, H.; Chen, X.; AghaKouchak, A. Determination of water required to recover from hydrological drought: Perspective from drought propagation and non-standardized indices. *J. Hydrol.* **2020**, *590*, 125227. [CrossRef]
13. Jehanzaib, M.; Sattar, M.N.; Lee, J.-H.; Kim, T.-W. Investigating effect of climate change on drought propagation from meteorological to hydrological drought using multi-model ensemble projections. *Stoch. Environ. Res. Risk Assess.* **2019**, *34*, 7–21. [CrossRef]
14. Gleick, P. Water, Drought, Climate Change, and Conflict in Syria. *Weather. Clim. Soc.* **2014**, *6*, 331–340. [CrossRef]
15. Craft, K.E.; Mahmood, R.; King, S.A.; Goodrich, G.; Yan, J. Twentieth century droughts and agriculture: Examples from impacts on soybean production in Kentucky, USA. *Ambio* **2015**, *44*, 557–568. [CrossRef]
16. Holman, I.; Hess, T.; Rey, D.; Knox, J. A Multi-Level Framework for Adaptation to Drought Within Temperate Agriculture. *Front. Environ. Sci.* **2021**, *8*. [CrossRef]
17. Spinoni, J.; Barbosa, P.; Bucchignani, E.; Cassano, J.; Cavazos, T.; Christensen, J.H.; Christensen, O.B.; Coppola, E.; Evans, J.; Geyer, B.; et al. Future Global Meteorological Drought Hot Spots: A Study Based on CORDEX Data. *J. Clim.* **2020**, *33*, 3635–3661. [CrossRef]
18. IPCC. Changes in Climate Extremes and Their Impacts on the Natural Physical Environment. In *Managing the Risks of Extreme Events and Disasters to Advance Climate Change Adaptation: Special Report of the Intergovernmental Panel on Climate Change*; Field, C.B., Barros, V., Stocker, T.F., Qin, D., Dokken, D.J., Ebi, K.L., Mastrandrea, M.D., Mach, K.J., Plattner, G.-K., Allen, S.K., et al., Eds.; Cambridge University Press: Cambridge, UK, 2012.
19. Vicente-Serrano, S.M.; Dominguez-Castro, F.; McVicar, T.R.; Tomas-Burguera, M.; Peña-Gallardo, M.; Noguera, I.; López-Moreno, J.I.; Peña, D.; El Kenawy, A. Global characterization of hydrological and meteorological droughts under future climate change: The importance of timescales, vegetation-CO_2 feedbacks and changes to distribution functions. *Int. J. Clim.* **2019**, *40*, 2557–2567. [CrossRef]
20. Chiang, F.; Mazdiyasni, O.; AghaKouchak, A. Evidence of anthropogenic impacts on global drought frequency, duration, and intensity. *Nat. Commun.* **2021**, *12*, 2754. [CrossRef]

21. Jehanzaib, M.; Kim, T.-W. Exploring the influence of climate change-induced drought propagation on wetlands. *Ecol. Eng.* **2020**, *149*, 105799. [CrossRef]
22. Gao, L.; Zhang, Y. Spatio-temporal variation of hydrological drought under climate change during the period 1960–2013 in the Hexi Corridor, China. *Arid Land* **2016**, *82*, 157–171. [CrossRef]
23. Ndehedehe, C.E.; Agutu, N.O.; Ferreira, V.G.; Getirana, A. Evolutionary drought patterns over the Sahel and their teleconnections with low frequency climate oscillations. *Atmos. Res.* **2019**, *233*, 104700. [CrossRef]
24. Bazrafshan, O.; Zamani, H.; Shekari, M. A copula-based index for drought analysis in arid and semi-arid regions in Iran. *Nat. Resour. Model.* **2020**, *33*, e12237. [CrossRef]
25. Dikici, M. Drought analysis with different indices for the Asi Basin (Turkey). *Sci. Rep.* **2020**, *10*, 1–12. [CrossRef]
26. Mahdavi, P.; Kharazi, H.G.; Eslami, H.; Zohrabi, N.; Razaz, M. Drought occurrence under future climate change scenarios in the Zard River basin, Iran. *Water Supply* **2020**, *21*, 899–917. [CrossRef]
27. Buras, A.; Rammig, A.; Zang, C.S. Quantifying impacts of the 2018 drought on European ecosystems in comparison to 2003. *Biogeosciences* **2020**, *17*, 1655–1672. [CrossRef]
28. Kubiak-Wójcicka, K.; Nagy, P.; Zeleňáková, M.; Hlavatá, H.; Abd-Elhamid, H. Identification of Extreme Weather Events Using Meteorological and Hydrological Indicators in the Laborec River Catchment, Slovakia. *Water* **2021**, *13*, 1413. [CrossRef]
29. Emadodin, I.; Corral, D.; Reinsch, T.; Kluß, C.; Taube, F. Climate Change Effects on Temperate Grassland and Its Implication for Forage Production: A Case Study from Northern Germany. *Agriculture* **2021**, *11*, 232. [CrossRef]
30. Koffi, B.; Kouadio, Z.A.; Kouassi, K.H.; Yao, A.B.; Sanchez, M. Impact of Meteorological Drought on Streamflows in the Lobo River Catchment at Nibéhibé, Côte d'Ivoire. *J. Water Resour. Prot.* **2020**, *12*, 495–511. [CrossRef]
31. Awange, J.L.; Mpelasoka, F.; Goncalves, R. When every drop counts: Analysis of Droughts in Brazil for the 1901-2013 period. *Sci. Total Environ.* **2016**, *566–567*, 1472–1488. [CrossRef] [PubMed]
32. Hari, V.; Rakovec, O.; Markonis, Y.; Hanel, M.; Kumar, R. Increased future occurrences of the exceptional 2018–2019 Central European drought under global warming. *Sci. Rep.* **2020**, *10*, 1–10. [CrossRef]
33. Kubiak-Wójcicka, K. Assessment of water resources in Poland. In *Quality in Water Resources in Poland*; Zeleňáková, M., Kubiak-Wójcicka, K., Negm, A.M., Eds.; Springer International Publishing: Cham, Switzerland, 2021; pp. 15–34. [CrossRef]
34. Somorowska, U. Changes in Drought Conditions in Poland over the Past 60 Years Evaluated by the Standardized Precipitation-Evapotranspiration Index. *Acta Geophys.* **2016**, *64*, 2530–2549. [CrossRef]
35. Szwed, M. Variability of precipitation in Poland under climate change. *Theor. Appl. Clim.* **2018**, *135*, 1003–1015. [CrossRef]
36. Kubiak-Wójcicka, K. Dynamics of meteorological and hydrological droughts in the agricultural catchments. *Res. Rural Dev.* **2019**, *1*, 111–117. [CrossRef]
37. Tomczyk, A.M.; Szyga-Pluta, K. Variability of thermal and precipitation conditions in the growing season in Poland in the years 1966–2015. *Theor. Appl. Clim.* **2018**, *135*. [CrossRef]
38. Osuch, M.; Romanowicz, R.J.; Booij, M.J. The influence of parametric uncertainty on the relationships between HBV model parameters and climatic characteristics. *Hydrol. Sci. J.* **2015**, *60*, 1299–1316. [CrossRef]
39. Łabędzki, L.; Bak, B. Indicator-based monitoring and forecasting water deficit and surplus in agriculture in Poland. *Ann. Wars. Univ. Life Sci. SGGW Land Reclam.* **2015**, *47*, 355–369. [CrossRef]
40. Kuśmierek-Tomaszewska, R.; Żarski, J.; Dudek, S. Assessment of Irrigation Needs in Sugar Beet (*Beta vulgaris* L.) in Temperate Climate of Kujawsko-Pomorskie Region (Poland). *Agronomy* **2019**, *9*, 814. [CrossRef]
41. Łabedzki, L. Estimation of local drought frequency in central Poland using the standardized precipitation index SPI. *Irrig. Drain.* **2007**, *56*, 67–77. [CrossRef]
42. Tokarczyk, T.; Szalińska, W. Combined analysis of precipitation and water deficit for drought hazard assessment. *Hydrol. Sci. J.* **2014**, *59*, 1675–1689. [CrossRef]
43. Kubiak-Wójcicka, K.; Bąk, B. Monitoring of meteorological and hydrological droughts in the Vistula basin (Poland). *Environ. Monit. Assess.* **2018**, *190*, 691. [CrossRef]
44. Tokarczyk, T.; Szalińska, W.; National Research Institute. Drought Hazard Assessment in the Process of Drought Risk Management. *Acta Sci. Pol. Form. Circumiectus* **2018**, *3*, 217–229. [CrossRef]
45. Łabędzki, L.; Bąk, B. Impact of meteorological drought on crop water deficit and crop yield reduction in Polish agriculture. *J. Water Land Dev.* **2017**, *34*, 181–190. [CrossRef]
46. Kubiak-Wójcicka, K.; Juśkiewicz, W. Relationships between meteorological and hydrological drought in a young-glacial zone (north-western Poland) based on Standardised Precipitation Index (SPI) and Standardized Runoff Index (SRI). *Acta Montan. Slovaca* **2020**, *25*, 517–531. [CrossRef]
47. Bąk, B.; Kubiak-Wójcicka, K. Impact of meteorological drought on hydrological drought in Toruń (central Poland) in the period of 1971–2015. *J. Water Land Dev.* **2017**, *32*, 3–12. [CrossRef]
48. Radzka, E. The assessment of atmospheric drought during vegetation season (according to Standardized Precipitation Index SPI) in central-eastern Poland. *J. Ecol. Eng.* **2015**, *16*, 87–91. [CrossRef]
49. Grzywna, A.; Bochniak, A.; Ziernicka-Wojtaszek, A.; Krużel, J.; Jóźwiakowski, K.; Wałega, A.; Cupak, A.; Mazur, A.; Obroślak, R.; Serafin, A. The analysis of spatial variability of precipitation in Poland in the multiyears 1981–2010. *J. Water Land Dev.* **2020**, *46*, 105–111. [CrossRef]
50. Bąk, B. Warunki klimatyczne Wielkopolski i Kujaw. *Woda-Środowisko-Obszary Wiejskie* **2003**, *3*, 11–39.

51. Natural Earth. Free Vector and Raster Map Data @ naturalearthdata.com. Available online: https://www.naturalearthdata.com/downloads/10m-cultural-vectors and https://www.naturalearthdata.com/downloads/10m-raster-data/10m-natural-earth-1 (accessed on 12 July 2021).
52. Wektorowe Warstwy Tematyczne aPGW. Available online: https://dane.gov.pl/pl/dataset/599/resource/672,wektorowe-warstwy-tematyczne-apgw/table (accessed on 12 July 2021).
53. SRTM 1 Arc-Second Global. Available online: https://earthexplorer.usgs.gov/ (accessed on 3 June 2021). [CrossRef]
54. Baza Danych Obiektów Ogólnogeograficznych. Available online: https://mapy.geoportal.gov.pl/imap/Imgp_2.html?locale=pl&gui=new&sessionID=5160004 (accessed on 15 July 2021).
55. Hydro IMGW-PIB. Available online: https://hydro.imgw.pl/#map/19.5,51.5,7,true,true,0 (accessed on 15 July 2021).
56. QGIS Quick Map Services Plugin. OSM Standard. Plugin Was Used on 13–15 July 2021. Author of the Plugin: NextGIS. Base Map from: OpenStreetMap and OpenStreetMap Foundation. Detailed Information about Copyright and License of the OpenStreetMap. Available online: https://www.openstreetmap.org/copyright (accessed on 15 July 2021).
57. Przybyłek, J. Aktualne problemy odwadniania złóż węgla brunatnego w Wielkopolsce. *Górnictwo Odkrywkowe* **2018**, *59*, 5–14.
58. Kubiak-Wójcicka, K.; Machula, S. Influence of Climate Changes on the State of Water Resources in Poland and Their Usage. *Geosciences* **2020**, *10*, 312. [CrossRef]
59. Tomaszewski, E.; Kubiak-Wójcicka, K. Low-flows in Polish rivers. In *Management of Water Resources in Poland*; Zeleňáková, M., Kubiak-Wójcicka, K., Negm, A.M., Eds.; Springer International Publishing: Cham, Switzerland, 2021; pp. 205–228. [CrossRef]
60. McKee, T.B.; Doesken, N.J.; Kleist, J. The relationship of drought frequency and duration to time scales. In Proceedings of the 8th Conference of Applied Climatology, Anaheim, CA, USA, 17–22 January 1993; pp. 179–184.
61. World Meteorological Organization (WMO); Global Water Partnership (GWP). *Handbook of Drought Indicators and Indices; Integrated Drought Management Tools and Guidelines Series 2*; Svoboda, M., Fuchs, B.A., Eds.; Integrated Drought Management Programme (IDMP): Geneva, Switzerland, 2016; ISBN 978-92-63-11173-9.
62. Guttman, N.B. Accepting the Standardized Precipitation Index: A calculation algorithm. *J. Am. Water Resour. Assoc.* **1999**, *35*, 311–322. [CrossRef]
63. Lloyd-Hughes, B.; Saunders, M. A drought climatology for Europe. *Int. J. Clim.* **2002**, *22*, 1571–1592. [CrossRef]
64. Fang, W.; Huang, S.; Huang, Q.; Huang, G.; Wang, H.; Leng, G.; Wang, L. Identifying drought propagation by simultaneously considering linear and nonlinear dependence in the Wei River basin of the Loess Plateau, China. *J. Hydrol.* **2020**, *591*, 125287. [CrossRef]
65. Łabędzki, L. Actions and measures for mitigation drought and water scarcity in agriculture. *J. Water Land Dev.* **2016**, *29*, 3–10. [CrossRef]
66. Lorenzo-Lacruz, J.; Morán-Tejeda, E.; Vicente-Serrano, S.M.; López-Moreno, J.I. Streamflow droughts in the Iberian Peninsula between 1945 and 2005: Spatial and temporal patterns. *Hydrol. Earth Syst. Sci.* **2013**, *17*, 119–134. [CrossRef]
67. Sahoo, R.N.; Dipanwita, D.; Khanna, M.; Kumar, N.; Bandyopadhyay, S.K. Drought assessment in the Dhar and Mewat Districs of India using meteorological and remote-sensing derived indices. *Nat. Hazards* **2015**, *77*, 733–751. [CrossRef]
68. Shukla, S.; Wood, A. Use of a standardized runoff index for characterizing hydrologic drought. *Geophys. Res. Lett.* **2008**, *35*. [CrossRef]
69. Bayissa, Y.; Maskey, S.; Tadesse, T.; Van Andel, S.J.; Moges, S.; Van Griensven, A.; Solomatine, D. Comparison of the Performance of Six Drought Indices in Characterizing Historical Drought for the Upper Blue Nile Basin, Ethiopia. *Geosciences* **2018**, *8*, 81. [CrossRef]
70. Vicente-Serrano, S.M.; Lopez-Moreno, I.; Beguería, S.; Lorenzo-Lacruz, J.; Azorin-Molina, C.; Morán-Tejeda, E. Accurate Computation of a Streamflow Drought Index. *J. Hydrol. Eng.* **2012**, *17*, 318–332. [CrossRef]
71. Mann, H.B. Nonparametric tests against trend. *Econometrica* **1945**, *13*, 245–259. [CrossRef]
72. Kendall, M.G. *Rank Correlation Methods*; Charles Grin: London, UK, 1975.
73. Silva, A.; Filho, M.; Menezes, R.; Stosic, T.; Stosic, B. Trends and Persistence of Dry–Wet Conditions in Northeast Brazil. *Atmosphere* **2020**, *11*, 1134. [CrossRef]
74. Bacanli, G. Trend analysis of precipitation and drought in the Aegean region, Turkey. *Meteorol. Appl.* **2017**, *24*, 239–249. [CrossRef]
75. Hu, M.; Dong, M.; Tian, X.; Wang, L.; Jiang, Y. Trends in Different Grades of Precipitation over the Yangtze River Basin from 1960 to 2017. *Atmosphere* **2021**, *12*, 413. [CrossRef]
76. Gocic, M.; Trajkovic, S. Analysis of changes in meteorological variables using Mann-Kendall and Sen's slope estimator statistical tests in Serbia. *Glob. Planet. Chang.* **2013**, *100*, 172–182. [CrossRef]
77. Tuan, N.H.; Canh, T.T. Analysis of Trends in Drought with the Non-Parametric Approach in Vietnam: A Case Study in Ninh Thuan Province. *Am. J. Clim. Chang.* **2021**, *10*, 51–84. [CrossRef]
78. Kuriqi, A.; Ali, R.; Quoc Bao, P.; Gambini, J.M.; Gupta, V.; Malik, A.; Nguyen Thi Thuy, L.; Joshi, Y.; Duong Tran, A.; Van Thai, N.; et al. Seasonality shift and streamflow flow variability trends in central India. *Acta Geophys.* **2020**, *68*, 1461–1475. [CrossRef]
79. Abeysingha, N.S.; Rajapaksha, U.R.L.N. SPI-Based Spatiotemporal Drought over Sri Lanka. *Hindawi Adv. Meteorol.* **2020**. [CrossRef]
80. Sen, P.K. Estimates of the regression coefficient based on Kendall's tau. *J. Am. Stat. Assoc.* **1968**, *63*, 1379–1389. [CrossRef]
81. Achite, M.; Krakauer, N.; Wałęga, A.; Caloiero, T. Spatial and Temporal Analysis of Dry and Wet Spells in the Wadi Cheliff Basin, Algeria. *Atmosphere* **2021**, *12*, 798. [CrossRef]

82. RStudio Team. *RStudio: Integrated Development Environment for R;* RStudio PBC: Boston, MA, USA, 2021.
83. Wickham, H.; Bryan, J. Readxl: Read Excel Files. R Package Version 1.3.1. 2019. Available online: https://CRAN.R-project.org/package=readxl (accessed on 17 July 2021).
84. Pohlert, T. Trend: Non-Parametric Trend Tests and Change-Point Detection. R package Version 1.1.4. 2020. Available online: https://CRAN.R-project.org/package=trend (accessed on 17 July 2021).
85. Reference Manual. Package "Trend". Non-Parametric Trend Tests and Change-Point Detection. Available online: https://cran.r-project.org/web/packages/trend/trend.pdf (accessed on 8 August 2021).
86. Portela, M.M.; Espinosa, L.A.; Zelenakova, M. Long-Term Rainfall Trends and Their Variability in Mainland Portugal in the Last 106 Years. *Climate* **2020**, *8*, 146. [CrossRef]
87. Ionita, M.; Tallaksen, L.M.; Kingston, D.G.; Stagge, J.H.; Laaha, G.; Van Lanen, H.A.J.; Chelcea, S.M.; Haslinger, K. The European 2015 drought from a climatological perspective. Hydrol. *Earth Syst. Sci. Discuss.* **2016**. [CrossRef]
88. Hanel, M.; Rakovec, O.; Markonis, Y.; Maca, P.; Samaniego, L.; Kyselý, J.; Kumar, R. Revisiting the recent European droughts from a long-term perspective. *Sci. Rep.* **2018**, *8*, 1–11. [CrossRef]
89. Pińskwar, I.; Choryński, A.; Kundzewicz, Z.W. Severe Drought in the Spring of 2020 in Poland—More of the Same? *Agronomy* **2020**, *10*, 1646. [CrossRef]
90. Barker, L.J.; Hannaford, J.; Parry, S.; Katie, A.; Smith, K.A.; Tanguy, M.; Prudhomme, C. Historic hydrological droughts 1891–2015: Systematic characterisation for a diverse set of catchments across the UK. *Hydrol. Earth Syst. Sci. Discuss.* **2019**. [CrossRef]
91. Solarczyk, A.; Kubiak-Wójcicka, K. The exhaustion of water resources in the Kuyavian-Pomeranian voivodship in drought condition in 2015. *Res. Rural Dev. Water Manag.* **2019**, *1*, 118–125. [CrossRef]
92. Ziernicka-Wojtaszek, A.; Kopcińska, J. Variation in Atmospheric Precipitation in Poland in the Years 2001–2018. *Atmosphere* **2020**, *11*, 794. [CrossRef]
93. Wachowiak, G. Wpływ zrzutów wód kopalnianych na wielkość przepływów wody w rzekach w początkowym okresie odwadniania odkrywki "Tomisławice" KWB "Konin". *Górnictwo i Geoinżynieria* **2011**, *35*, 153–166.
94. Bewick, V.; Cheek, L.; Ball, J. Statistics review 7: Correlation and regression. *Crit. Care* **2003**, *7*, 451–459. [CrossRef] [PubMed]
95. Salimi, H.; Asadi, E.; Darbandi, S. Meteorological and hydrological drought monitoring using several drought indices. *Appl. Water Sci.* **2021**. [CrossRef]
96. El-Afandi, G.; Morsy, M.; Kamel, A. Estimation of Drought Index over the Northern Coast of Egypt. *IJSRSET* **2016**, *2*, 335–344.

MDPI

Article

The Role of Aquatic Refuge Habitats for Fish, and Threats in the Context of Climate Change and Human Impact, during Seasonal Hydrological Drought in the Saxon Villages Area (Transylvania, Romania)

Doru Bănăduc [1,*,†], Alexandru Sas [2], Kevin Cianfaglione [3], Sophia Barinova [4] and Angela Curtean-Bănăduc [1,†]

1 Applied Ecology Research Center, Lucian Blaga University of Sibiu, I. Raţiu Street 5-7, RO-550012 Sibiu, Romania; angela.banaduc@ulbsibiu.ro
2 Ecotur Sibiu NGO, Rahova Street, 43, RO-550345 Sibiu, Romania; alexandru.sas@europe.com
3 UMR UL/AgroParisTech/INRAE 1434 Silva, Université de Lorraine, Faculté des Sciences et Techniques, BP 70239, 54506 Vandoeuvre-lés-Nancy, France; kevin.cianfaglione@univ-lorraine.fr
4 Institute of Evolution, University of Haifa, Mount Carmel, Abba Khoushi Avenue 199, Haifa IL-3498838, Israel; sophia@evo.haifa.ac.il
* Correspondence: ad.banaduc@yahoo.com; Tel.: +40-722604338
† The first and the last authors have equal contributions.

Citation: Bănăduc, D.; Sas, A.; Cianfaglione, K.; Barinova, S.; Curtean-Bănăduc, A. The Role of Aquatic Refuge Habitats for Fish, and Threats in the Context of Climate Change and Human Impact, during Seasonal Hydrological Drought in the Saxon Villages Area (Transylvania, Romania). *Atmosphere* **2021**, *12*, 1209. https://doi.org/10.3390/atmos12091209

Academic Editors: Andrzej Walega and Agnieszka Ziernicka-Wojtaszek

Received: 31 July 2021
Accepted: 10 September 2021
Published: 16 September 2021

Publisher's Note: MDPI stays neutral with regard to jurisdictional claims in published maps and institutional affiliations.

Abstract: In spite of the obvious climate changes effects on the Carpathian Basin hydrographic nets fish fauna, studies on their potential refuge habitats in drought periods are scarce. Multiannual (2016–2021) research of fish in some streams located in the Saxon Villages area during hydrological drought periods identified, mapped, and revealed the refuge aquatic habitats presence, management needs, and importance for fish diversity and abundance for small rivers. The impact of increasing global temperature and other human activities induced hydrologic net and habitats alteration, decreased the refuge habitats needed by freshwater fish, diminished the fish abundance, and influenced the spatial and temporal variation in fish assemblage structure in the studied area. The sites more than one meter in depth in the studied lotic system were inventoried and all 500 m of these lotic systems were also checked to see what species and how many individuals were present, and if there is was difference in their abundance between refuge and non-refuge 500 m sectors. The scarce number of these refuges due to relatively high soil erosion and clogging in those basins and the cumulative effects of other human types of impact induced a high degree of pressure on the fish fauna. Overall, it reduced the role of these lotic systems as a refuge and for reproduction for the fish of downstream Târnava Mare River, into which all of them flow. Management elements were proposed to maintain and improve these refuges' ecological support capacity.

Keywords: drought; lotic systems; refuge habitats; fish; risk management

1. Introduction and Background

Water, the fundamental factor of initiation, persistence, and evolution of life on Earth is ever-present; it influences everything and is a key to comprehending the universe in general [1,2], including the biodiversity structure, distribution, and ecological state [3–11]. In this framework, the relevance of small water bodies for biodiversity and ecosystem services is not negligible [12,13].

Climate change is one of the most known crises to all-encompassing environmental, economic, social, and human health conditions [14–17] modern human have ever challenged [18]. The simulations conducted using global climate models uncover that the most important factors that induce this planetary phenomena are natural (variations in solar radiation, volcanic activity, and aerosol concentrations) and anthropogenic (fluctuation in the content of the atmosphere due to human actions); only the amassing effect of the two factors can provide a reason for the transformations noticed in the world average temperature in the last century and a half [19,20], despite the fact that this is only a small part of the extent of fluctuations in climatic parameters, even including the planet ocean level [21].

It is accepted that inland water conditions are closely connected with weather and atmospheric temperature fluctuations, so climate change may have forceful direct and indirect effects on freshwater biota [22].

The most recent IPCC Climate Report, "Code Red for Humanity", emphasizes undeniable proofs accompanying the reality that warming has sped up in recent decades; the planet warming is affecting all regions on Earth, and additional heating is anticipated for the next century. Many of the modifications becoming irreversible climate impacts will certainly exacerbate [23]. Another effect of climate change is the alteration of hydrologic cycles; with rising intensity and frequency of extreme events such as droughts, this scenario could influence freshwater biota, generating changes in phenology, life cycles, and dispersion areas, and even the extinction of sensitive species [24]. Accelerated climate change is estimated to influence the biodiversity of huge areas forcefully, with changes in the presence and distribution of numerous species, the decline of the taxonomic richness, and the vanishing of entire ecosystems [25]. There is much proof that global warming is threatening the biodiversity of our planet, including fish [26–32], a significant taxonomic group under permanent high global human-impact threats and risks [33–45].

In this global warming scenario, freshwater ecosystems are highly vulnerable and their communities could experience significant impacts. Some new research highlights that freshwater biodiversity has declined quicker than both marine and terrestrial [46].

Due to the fact that in the above-mentioned recent United Nations report that the researched Carpathian area will be characterized by heat waves and severe drought periods [46], and the freshwater biota is under a accentuated risk in general, the aim of this study was to provide information about the state of the local fish species refuge habitats in the context of climate change and human impact effects.

Climate change forecasts for Europe are not an exception and it is clear that air temperatures will rise because of the impact of human activities on the atmosphere [47–49]. The predictions are also clear that the contemporary precipitation regime will be modified and altered and will vary from one area to another beyond the normal known seasonal patterns, with drought episodes becoming accentuated [50], and severe and even extreme climatic crisis are anticipated [51].

The climate system heating is unquestionable due to the relatively uninterrupted long-term warming trend since the mid-20th century, which may be correlated with anthropogenic influence [52–54].

Drought is an effect-dependent phenomenon [55], and considerable human impact is at least partly culpable for the harshness of the contemporary frequent and persistent drought episodes [56] generating a very intricate hydro-climatic risk affecting the natural and anthropogenic systems [57]. More than that, the heat wave magnitude is projected to rise everywhere in future [58].

Climate change-associated issues are some of the most contentious scientific issues of the present day. At this time of the climate change situation, the temperature increases all around [54], even in unanticipated ranges on Earth [59–63], and drought, decreasing altered flows on streams, which is a significant driver for aquatic ecosystems' ecological status [64], aquatic biodiversity [65], and even their potential economic use [66,67] appear and remains even in what are considered "safe" geographical areas. From this viewpoint, the Carpathian Basin was considered, bringing their synergic effects together with other human impact types that could also have consequences on functional traits of aquatic assemblages in terms of the abundance and distribution of their species [68–75].

Carpathian Basin climate change trends exceed the Earth warming rate since the 1950s of the last century through rising temperature and precipitation and drier spring, summer, and autumn seasons [76–79]. For example, in the last century, a 0.8 °C increase in surface temperature and a 60–80 mm decrease in precipitation were registered [80].

In the last decades, the Carpathian Basin has experienced very persistent and accentuated droughts, a trend connected with climate change, inducing remarkable drying in this region, particularly in the summer, in the chiefly exposed south-eastern sectors that have had environmental and socio-economic effects [81].

In the general climate change circumstances, the Carpathian Basin and its valuable biosphere are considered to be very sensitive. Drought is one of the important climate-related detrimental natural phenomena, and it has been appearing with increasing frequency, severity, and extent in the last decades [82,83].

The causes of fast expansion of the drought in the South-Eastern Carpathian basin are: the rise of the yearly average temperature by 0.3 °C, the rise in number of tropical days (>30 °C), the diminishing of winter days (<0 °C), the diminishing of precipitation, and the lessening of runoff. As results: reduced inflows to aquatic ecosystems, decreased stream-flows in a most of basins, reduced recharge of groundwater, high repetitiveness and period of the drying up of lotic systems (principally those smaller than 500 km^2), and the stream flow drought which has arisen more often since 2000 [84,85].

Climate changes disturb ecosystems and cause potential threats and risks regarding their natural products and services; in these circumstances, human society should anticipate and adapt in time to these major global challenges [60].

The ephemeral and small lotic systems are the most abundant and hydrologically dynamic of all freshwater ecosystems, existing across most of our planet, including rivers in alpine and hilly zones as well as temperate regions [86–88].

The Carpathian Mountains' geographic typical features (i.e., form, orientation, latitude, and altitude) were and are essential elements which played a key driving role concerning the fish species' presence, dispersion, evolution, and their populations' ecological status [89]. In the recent climate change pattern for the hydrological nets, is the drought a new game changer in the Carpathian area? What will be the effect of this type of change on small lotic low-flow systems' habitats and their fish populations, which are already under high stress from human impact? How can we be proactive and identify such risk hotspots and how can we design proper management plans for these threatened lotic systems in order to diminish these climate-induced negative effects?

This research intended to deal with several local case studies of small lotic systems based on which similar areas can benefit by the proposed habitats and fish population risk assessment, monitoring, and management elements. Such a research-targeted area, the so-called "Saxon Villages" area/Southern Transylvania Tableland, in South-East Transylvania (Romania), is one of these types of identified areas where climate change can lead to great pressures both on the water-related habitats and biodiversity.

The research area is located in the arena-like Transylvanian Depression. Encircled by the South-Eastern Carpathians, the middle Târnava Mare River basin sector is located in the central-south part of the Transylvanian Plateau, particularly in the central Târnave Plateau, populated by more than seven million people [90], and its lotic systems are under both historical and modern diverse human-impact negative effects [91–93].

The geological basis is under a deep Neogene bed, formed by the southern segment of the Central Transylvanian massif, composed of crystalline schists, over which were accumulated Miocene, Pliocene, and Quaternary soft structures. The interfluvial areas are topped by dispersed Sarmatian marls-clays, sands, and tuffs. The most frequent is the Pliocene (Pannonian) structure, characterized by marl-clays and sands [94–98].

From the geomorphologic viewpoint, the base of the researched area was the river meadow suspended above the Târnava Mare riverbed and the lower terraces, creating transversely fragmented hilly surfaces. The energy relief is lessened, with a maximum of 100 m, and the regular fragmentation degree is 05–07 km/km^2 [94–98].

The climatic regime, connected with the sector of the hilly area with a moderate-continental climate, is defined by warm summers with rather low precipitation, and winters with warmer periods. The circulation of the atmosphere is characterized by the high frequency of western and north-western temperate-oceanic air masses, chiefly in summer. Less often, south-west and south Mediterranean air masses appear, and northern Arctic masses less often. The annual average temperatures lie between +8 °C and +9 °C, with absolute values of +37 °C and −32 °C. The average July temperatures vary between +18 °C and +20 °C, and those of January between −3 °C and −4 °C. The atmospheric precipitation has an annual quantity of 500–700 mm [94–98].

In the study area of the Târnava River basin, the torrential character of the superficial flow has high maximum flows in rainy periods and common minimum low-flows, with run-dry during drought [94–98]. In the field activities, in the August months of the 2016–2021 study period, the drought and heat waves kept the general trend of the last decades; moreover, the years 2019 and 2020 were the hottest years since 1961 [99–103].

The studied lotic systems, Dupuş (5 km length, 10 km^2 basin surface, 0.024 m^3/s multiannual flow—2016–2021), Biertan (17 km, 58 km^2, 0.124 m^3/s), Valchid (16 km, 56 km^2, 0.120 m^3/s), Laslea (22 km, 111 km^2, 0.345 m^3/s), Mălâncrav (14 km, 41 km^2, 0.100 m^3/s), and Felţa/Florești (9.6 km, 17 km^2 0.041 m^3/s) belong to the southern Târnava Mare Basin. This basin's aquatic and riverine habitats and associated biodiversity are under the influence of the impacts of multiple and diverse human activities [104–113].

The rather new extended-heat summers and warmer winters increased the temperature of rivers, decline snowpack, and the accessibility water and its related resources to riverine human communities. In addition, the present higher human pressure, contrasting with the traditional past environmentally friendly practices of natural resources use, on water resources finally induced a decrease in water quality and quantity.

Recurrently, the accent of drought-connected effects are pushed onto human-centric water resources and agricultural, socioeconomic, and migration aspects due to the related economic losses and social tension [114–117], and avoiding the related and primary triggering features of droughts, namely the meteorological and hydrological elements.

This study addressed some of the ecological dimensions of frequent and prolonged droughts, and their lasting structural effects on some aquatic habitats and their fish communities, which are communities with high relevance for the studied lotic systems' ecological status under climate change/drought seasons' constant pressure.

The human indirect (climate changes/drought) and direct influence (water overuse, water pollution, habitats fragmentation and destruction, scarcity of refuge habitats, wetland and riverine areas mismanagement, stimulating over sedimentation, etc.) are key aspects of understanding many of the drought effects on the researched lotic systems.

Other objectives of this study were to reveal some landscape (geographical, cultural, and natural heritage) values and traditional best-practices loss effects in the new climatic change situation and to identify new threats and the different types of human impact affecting some lotic system fish fauna, based upon specially designed scientific research. In the beginning of the twenty-first century, the human-nature relationship is far from balanced and with significant and variable negative effects on the biodiversity, including the whole Danube Basin [118–133], to which the studied area belongs.

It is very likely that the tendency of climate change in the twenty-first century will be very much alike that of the end of the twentieth century, manifested by rising values of extreme maximum temperatures and heat waves [76–85]; this reiterates the need for applied studies by identification of problems and proposals of integrated management plans for the ecological support systems for human society and its enterprises.

This study had as a main aim to identify and map the lotic refuges in drought periods, and also the sectors where these should be rehabilitated or made. Adjacent monitoring and management elements were proposed for the studied lotic systems' natural processes recovery. It must be highlighted that no such research approach regarding the south-east Carpathians small rivers fish refuge habitats has previously been realized.

2. Materials and Methods

This study was based on a fish samples assessment, from 2016 to 2021, in summer drought periods, at five rivers, on every 500 m sector, in the west of the Saxon Villages area (Figure 1). The sites more than one meter in depth in the studied lotic systems were measured and inventoried, walking in all the riverbeds length. The near downstream 500 m-long stretches of these lotic systems were also checked to see what fish species and how many individuals thereof are present, and if there is a difference in their abundance between refuge and non-refuge 500 m sectors. The fish numbers in the identified refuge habitats (lotic habitats with a depth of minimum one meter and a length of maxim 10 m) were compared with the near downstream 500 m-long lotic sector fish number of individuals and presented. The major method limitation is that it is relatively cronophagous; the river's entire length should be walked by the researchers through the riverbed.

Figure 1. The study area location on the South-Eastern Carpathians basin (Romania).

The sampling highlights the presence of aquatic refuge habitats and fish communities' richness in and near the refuge habitats.

During the drought season, field studies concerning the habitats were carried out on Dupuş, Biertan, Valchid, Laslea, Mălâncrav, and Felţa/Florești rivers (Figure 2).

The fish assemblages' survey presented, through time/on effort (one hour/500 m), these quantitative samplings, which were gathered with a hand-net.

For the fish communities' quantitative structure, the used description was: the individuals' number in the unit of time/effort unit—average value for the samples of the same station, for six years of the study, on each identified refuge habitat and its downstream near-500-m sector.

The sampled fish were identified, counted, and released immediately back into their natural habitats. Different habitat characteristics data (refuge depth, banks description, land use, substrate, banks height, minor riverbed width, GPS coordinates, vegetation, and human impact) were collected (Tables A1–A6).

This research proposed some in situ adapted management elements for the recovery, at least partially, of the previous ecologic status of the lotic systems' habitats in the area and of their associated fish species communities.

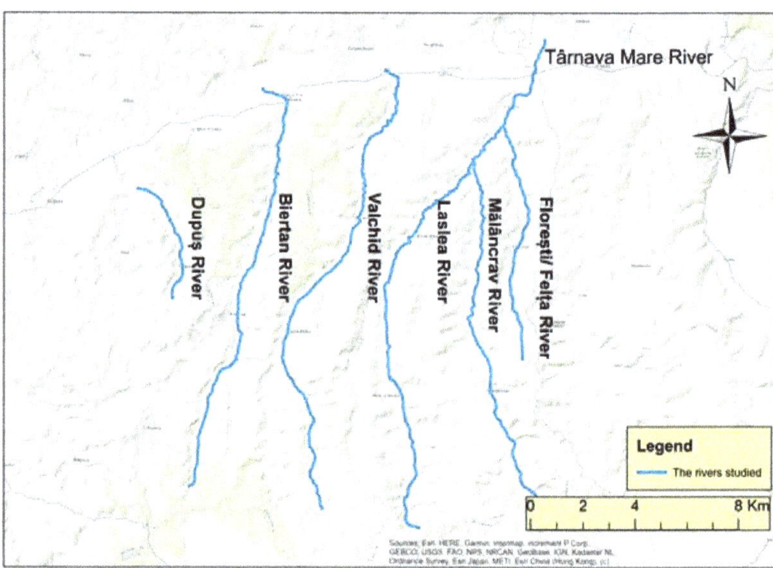

Figure 2. The studied lotic systems.

3. Results

3.1. Dupuş River

Based on the field inventory, on the Dupuș River, only one refuge was identified with a depth of 150 cm (i.e., Figure 3), and the other non-refuge sectors (i.e., Figure 4) had different depths between 10 to 80 cm.

In the Dupuș River, three fish species have permanent populations: *Gobio gobio* (Linnaeus, 1758) (9 total caught individuals), *Barbus meridionalis* Risso 1827 (6), and *Squalius cephalus* (Linnaeus, 1758) (3).

The ratio between the total number of fish between the refuge habitat of maximum 10 m length and the adjacent/downstream non-refuge sector of 500 m length was two to one.

Figure 3. Dupuş River refuge habitat.

Figure 4. Dupuş River non-refuge sector.

The single identified refuge habitat is present in the lower third of the river (Scheme 1) revealing the drought-related risks for the biodiversity in at least two thirds of the river sectors.

Given the shallow depth of the Dupuș River and the vulnerabilities that occur for fish during periods of drought, the building of artificial refuges for fish from about 500 to 500 m is proposed. In total, it would be necessary to build a number of 12 such refuges with a depth of at least one meter (Scheme 1).

3.2. Biertan River

Based on the field inventory, on the Biertan River, only three refuges were identified with a depth of 100–120 cm (i.e., Figure 5); the other non-refuge sectors (i.e., Figure 6) had different depths between 5 and 100 cm.

Figure 5. Biertan River refuge habitat.

Figure 6. Biertan River non-refuge sector.

In the Biertan River, 11 fish species have permanent populations: *Squalius cephalus* (Linnaeus 1758) (total caught individuals 21), *Alburnus alburnus* (Linnaeus 1758) (7), *Alburnoides bipunctatus* (Bloch 1782) (20), *Chondrostoma nasus* (Linnaeus 1758) (8), *Gobio gobio* (Linnaeus 1758) (19), *Barbus barbus* (Linnaeus 1758) (9), *Barbus meridionalis* Risso 1827 (23), *Carassius gibelio* (Bloch 1782) (5), *Barbatula barbatula* (Linnaeus 1758) (11), *Sabanejewia romanica* (Băcescu 1943) (31), and *Sabanejewia aurata* (De Filippi 1863) (14).

The ratio between the number of fish between the refuge habitats of maximum 10 m in length and the adjacent/downstream non-refuge sectors of 500 m length was six to one.

The identified refuge habitats are present only in the lower one-fifth sector of the river (Scheme 2), revealing the drought-related risks for the biodiversity in four-fifths of the river sectors.

Given the shallow depth of the Biertan River and the vulnerabilities that occur for fish during periods of drought, it is proposed to build artificial refuges for fish from about 500 to 500 m. In total, it would be necessary to build a number of 34 such refuges with a depth of at least one meter (Scheme 2).

3.3. Valchid River

Based on the field inventory, on the Valchid River, 27 refuge habitats were identified with a depth of 100–160 cm (i.e., Figure 7); the other non-refuge sectors (i.e., Figure 8) had different depths between 5 and 100 cm.

In the Valchid River, 10 fish species have permanent populations: *Squalius cephalus* (Linnaeus 1758) (total caught individuals 28), *Alburnus alburnus* (Linnaeus 1758) (6), *Alburnoides bipunctatus* (Bloch 1782) (22), *Chondrostoma nasus* (Linnaeus 1758) (3), *Gobio gobio* (Linnaeus 1758) (26), *Barbus barbus* (Linnaeus 1758) (3), *Barbus meridionalis* Risso 1827 (30), *Barbatula barbatula* (Linnaeus 1758) (19), *Sabanejewia romanica* (Băcescu 1943) (20), and *Sabanejewia aurata* (De Filippi 1863) (19).

The ratio between the number of fish and the refuge habitats of maximum 10 m in length and the adjacent/downstream non-refuge sectors of 500 m length was seven to one.

Refuge habitats are present in four-fifths of the lower and middle parts of the river, and should be extended to the upper part and multiplied so that at every 500 m there is at least one.

Given the general depth of the Valchid River and the vulnerabilities that occur for fish during periods of drought, the building of artificial refuges for fish from about 500 to 500 m is proposed. In total, it would be necessary to build a number of 17 such refuges with a depth of at least one meter (Scheme 3).

Figure 7. Valchid River refuge habitat.

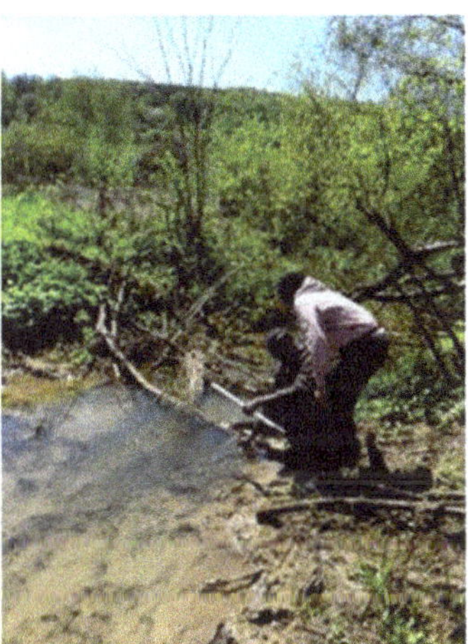

Figure 8. Valchid River non-refuge sector.

3.4. Laslea River

Based on the field inventory, on the Laslea River, only four refuge habitats were identified with a depth of 100–120 cm (i.e., Figure 9); the other non-refuge sectors (i.e., Figure 10) had different depths between 5 and 120 cm.

In the Laslea River, seven fish species have permanent populations: *Alburnus alburnus* (Linnaeus 1758) (total caught individuals 9), *Chondrostoma nasus* (Linnaeus 1758) (8), *Gobio gobio* (Linnaeus, 1758) (40), *Barbus meridionalis* Risso 1827 (42), *Barbatula barbatula* (Linnaeus 1758) (44), *Sabanejewia romanica* (Băcescu 1943) (60), and *Sabanejewia aurata* (De Filippi, 1863) (31).

The ratio between the number of fish between the refuge habitats of maximum 10 m in length and the adjacent/downstream non-refuge sectors of 500 m length was 12 to 1.

Figure 9. Laslea River refuge habitat.

Figure 10. Laslea River non-refuge sector.

Refuge habitats are rarely present on two-thirds of the lower and middle parts of the river, and should be extended to the upper part of it and also multiplied so that at every 500 m there is at least one.

Given the general depth of the Laslea River and the vulnerabilities that occur for fish during periods of drought, it is proposed to build artificial refuges for fish from about 500 to 500 m. In total, it would be necessary to build a number of 37 such refuges with a depth of at least one meter (Scheme 4).

3.5. *Mălâncrav River*

Based on the field inventory on the Mălâncrav River, only three refuge habitats were identified with a depth of 100–110 cm (i.e., Figure 11); the other non-refuge sectors (i.e., Figure 12) had different depths between 10 and 100 cm.

Figure 11. Mălâncrav River refuge habitat.

Figure 12. Mălâncrav River non-refuge sector.

In the Mălâncrav River, four fish species have permanent populations: *Squalius cephalus* (Linnaeus 1758) (total caught individuals 12), *Gobio gobio* (Linnaeus, 1758) (8), *Barbus meridionalis* Risso 1827 (22), and *Barbatula barbatula* (Linnaeus 1758) (13).

The ratio between the number of fish between the refuge habitats of maximum 10 m in length and the adjacent/downstream non-refuge sectors of 500 m length was four to one.

Refuge habitats are rarely present throughout the length of the river, but should be more numerous, with at least one every 500 m.

Given the general depth of the Mălâncrav River and the vulnerabilities that occur for fish during periods of drought, the building of artificial refuges for fish from about 500 to 500 m is proposed. In total, it would be necessary to build a number of 21 such refuges with a depth of at least one meter (Scheme 5).

3.6. Felţa River

Based on the field inventory on the Felţa/Floreşti River, only two refuge habitats were identified with a depth of 100 cm (i.e., Figure 13)' the other non-refuge sectors (i.e., Figure 14) had different depths between 10 and 100 cm.

Figure 13. Felţa River refuge habitat.

Figure 14. Felţa River non-refuge sector.

In Felţa/Floreşti River 3 fish species have permanent populations: *Squalius cephalus* (Linnaeus 1758) (total caught individuals 14), *Gobio gobio* (Linnaeus, 1758) (24), and *Barbus meridionalis* Risso 1827 (10).

The ratio between the number of fish between the refuge habitats of maximum 10 m length and the adjacent/downstream non-refuge sectors of 500 m in length was three to one.

The only two identified refuge habitats were in the lower third of the river (Scheme 6), revealing the drought related risks for the biodiversity in at least two middle and upper-third sectors of the river.

Given the general depth of the Felţa River and the vulnerabilities that occur for fish during periods of drought, the building of artificial refuges for fish from about 500 to 500 m is proposed. In total, it would be necessary to build a number of 12 such refuges with a depth of at least one meter (Scheme 6).

4. Discussion

4.1. Identified Ecological State Elements

Practically all fish species have a definite range of habitat preference delineated by abiotic attributes such as substrate, depth, velocity, floods, temperature, dissolved oxygen, etc. [134–141]. The spawning periods of fish also differ with respect to such distinct ecological elements as stagnant or running water, as well as altitude, temperature, quality of water, and etc. A fundamental property of fecundity is its raise during the development of the fish; a big fish produces more eggs than a small one [142,143]. If no relative large habitats, including refuge-type habitats, exist, no large fish can exist, and consequently reduced fecundity of fish populations can appear in the rivers.

As stream discharge lessens, decreasing water quantitative and qualitative characteristics may exceed resilience limits, pushing fish to search for and use refuge habitats. These can be characterized as sites where the adverse effects of disruption are absent or reduced than in more damaged sectors. For example, as discharge diminishes, the fish species which prefer riffles leave and search for refuge in pools. In these general circumstances, it is self-evident that the refuge habitats play a critical role in reducing the disruptive effects of drought in terms of individual endurance of fish and population survival. However, the quality of the pools varies in relation with many elements such as abiotic suitability, trophic opportunities, predation risks, etc.

Drought drying is a critical disruption circumstance in many small streams creating irregular or isolated lotic habitats [144]. We assessed and found that, based on the ratio between the fish numbers in refuge habitats and in the near lotic sectors, the remaining pools act as needed refuge habitats, especially during drought episodes.

In the researched area, there are two fish species of conservative interest, namely *Barbus meridionalis* Riso, 1827 and *Sabanejewia aurata* (De Filipi, 1863). The barbels, Ord. Cypriniformes; Fam. Cyprinidae are influenced by stream habitat quality such as, for example, drought periods [145]. *Barbus meridionalis* species terra typica is the Mureş River basin to which the studied Târnava River tributaries belong. This fish is a bottom-dweller, relatively fresh, cool, and well saturated with oxygen water species. It prefers hard substrate [134]. The loaches, Ord. Cypriniformes; Fam. Cobitidae are also influenced by the streams' habitat quality, inclusive of drought effects [146]. *Sabanejewia aurata* is a mainly nocturnal, demersal, freshwater, bottom-feeder species. The sand presence in the riverbed is an important habitat precondition, with individuals spending extended periods in the sand. Its presence in muddy or silted areas is rare. It needs a quite warm water temperature especially in the summer season, but not over 20 °C [134].

When habitat characteristics match the favored range, anticipated fish species will appear in good abundance [147]. The first fact, which was observed in the research in the field, was that in hydrological drought periods, the number of fish was higher in the identified refuge habitats, a critical aspect for fish conservation, in contrast with transitional lotic sectors and especially in riffles, which are usually the first habitats to dewater at low flows. It is clear that the pools offer refuge from the most negative effects of drought (i.e., isolations and stranding of fish). Specifying a niche is equivalent to defining habitat conditions [148,149] that allow a species to remain in space and time. That is why proper habitats in general and especially the highly needed refuge habitats should be identified, characterized, and inventoried for considering present-future climate changes.

Together with flow fluctuations and low flow prolonged periods, a second stressor which can be important for the fish communities' deteriorated ecological status is the temperature, which can differ among river sectors with refuge habitats bordered by dense riparian vegetation and sectors with no such refuge habitats and no or rare vegetation. These sectors are exposed to the sun heat and have a higher temperature and evaporation rate; in consequence, both the water's quality and quantity decrease.

The rise in frequency and magnitude of hydrological droughts has quantitative and qualitative negative effects, lowering the refuge habitats' surface and volume, the lateral connectedness between rivers and floodplains, the longitudinal connectedness, and the quality of water. Drought also impacts the relative habitat availability (i.e., ratio of pools to riffles), reproduction sectors and activities, and diminishes the trophic base support surfaces. The shallow habitats may dry pools, decreasing them in dimension, but still hold water for longer after the surface flow ends. Drought also intensifies density-dependent biotic interactions such as competition and predation, as fish are congested into a smaller volume of water.

A fall in fish numbers among refuge-sampled habitats goes along with descending discharge and growth in the sampled refuge habitat areas.

Over the course of the summer low-flow period, numbers of fish in the sampled non-refuge habitats were lower in comparison with the refuge habitat, indicating degradation in abiotic and biotic conditions linked with drought.

Interest and struggle are increasing on improving environments that are human-dominated [150]. In this study, the skewed natural fish communities' structure, or even disappearance/local extinction of some fish species on some studied lotic systems sectors, had complex causes, among which climate change and human impact synergy in these basins can play a central role, inducing as a mitigation measure the completion of the needed refuge habitats proposed herein.

The traditional land and water use best practices in the studied area, i.e., lotic systems with natural courses, using the phreatic rather than the river water for household needs, thick riverine natural vegetation (especially *Alnus incana*, *Telekia speciosa*, and *Chrysosplenium alternifolium* Erika Schneider-Binder *in verbis*), terraces with vineyards with anti-erosional effects on slopes, full coverage with forests on the tops of the hills, soft agricultural practices in small family farms in lowland areas, raising of cows, etc., were replaced by new, modern activities with negative effects on the environment.

The traditional best practices related with the land and water use were replaced by new ones which are aggressive with the aquatic environment: habitat fragmentation with riverbed barriers with no fish ladders or passages; aquatic lotic habitats over sedimentation due to improper riverine forestry and agricultural area use; the riverine vegetation destruction on some sectors reduces their capacity to retain sediments carried down by floods from the basin; overgrazing mainly with sheep and, to a lesser degree, with cows also induces sediment mobilization in the basin; the relatively big riverine industrial farms and the small family farms overflowing their animal manure direct into the river without any cleaning treatment; household water and solid waste overflows without any treatment; the lotic natural courses being affected in some sectors by modifications of the course, banks, and riverine habitats; the escape of alien fish from adjacent fish farms/ponds, etc.

All these quantitative shortages in refuge habitats and the qualitative problems induce significant pressure on fish fauna in the context of climate change and should be addressed with specific in situ adapted management measures.

The shortage of refuge habitats did not allow the presence of proper conditions for fish related first to the needed hard and coarse sand substrata due to fine sediments and mud over sedimentation and clogging, and second with water temperature and oxygenation due to the rising of water temperature and decreasing of oxygenation in lotic sectors not protected by riverine vegetation, exposed to direct sun heat, without refuges, and which are shallow.

Human activities in drought conditions, which synergically induces erosion regularly, deliver massive quantities of fine sediments into streams and rivers, forming large static bodies of sediment known as sand slugs, which smother in-stream habitat, alter community structures, and decrease biodiversity [151–153]. Based on our findings of such droughts plus a high variety of human impacts in the studied area, a strategic basin management should be designed and put into action to accomplish effective and viable aquatic refuge habitats used for the associated fish communities. Correct protection of the fish populations should rely on the protection of their environment through integrated management, which should solve the following identified problems: riparian zones were reduced on the surface or removed by the agriculture expansion and destructive agriculture works, lessening river shading and raising aquatic habitat temperatures and decreasing water oxygenation; severe sedimentation problems due to basin erosion, channel cuttings and strong weather water runoff raised by the insufficiency or absence in some sectors of riparian vegetation; accentuated erosion and sedimentation issues that develop from a lack of riparian vegetation along long sectors of river corridors and which can lead to gravel bed siltation, critical to the insectivorous fish species; disturbed hydrologic regimes in some sectors by water over extraction and use; the man-made stream barriers which have repercussions on the ability of fish to move among refuge habitats; lasting inputs of pollutants; habitat loss-induced native fish species decline; stream sectors channelization and isolation from their alluvial plain, important for cyprinids; habitat deficit easing fish overcrowding circumstances, which lead to occurrences and outbreaks of diseases; etc.

Based on the biological and ecological characteristics of the two sampled fish species of conservative interest and on the identified problems in relation with this species throughout the six years of monitoring activities, some management directions can be highlighted so that basin managers can have the key information and recommendations needed to preserve a good quantitative and qualitative conservation status of the local water resources and fish populations in synergic climate change and anthropogenic impact conditions.

Drought is a natural disruption including aquatic ecosystems in terms of physical and chemical water quality, increased fish death rates, and decreased fish birth rates and/or expanding migration rates, and can be an important factor in destruction of aquatic communities; for fish to remain in affected lotic systems, they must have refuge from interfering factors. Refuges' dimension, commonness, and connectedness play an important role in fish populations' ecological status and persistence. Population dynamics of fish using refuges during drought are best modeled by modified source-sink dynamics, but such dynamics are likely to change with spatial scale [154–158].

For the studied lotic systems, habitats, and their fish populations ecological status recovery, as a part of "a best practice" management of the researched area, relatively numerous new refuge habitats should be created and properly managed: nine times more than the existing number on the Dupuş River, 10 times more on the Biertan River, 0.4 times more on the Valchid River, 10 times more on the Laslea River, eight times more on the Mălâncrav River, and six times more on the Felţa River. Among all of these rivers, the Valchid River state reveals a potential good status to be reached by all lotic systems in the area. We can understand now how this area's conditions were in the past, and why, in some localities, coat of arms waterfowls were present (Erika Schneider-Binder *in verbis*).

A complex, integrated, and permanent monitoring system should be implemented in the studied area, based on the fact that the physic-chemical and biological variables can show and anticipate the human impact actions and climate change effects on lotic systems' basin ecologic status and perspectives [159]. For the efficiency of the proposed monitoring activities, people with appropriate fish-related taxonomic skills should be involved [160].

The torrential nature of the superficial flow in the studied lotic ecosystem basins set off high maximum flows especially in rainy periods and minor flows, with very low-flow episodes, during drought periods of the year [98], periods which are more numerous and longer under the present climate changes. The researched area's geomorphologic processes are of high amplitude, have a high frequency, and an intensity of manifestation that confers on them a risk character on the local geomorphosystems [151], especially in the relatively new appearance of climate change effects, including on the refuge habitats of fish.

Sediment motion, flow, amassing, and clogging are influential phenomena in lotic habitats and their associated fish communities due to their significance for: riverbed features, river channel morphology and stability, and refuge habitats availability and quality. In the studied area, the sediment entrainment, transport, and deposition induced a significant refuge habitat loss with associated fish communities, decreasing living conditions. This situation is stimulated especially in the climate change context of change in flow regime, respectively, the diminishing of the water flow in the drought periods, through affecting entrainment, transport, and deposition processes. Changes in the frequency, duration, and magnitude of flows competent to move the pre-existent bed materials have significance for substrate character formation and maintenance. For example, if riffles are not sufficiently flushed on a regular enough basis, fine sediment deposits amass in the interstices among the gravels and rocks, lowering or even removing the interstitial micro-habitats. Over the longer term, the hydrographical net substrate is buried, depth sectors/refuge habitats are clogged, and some disappear, and all of this leads to the loss of a natural series of river sectors along with the refuge habitats. Reductions in the frequency, duration, and magnitude of flows sufficient for transporting tributary sediment inputs further downstream have significance for river channels sedimentology and morphology, as sediment bars tend to expand, including in the confluence areas with tributaries, reducing fish access and mobility to long sectors of habitats and their resources.

The geomorphologic balance adjustment and sustenance in the studied area can be dealt with in an anthropic way by a set of measures for the prevention and diminishing of the effects of geomorphologic risk processes, but the witnessing of their manifestation remains for long periods, and needs a reconversion of the lands and a realistic management of the soil and sediments in the resilience limits of the area [151].

4.2. Fish Fauna Refuge Habitats Management Elements Proposals

It must be highlighted that no such research approach regarding the south-east Carpathians small rivers fish refuge habitats was previously realized.

Appropriate basin management measures should begin from the following elements: water is part of the environment and for that reason, different users can adapt only to the surplus quantity that is naturally available from rivers. In nature, there is more water in riverine systems than is needed for support of the riverine ecosystems; if the main features of the natural flow regime can be identified and adequately maintained in a changed flow regime, then the valuable biota should be preserved.

The management measures should mostly include: basin soil stabilization and conservation, slopes and tributaries erosion diminishing measures, treatment of point sources, etc.

The silting of deep stream habitats is a natural physical phenomenon which cannot be avoided, but by understanding the effects of hydro-morphological pressure due to a lack of basin sediment management, mitigation of this situation's effects on fish refuge habitats can be obtained based on basin sediment management measures: buildups of sediments could be managed by physical intervention like mechanical raking or excavation; restoration of more natural levels of sediment inputs reduce the demand imposed on streams to transport and rework elevated sediment loads; creating outside-of-river-beds storage ponds in which, through some mobile diversion dams, the high-flow period water will be channeled for the deposition of sediments before they can clog the refuge habitats; and all the torrent and soil erosion control works in the basin should be carried out.

Numerous forestry management measures can be useful for basin sediment management: the forest stands' vertical and horizontal structure should be as close as possible to its natural model (multi-layers, groups, and patches) in order to reduce soils being destabilized because of slope characteristics, loss of moisture, loss of large canopy cover, loss of root strength, and etc.; permanent watercourses will be protected by commercial intervention with a minimum 50-m-wide stripe on both sides of the watercourse. It is important to keep and respect the natural dynamics of habitats (i.e., reed beds, sedges and forests), together with their necromass (i.e., died wood), in banks (i.e., major/minor riverbed and alluvial terraces) and in water. The cross-over, in accidental or extraordinary cases when it cannot be avoided, will be made over the shortest distance on installations adopted to keep the water clean and soil erosion at a minimum; riparian forest and the natural bushy and grassy, local vegetation and flora will be strictly protected; any timber harvesting or transport activity is recommended to take place on a frozen substrate, covered by snow and/or ice and on well-maintained forestry roads with a layer of mineral aggregates on top of the road; the network of forestry roads must be optimally dimensioned for maximum efficiency with the minimum amount of damage; existing roads should be preserved and maintained in order to avoid or minimize erosion, sediment transport, and accumulation in the area; building of new forestry roads should be avoided; ecologically friendly, low-impact logging technologies (oxen, horses, cableway-skidders) are always preferred; timber will not be removed from forests during or after precipitation while the soil is moist; skid-roads will always be designed, built, and monitored in order to avoid soil-erosion as much as possible, ensure the protection of permanent and temporary watercourses, and protect remaining trees; sensitive areas like potential land-slide zones, talus, cliffs, steep slopes, etc., will be identified, protected (excepted from logging), and monitored; it will be prohibited to store timber (even for short periods of time) in the riverbed, on its banks, and on the minimum 50-m-wide protection stripes adjacent to both perennial and ephemeral waterways. All the vegetation within the protection belts shall remain intact; intervention cuttings will be applied in openings with a diameter up to a tree height, from which the old trees can be completely removed; timber harvesting will not exceed 10% of the volume of the stand (exceptions can be allowed in special accidental situations, for example windfalls and/or snowfalls, etc.), and the harvest volume will be correlated with the condition of the stand, the dynamics of natural regeneration, and with assigned conservation requirements; logging techniques will be adopted to minimize the level of injuries to the remaining trees and soil; if there are stands in which the natural regeneration is very difficult or stands are affected by calamities, replanting or direct sowings will be carried out using only seminological material of local origin or, if not possible, from identical ecotypes; timber harvesting access corridors should run parallel to the lotic and lentic ecosystems; pioneer species will not be extracted as they are important for soil improvement; illegal logging will be controlled with the aim of eradicating illegitimate logging activities; during logging and timber transport activities, sediment traps will be installed on the main watercourses (with around 500 m distance between them), and will be cleaned as often as necessary, with their evacuation/transport in areas that do not influence the degree of sedimentation in waterways; construction/monitoring/maintenance of drainage ditches for liquid and sediment flows from the transport routes designed to manage excessive and rapid precipitation events that are characteristic of the mountain area; installation of sediment traps (with around 500 m distance between them) on ditches used to evacuate liquids from the transport routes, cleaned as often as necessary, with the anticipation that the proposed sediment traps will stop, or at least diminish, the potential of these channels' networks to be an unwanted sediment delivery system directly to downstream water bodies; leafy branches and debris left over from the logging will be placed on remaining stumps; full reforestation of the watershed with canopy projection over 0.8 (to reduce the kinetic energy of raindrops on the soil surface, surface runoff from precipitation, air currents, solar radiation, and minimize large temperature differences, etc.) and a subsequent forest management regime; excluding human activity in the riverbed of watercourses and preserving permanent or temporary riparian wetlands as natural sediment traps during periods of increased precipitations and high flows; the construction of new bridges, if needed, should not narrow the waterway to avoid increasing water speed and its capacity for erosion and sediment transport to downstream sectors; ecological restoration of sediment deposits formed at high waters in the riparian areas by fixing sediments or planting on sedimentary areas will prevent sediments from moving downstream in subsequent episodes of high flows; efforts should be made to favor existing flora, undergrowth, shrubs, and the herbaceous bed; it is acceptable to intervene with sowing or planting in critical areas; leaving the stumps in situ; and controlling waste management [152,153].

Sediment management in watersheds is necessary for their water quality, but indirectly, sediments are also the sources of other problems. Sediments are not unmixed and can be adhered to or can bear pollutants. To prevent such situations, various measures can be enforced: monitoring tributary torrents, slopes, and banks, etc., specifically those prone to accidental or permanent erosion; in sectors with accidental erosion, its effects can be drop off with blankets, sandbags, gravel bags, rugs, plastic materials barricades, etc., until the ecological state is secured through ecological rehabilitation and reconstruction;

development of dense, vegetated fencerows, due to their function as sediment traps; banning the extraction of mineral resources in the basin; banning damming or regulating riverbeds, with the exception of debris basins, settling ponds, and other similar structures which can catch and retain sediments, which should then be frequently cleaned up; banning burning vegetation and trimmings; banning grazing and watering domestic animals in the forest; in pastoral lands, grazing will be organized to prevent the destruction of flora, soil compaction and the onset of erosion phenomena and will even be banned in sensitive seasons or sectors; a ban on the substitution forests and pastures with intensive agriculture lands; small-scale family farms are preferable to big-scale, industrial farms; and natural water filtration sectors and retention ponds should be protected and created and largely used in localities and in opposition solid sectors, growth runoff, and the transportation of sediments should be reduced, particularly hard sectors linked with the construction of buildings and roads infrastructure [152,153].

The general indiscipline with the dispositions commanding the wastewater management and the deficiency of dilution is one of the main causes for some of the researched sectors' environmental bad situations. In this circumstance some management elements are suggested: rising water use performance through general use of contour meters and trustworthy transport pipelines; the development of a hazardous waste site assessment and monitoring unit; the protected lotic sectors must be big and abundant enough to admit the river natural self-cleaning processes to work; the lotic ecosystems must be used like ecological capital seeking to decrease the costs for water cleaning technologies and low-cost fish protein for local consumers.

The present hydrotechnical works' impact should be reduced following some main management directions: increasing the river assimilative capacity thorough its restoration activities; revitalization of the traditions for land protection and use; avoiding the impact of wetland loss; and restoration of sectors of typical lotic ecosystems.

The riverine land exploitation should follow some main directions in this respect: determining the policies for cultivation of multi-year cultures; rehabilitation of the riverine forest corridor, hoping also in areas where the forest and other ecosystems in contact can be allowed to evolve according to natural dynamics; prohibiting access to the upper parts of the catchment areas so that spontaneous perennial vegetation could regenerate in good conditions and limit erosion damages; and rotating sylviculture and grazing (diminishing forestry, agriculture, and livestock impacts) with regard to seasonal conditions, especially on the river adjacent areas.

To sustain the protection of the conservative interest fish species and their habitats, their shelters should to be protected from all man-related aggression; the fact that proper protection is a useful help to the economic development of the rural communities should be highlighted; and complementarily should exist between human society's development and conservation.

The collector river, The Târnava Mare River, which belongs, together with some of its tributaries, to the Natura 2000 Sighişoara-Târnava Mare protected area, cannot be properly managed if its tributaries are not in the ecological state to offer permanent refuge habitats for fish reproduction, development at early growing stages, feeding, and refuge in periods of accidental pollution and high flows.

The elements of the design used for the management of these river basins management can be followed as a general suggested model for other similar Carpathian Basin river basins of conservation interest facing similar environmental issues.

5. Conclusions

The global increasing temperature and other human activities impact induced the hydrologic net and habitats alteration, with a decrease of needed refuge habitats for freshwater fish, diminishing the fish distribution and abundance in the studied area.

In the new climate change scenario, an ongoing pattern in the context of the already-present human impact stress, especially the small hydrological nets, their lotic refuge habitats, and associated fish fauna are under a significant increasing environmental and anthropogenic risk due to an accentuated, new game changer, the drought/low flows, even in what have been, until recently, considered safe geographical areas like the Carpathian Basin.

New proactive special in situ adapted assessment, monitoring, and management integrated systems should be designed and implemented in such risky, potential hot spots to prevent and mitigate these present and future negative effects for the conservation and restoration of fish communities.

Poor water and land use imposed long-term effects on natural lotic systems, namely a change in the physical structure and substratum cover of fish habitat and the fish themselves. This is due to the human-induced, increase in fine sediment clogging in the refuge habitats of fish, which had an adverse effect on them, a process in a continuous acceleration due to climate change-induced diminishing of liquid flow on streams.

A composite model of climatic and anthropogenic induced threats and pressures of the researched river basins has significantly jeopardized the ecological status of its fish. As indirect (climate changes) and direct (various human activities in the basin) human impact affects the lotic habitats, they have become, in some sectors, critically altered or degraded, and the fish populations of conservative interest, and others, have been affected.

For the studied lotic systems' fish populations ecological status recovery, relatively numerous new refuge habitats for fish should be created: nine times more than the existing number on the Dupuş River, 10 times more on the Biertan River, 0.4 times more on the Valchid River, 10 times more on the Laslea River, eight times more on the Mălâncrav River, and six times more on the Felţa River.

The habitat managers are required to specifically monitor the extent to which the changes in physical structures and cover for fish refuge habitats will affect these fish populations in the future, and a key management element should be the refuge habitats' proper management and the creation of others.

The proposed fish refuge habitat monitoring and the creation of new ones should increase the fish survival rate and the recovery of species populations experiencing climate change-induced environmental disturbance.

Author Contributions: Conceptualization, D.B. and A.C.-B.; methodology, D.B., A.S., K.C., S.B. and A.C.-B.; software, A.S.; validation, D.B. and A.C.-B.; formal analysis, D.B., A.S., K.C., S.B. and A.C.-B.; investigation, D.B., A.S. and A.C.-B.; resources, D.B., A.S. and A.C.-B.; data curation, D.B. and A.S.; writing—original draft preparation, D.B., A.S., K.C., S.B. and A.C.-B.; writing—review and editing, D.B. and A.C.-B.; visualization, D.B., A.S., K.C., S.B. and A.C.-B.; supervision, A.C.-B.; project administration, D.B.; funding acquisition, A.C.-B. All authors have read and agreed to the published version of the manuscript. The first and the last authors have equal contributions.

Funding: The APC was funded by Lucian Blaga University of Sibiu, Applied Ecology Research Center, and Hasso Platner Foundation research action LBUS-RC 2020-01-12.

Institutional Review Board Statement: Not applicable.

Informed Consent Statement: Not applicable.

Data Availability Statement: The data presented in this study are available within the article.

Acknowledgments: The project was financed by Lucian Blaga University of Sibiu, Applied Ecology Research Center, and Hasso Platner Foundation research action LBUS-RC 2020-01-12. The first and the last authors thank to their students (Ceauşu M., Tchoumta R. and Nagapin S.) for their active involvement in the field activities. Last but not least the authors thank for the editorial English language support to Baumgartel D., writer/editor in Lausanne, Switzerland, and to Atmosphere journal editors and reviewers support.

Conflicts of Interest: The authors declare no conflict of interest. The funders had no role in the design of the study; in the collection, analyses, or interpretation of data; in the writing of the manuscript, or in the decision to publish the results.

Appendix A

Table A1. Dupuş River Habitat Characteristics.

Station Code	R-Refuge/M-Downstream Refuge Sector	Refuge Depth (cm)	Banks Description	Land Use	Substrate	Banks Height (m)	Fish Presence	Minor Riverbed Width (cm)	GPS Coordinates	Others
A1D0	R	35	Arboreal layer	Arable	Oozy	3		70	46°10′46.0″ N 24°28′20.6″ E	Lush vegetation and branches in the riverbed
AD1	R	80	Arboreal layer	Arable	Oozy	0.5		70	46°10′41.6″ N 24°28′39.2″ E	100 m upstream accumulation lake with fish, probably holiday house, rush, floodplain
A1D2	R	50	Arboreal layer	Arable	Oozy	1	+	100	46°10′36.6″ N 24°28′57.3″ E	Downstream 100 m from the lake, lush vegetation, plastic wastes
A1D3	R	150	Arboreal layer	Arable	Oozy	0.5		120	46°10′24.9″ N 24°29′06.6″ E	Meander, sheep farm, metal, plastic, textile waste
A1D4	R	30	Arboreal layer	Grassland	Oozy	0.5		60	46°10′24.9″ N 24°29′06.7″ E	Lush vegetation
A1D5	R	60	Arboreal layer	Arable	Oozy	3		60	46°10′11.0″ N 24°29′18.1″ E	Bridge, furniture waste
A1D6	R	30	Arboreal layer	Arable	Oozy	3		50	46°10′00.0″ N 24°29′27.0″ E	Rush
A1D7	M	20	Arboreal layer	Arable	Sandy	2	+	60	46°09′51.5″ N 24°29′32.6″ E	Sheep farm
A1D8	M	20	Shrub layer	Arable	Oozy	2		60	46°09′38.8″ N 24°29′41.1″ E	Rush, Plastic wastes
A1D9	M	20	Shrub layer	Arable	Oozy	2		40	46°09′25.5″ N 24°29′37.6″ E	Wool waste, construction (brick, tile), at the entrance in Dupus village
A1D10	M	15	Shrub layer	In locality	Oozy	1.5		30	46°09′15.7″ N 24°29′41.7″ E	Minor riverbed with grass vegetation
A1D11	M	15	Shrub layer	In locality	Sandy	2		35	46°09′06.3″ N 24°29′43.7″ E	
A1D12	M	15	Shrub layer	In locality	Sandy	1.5		35	46°08′54.9″ N 24°29′36.7″ E	
A1D13	M	15	Shrub layer	Arable	Oozy	2		35	46°08′47.3″ N 24°29′25.8″ E	At the exit of the village, branches in the riverbed
A1D14	M	10	Shrub layer and arboreal	Pasture	Oozy	0.5		20	46°08′38.3″ N 24°29′23.3″ E	Rush, cattle farm, close to the river source

Table A2. Biertan River Habitat Characteristics.

Station Code	R-Refuge/M-Downstream Refuge Sector	Refuge Depth (cm)	Banks Description	Land Use	Substrate	Banks Height (m)	Fish Presence	Minor Riverbed Width (cm)	GPS Coordinates	Others
A2B0	R	75	Arboreal layer and grass	Arable	Oozy	5	+	200	46°12′50.3″ N 24°32′04.7″ E	Confluence with Tarnava Mare (close to the road and railway
A2B1	R	35	Shrub and arboreal layer	Arable	Oozy	4		150	46°12′51.3″ N 24°32′04.1″ E	
A2B2	R	50	Arboreal layer	Arable	Oozy	4		200	46°12′45.4″ N 24°32′08.0″ E	Plastic waste
A2B3	R	60	Shrub layer	Arable	Oozy	2		150	46°12′46.3″ N 24°32′17.4″ E	Waste metal, plastic, textiles
A2B4	R	55	Shrub layer	Arable	Oozy	2		150	46°12′46.0″ N 24°32′17.5″ E	Rush, polystyrene, plastic waste
A2B5	R	55	Grass layer	Arable	Oozy	3		300	46°12′44.4″ N 24°32′23.9″ E	Concreted banks, near the railway bridge
A2B6	M	25	Grass layer	Pasture	Oozy, concreted	3		400	46°12′42.3″ N 24°32′30.1″ E	Banks and bed of the river concreted on a length of 30 m
A2B7	R	40	Grass layer	Arable and pasture	Oozy	2		100	46°12′42.3″ N 24°32′30.8″ E	Reeds, gardens
A2B8	R	60	Grass layer	Arable and In locality	Oozy	2		150	46°12′40.5″ N 24°32′40.3″ E	Entrance to Saros on Tarnave
A2B9	R	35	Grass layer	In locality	Oozy	2		150	46°12′40.5″ N 24°32′40.5″ E	In the village, Saros on Tarnave, drenate banks, stone nets
A2B10	R	65	Grass layer	In locality	Oozy	2		150	46°12′32.7″ N 24°32′46.8″ E	Sewer holes, untreated wastewater, plastic waste, manure dumps
A2B11	R	60	Grass layer	In locality	Oozy	2	+	100	46°12′27.9″ N 24°32′45.7″ E	Sewer holes, untreated wastewater, plastic waste, manure dumps
A2B12	R	50	Grass layer	In locality	Oozy	2		100	46°12′22.5″ N 24°32′44.6″ E	Discharge of wastewater, traces and smell of detergents

Table A2. *Cont.*

Station Code	R-Refuge/M-Downstream Refuge Sector	Refuge Depth (cm)	Banks Description	Land Use	Substrate	Banks Height (m)	Fish Presence	Minor Riverbed Width (cm)	GPS Coordinates	Others
A2B13	R	100	Grass layer	In locality	Oozy	1.5		100	46°12′18.2″ N 24°32′41.8″ E	Discharge of wastewater
A2B14	R	45	Grass layer and shrub	In locality	Oozy	1.5		100	46°12′18.2″ N 24°32′41.8″ E	Plastic waste, rush
A2B15	R	40	Grass layer	In locality	Oozy	1.5	+	100	46°12′09.0″ N 24°32′39.7″ E	
A2B16	R	60	Grass layer	In locality and pasture	Oozy	1.5	+	100	46°12′07.7″ N 24°32′39.2″ E	Pumping from the river for irrigation
A2B17	R	55	Grass layer	Pasture	Oozy	2		150	46°11′58.0″ N 24°32′40.9″ E	Plastic waste
A2B18	R	40	Shrub layer	Pasture	Oozy	2		150	46°11′52.7″ N 24°32′37.9″ E	
A2B19	R	70	Shrub layer	Pasture	Oozy	2	+	200	46°11′52.1″ N 24°32′38.6″ E	
A2B20	R	90	Arboreal layer	Arable	Oozy	1.5		150	46°11′45.2″ N 24°32′39.0″ E	Sector with meanders and fallen trees in the minor riverbed
A2B21	R	100	Arboreal layer	Arable	Oozy	1.5		150	46°11′42.0″ N 24°32′31.3″ E	Plastic waste
A2B22	R	60	Arboreal layer	Arable	Oozy	2		150	46°11′42.2″ N 24°32′31.1″ E	Hard to reach, twigs
A2B23	R	65	Arboreal layer	Arable	Oozy	2		150	46°11′36.1″ N 24°32′32.7″ E	
A2B24	M	25	Arboreal layer	Arable	Oozy	2		150	46°11′28.4″ N 24°32′30.9″ E	Plastic waste
A2B25	R	95	Arboreal layer	Arable	Oozy	2		150	46°11′28.2″ N 24°32′30.7″ E	Fallen branches in the minor riverbed, many nettles, traces of deer crossing

Table A2. *Cont.*

Station Code	R-Refuge/M-Downstream Refuge Sector	Refuge Depth (cm)	Banks Description	Land Use	Substrate	Banks Height (m)	Fish Presence	Minor Riverbed Width (cm)	GPS Coordinates	Others
A2B26	R	80	Arboreal layer	Arable	Oozy	1.5		150	46°11′25.6″ N 24°32′31.3″ E	Meandered sector, with high meanders, low flow speed
A2B27	R	55	Arboreal layer	Arable	Oozy	2		150	46°11′21.4″ N 24°32′35.1″ E	Meandering, bridge crossing
A2B28	R	25	Shrub layer	Arable and grass	Oozy	2		200	46°11′14.2″ N 24°32′30.0″ E	
A2B29	R	80	Shrub layer	Arable and grass	Oozy	2		200	46°13′16.6″ N 24°33′01.7″ E	Plastic waste
A2B30	R	40	Arboreal layer	Arable and grass	Oozy	1.5		200	46°13′16.6″ N 24°33′01.7″ E	
A2B31	R	120	Arboreal layer	Arable and grass	Oozy	1.5		200	46°11′06.7″ N 24°32′32.4″ E	Large meanders, sheepfold without animals probably present by cows
A2B32	R	45	Grass layer and arboreal	Arable	Oozy	1.5		150	46°11′09.3″ N 24°32′27.3″ E	Floodplain
A2B33	R	40	Grass layers and arboreal	Arable	Oozy	5.5		150	46°13′16.9″ N 24°33′01.8″ E	
A2B34	M	25	Arboreal layer	Arable	Oozy	3		150	46°13′16.9″ N 24°33′01.8″ E	
A2B35	R	55	Shrub layer	Arable	Oozy	2		150	46°10′55.1″ N 24°32′26.5″ E	
A2B36	R	70	Shrub layer	Arable and grassland	Oozy	2		200	46°10′44.7″ N 24°32′27.1″ E	
A2B37	R	80	Arboreal layer	Arable and grassland	Oozy	3		200	46°10′44.7″ N 24°32′27.1″ E	
A2B38	R	30	Arboreal layer	Arable	Oozy	5		200	46°10′31.2″ N 24°32′21.1″ E	

Table A2. *Cont.*

Station Code	R-Refuge/M-Downstream Refuge Sector	Refuge Depth (cm)	Banks Description	Land Use	Substrate	Banks Height (m)	Fish Presence	Minor Riverbed Width (cm)	GPS Coordinates	Others
A2B39	R	45	Arboreal layer	Arable	Oozy	6		150	46°10′31.2″ N 24°32′21.1″ E	Landslide, denuded shore, meander, poultry farm across the street
A2B40	R	70	Arboreal layer	Arable	Oozy	5		200	46°13′05.2″ N 24°32′42.2″ E	Wildly
A2B41	R	70	Arboreal layer	Arable and grassland	Oozy	4		150	46°13′04.3″ N 24°32′58.4″ E	
A2B42	R	40	Shrub layer	Arable	Oozy	4		150	46°09′51.7″ N 24°32′06.1″ E	Wildly
A2B43	R	60	Grass layer and arboreal	Arable	Oozy	4		150	46°09′58.3″ N 24°32′07.8″ E	Plastic waste, fallen trees in the minor riverbed
A2B44	R	60	Grass layer and arboreal	Arable	Oozy	4		150	46°09′49.0″ N 24°32′04.1″ E	
A2B45	M	20	Shrub layer and arboreal	Arable and pasture	Rocky	2		150	46°09′47.4″ N 24°32′04.6″ E	Construction and plastic waste, possible sheep farm, traces of wool on branches
A2B46	R	40	Arboreal layer	Arable	Oozy	4		150	46°09′33.3″ N 24°31′56.4″ E	
A2B47	R	45	Shrub layer and arboreal	Arable	Oozy	4.5		200	46°09′16.1″ N 24°31′47.1″ E	Wild, hard to reach, closed by shrub vegetation
A2B48	R	65	Shrub layer	Arable	Oozy	4		200	46°09′10.1″ N 24°31′43.9″ E	
A2B49	R	45	Shrub layer	Arable	Oozy	4		200	46°09′04.2″ N 24°31′40.7″ E	Steep banks
A2B50	R	120	Shrub layer	Arable	Oozy	1.5		200	46°08′58.6″ N 24°31′37.1″ E	Many plastics waste, twig dam, cloudy water
A2B51	R	40	Shrub layer	Arable	Oozy	3		150	46°08′39.2″ N 24°31′24.6″ E	

Table A2. *Cont.*

Station Code	R-Refuge/M-Downstream Refuge Sector	Refuge Depth (cm)	Banks Description	Land Use	Substrate	Banks Height (m)	Fish Presence	Minor Riverbed Width (cm)	GPS Coordinates	Others
A2B52	M	25	Shrub layer	Ir locality	Oozy	2		150	46°08′32.7″ N 24°31′20.5″ E	
A2B53	R	35	Grass layer	Ir locality	Oozy	2		150	46°08′34.5″ N 24°31′21.8″ E	
A2B54	M	25	Shrub layer	Ir locality	Oozy	2		150	46°08′32.7″ N 24°31′20.5″ E	
A2B55	M	25	Shrub layer	Ir locality	Oozy	2		150	46°08′18.4″ N 24°31′26.8″ E	In the village, bridge under construction, river channeled through pipes
A2B56	M	20	Shrub layer	Ir locality	Oozy	2		150	46°08′05.3″ N 24°31′14.8″ E	
A2B57	M	25	Shrub layer	Ir locality	Oozy	1.5		150	46°08′05.3″ N 24°31′14.8″ E	
A2B58	R	35	Shrub layer	Arable	Oozy	2		150	46°08′04.3″ N 24°31′09.2″ E	
A2B59	R	35	Arboreal layer	Arable	Oozy	2		150	46°08′02.8″ N 24°31′05.1″ E	
A2B60	R	35	Arboreal layer	Arable	Oozy	1.5		150	46°07′59.3″ N 24°31′01.5″ E	
A2B61	R	35	Shrub layer	Arable and grassland	Oozy	1.5		100	46°07′13.8″ N 24°29′59.2″ E	
A2B62	R	30	Shrub layer	Arable	Oozy	2		100	46°06′59.2″ N 24°29′41.8″ E	
A2B63	R	55	Arboreal layer	Pasture	Oozy	2		150	46°06′46.8″ N 24°29′30.3″ E	Twigs in the minor riverbed
A2B64	R	30	Arboreal layer	Arable	Oozy	2		150	46°06′47.4″ N 24°29′29.7″ E	Underarm clay
A2B65	R	40	Arboreal layer	Arable	Oozy	2		100	46°06′31.7″ N 24°29′06.8″ E	Cow farm

Table A2. *Cont.*

Station Code	R-Refuge/M-Downstream Refuge Sector	Refuge Depth (cm)	Banks Description	Land Use	Substrate	Banks Height (m)	Fish Presence	Minor Riverbed Width (cm)	GPS Coordinates	Others
A2B66	R	35	Grassy layer	Arable	Oozy	2		55	46°06′29.2″ N 24°29′05.4″ E	
A2B67	R	35	Grass layer	Arable	Oozy	2		60	46°06′23.8″ N 24°28′57.0″ E	
A2B68	R	30	Grass layer	Arable	Oozy	2		80	46°06′13.2″ N 24°29′00.5″ E	
A2B69	M	25	Grass layer	In locality and Arable	Oozy	1.5		55	46°06′10.2″ N 24°29′00.3″ E	
A2B70	M	25	Grass layer	In locality and Arable	Oozy	1	-	55	46°05′56.7″ N 24°28′48.7″ E	Green water, eutrophication
A2B71	M	25	Grass layer	In locality and Arable	Oozy	1		55	46°05′59.0″ N 24°28′52.5″ E	Plastic waste
A2B72	M	15	Grass layer	In locality and Arable	Oozy			55	46°05′58.1″ N 24°28′52.3″ E	Discharges of domestic water directly into the river
A2B73	M	20	Grass layer and arboreal	In locality	Oozy	1.5		55	46°05′37.7″ N 24°28′35.9″ E	Discharge of domestic water directly into the river
A2B74	M	15	Grass layer	In locality and arable	Oozy	1.5		55	46°05′35.5″ N 24°28′35.4″ E	
A2B75	M	15	Grass layer and shrub	Pasture	Oozy	2		55	46°05′14.4″ N 24°28′37.7″ E	
A2B76	R	35	Grass layer	Pasture	Oozy	2		70	46°04′60.0″ N 24°28′34.8″ E	Plastic and electronic waste
A2B77	M	15	Grass layer	Pasture	Oozy	0.5		50	46°05′00.7″ N 24°28′31.3″ E	Wetland, rush
A2B78	M	5	Arboreal layer	Pasture and Forest		0.5		-	46°04′27.5″ N 24°28′40.2″ E	Spring area

Table A3. Valchid River habitat characteristics.

Station Code	R-Refuge/M-Downstream Refuge Sector	Refuge Depth (cm)	Banks Description	Land Use	Substrate	Banks Height (m)	Fish Presence	Minor Riverbed Width (cm)	GPS Coordinates	Others
A3V0	R	60	Shrub and arboreal layer	Pasture and arable	Oozy	5	+	300	46°13′10.7″ N 24°35′52.9″ E	Valchid confluence with Tarnava Mare, near to the road
A3V1	R	35	Shrub and arboreal layer	Pasture and arable	Oozy	1	+	200	46°13′13.0″ N 24°35′48.9″ E	
A3V2	R	51	Concreted banks	Pasture and arable		3		200	46°13′12.1″ N 24°35′51.9″ E	River channeled on a portion of 5 m, upstream banks and concrete riverbed, road, railway
A3V3	R	27	Concreted banks	Arable		4		800	46°13′12.0″ N 24°35′52.1″ E	Succession of concrete thresholds of about 60 cm
A3V4	R	45	Grassy layer	Arable	Oozy	4		250	46°13′08.5″ N 24°35′56.7″ E	Boulder Dam
A3V5	R	100	Shrub layer	Arable	Oozy	4		200	46°13′08.2″ N 24°35′59.1″ E	
A3V6	R	80	Arboreal layer	Arable	Oozy	6	+	200	46°13′07.1″ N 24°36′01.3″ E	
A3V7	R	100	Grassy and arboreal layer	Arable	Oozy	4.5	+	200	46°13′06.6″ N 24°36′05.0″ E	
A3V8	R	120	Grassy layer, Exposed shore, without trees	Arable	Oozy	1.5		200	46°13′06.3″ N 24°36′08.8″ E	Low flow rate
A3V9	R	130	Shrub layer	Arable	Oozy	2		200	46°13′02.8″ N 24°36′07.0″ E	
A3V10	M	no refuges, fleeting	Grassy layer and denuded banks	Arable	Oozy	1.5		200	46°12′58.2″ N 24°36′07.2″ E	
A3V11	R	110	Arboreal layer	Arable	Oozy	1		200	46°13′34.4″ N 24°37′52.6″ E	
A3V12	R	35	Arboreal layer	Arable	Oozy	1.5		200	46°12′53.9″ N 24°36′04.9″ E	

Table A3. *Cont.*

Station Code	R-Refuge/M-Downstream Refuge Sector	Refuge Depth (cm)	Banks Description	Land Use	Substrate	Banks Height (m)	Fish Presence	Minor Riverbed Width (cm)	GPS Coordinates	Others
A3V13	R	85	Arboreal layer	Arable	Oozy	2		200	46°12′53.6″ N 24°36′05.9″ E	
A3V14	R	40	Arboreal layer	Arable	Oozy	2		250	46°12′47.6″ N 24°36′04.9″ E	
A3V15	R	150	Arboreal layer	Arable	Oozy	2		250	46°12′45.5″ N 24°36′03.8″ E	Cloudy water
A3V16	R	100	Arboreal layer	Arable	Oozy	3		200	46°12′45.6″ N 24°36′03.8″ E	
A3V17	R	120	Arboreal layer	Arable	Oozy	2		200	46°12′43.3″ N 24°36′03.4″ E	Beaver dam
A3V18	R	102	Arboreal layer	Pasture	Oozy	2		150	46°12′43.0″ N 24°36′03.4″ E	
A3V19	R	120	Arboreal layer	Arable	Oozy			200	46°12′42.3″ N 24°36′04.1″ E	
A3V20	R	35	Arboreal layer	Pasture	Oozy	2		150	46°12′37.1″ N 24°36′03.9″ E	Household waste (plastic, pet, cardboard, textiles)
A3V21	R	50	Arboreal layer	Pasture	Oozy	3		200	46°13′25.7″ N 24°36′36.1″ E	
A3V22	R	30	Arboreal layer	Pasture	Gravel, ballast	2		150	46°12′39.4″ N 24°36′10.3″ E	
A3V23	M	Shallow depth	Arboreal layer	Pasture	Gravel, ballast	2		150	46°12′39.4″ N 24°36′10.3″ E	
A3V24	R	55	Arboreal layer	Pasture	Oozy	3		100	46°12′24.3″ N 24°36′01.0″ E	
A3V25	R	65	Arboreal layer	Pasture	Oozy	3		100	46°12′20.4″ N 24°36′03.3″ E	
A3V26	R	45	Arboreal layer	Pasture	Oozy	1.5		100	46°13′06.9″ N 24°36′26.9″ E	Secondary arm

Table A3. *Cont.*

Station Code	R-Refuge/M-Downstream Refuge Sector	Refuge Depth (cm)	Banks Description	Land Use	Substrate	Banks Height (m)	Fish Pres-ence	Minor Riverbed Width (cm)	GPS Coordinates	Others
A3V27	R	70	Arboreal layer	Pasture	Oozy	2		100	46°13′16.9″ N 24°36′32.1″ E	
A3V28	R	50	Arboreal layer	Pasture	Oozy	2		200	46°13′22.3″ N 24°36′34.8″ E	
A3V29	R	100	Arboreal layer	Arable	Oozy	2		150	46°13′29.4″ N 24°36′38.5″ E	Foam stain of 1 m^2, waste (pet, bran), fallen trees
A3V30	R	160	Arboreal layer	Arable	Oozy	2		150	46°12′10.1″ N 24°35′48.5″ E	
A3V31	R	120	Arboreal layer	Arable	Oozy	4		150	46°12′06.2″ N 24°35′47.1″ E	Low flow rate, 50 cm course width
A3V32	R	60	Arboreal layer	Pasture	Oozy	2		100	46°12′05.3″ N 24°35′43.7″ E	Low flow rate, course width of 50 cm, forest
A3V33	R	80	Arboreal layer	Arable	Oozy	2		150	46°12′03.3″ N 24°35′41.9″ E	
A3V34	R	90	Arboreal layer	Arable	Oozy	3		150	46°12′01.7″ N 24°35′40.3″ E	
A3V35	R	70	Arboreal layer	Arable	Oozy	3		180	46°11′59.9″ N 24°35′38.7″ E	
A3V36	R	120	Arboreal layer	Arable	Oozy	2		220	46°11′57.6″ N 24°35′34.7″ E	
A3V37	M	25	Arboreal layer	Arable				200	46°11′55.7″ N 24°35′30.0″ E	Sheepfold
A3V38	R	130	Arboreal layer	Arable	Oozy	3		150	46°11′55.7″ N 24°35′29.7″ E	Sheepfold
A3V39	R	130	Arboreal layer	Arable	Oozy	3	+	180	46°11′56.6″ N 24°35′30.9″ E	

Table A3. *Cont.*

Station Code	R-Refuge/M-Downstream Refuge Sector	Refuge Depth (cm)	Banks Description	Land Use	Substrate	Banks Height (m)	Fish Presence	Minor Riverbed Width (cm)	GPS Coordinates	Others
A3V40	R	75	Arboreal layer	Arable	Oozy	1.5	+	150	46°11′52.7″ N 24°35′20.9″ E	Sheepfold, culvert, in the summer the level drops by half, minimum flow provided almost all the time, in 2020 much weaker flow than so far in 2021
A3V41	R	70	Arboreal layer	Arable	Oozy	2	+		46°11′49.5″ N 24°35′17.9″ E	
A3V42	R	35	Arboreal layer	Arable and Pasture	Oozy	2	+	150	46°11′49.5″ N 24°35′18″ E	
A3V43	R	100	Arboreal layer	Arable and Pasture	Oozy	2	+	180	46°11′46.8″ N 24°35′15.6″ E	
A3V44	R	85	Arboreal layer	Arable and Pasture	Oozy	2	+	50	46°11′38.3″ N 24°35′09.4″ E	
A3V45	R	75	Arboreal layer	Arable and Pasture	Oozy	1.5		150	46°11′38.3″ N 24°35′10″ E	No human settlements, only three sheep farms
A3V46	R	60	Arboreal layer	Arable and Pasture	Oozy	2			46°11′34.0″ N 24°34′58.6″ E	Many twigs in the water
A3V47	R	70	Arboreal layer	Arable and Pasture	Oozy	2		100	46°11′29.8″ N 24°35′04.7″ E	Waste (plastic, cardboard, electro-household)
A3V48	R	45	Arboreal layer	Arable	Oozy	2		150	46°11′19.7″ N 24°35′02.1″ E	
A3V49	R	100	Arboreal layer	Arable	Oozy	2		150	46°11′14.8″ N 24°35′04.7″ E	Waste (plastic)
A3V50	R	100	Arboreal layer	Arable	Oozy	2		100	46°11′14.8″ N 24°35′04.7″ E	
A3V51	R	180	Arboreal layer	Arable, orchard	Oozy	3		150	46°11′14.8″ N 24°35′04.7″ E	
A3V52	R	175	Arboreal layer	Arable	Oozy	2			46°11′14.0″ N 24°35′08.4″ E	

Table A3. *Cont.*

Station Code	R-Refuge/M-Downstream Refuge Sector	Refuge Depth (cm)	Banks Description	Land Use	Substrate	Banks Height (m)	Fish Presence	Minor Riverbed Width (cm)	GPS Coordinates	Others
A3V53	R	150	Arboreal layer	Arable	Oozy	2		150	46°10′57.6″ N 24°35′09.5″ E	
A3V54	R	160	Arboreal layer	Arable	Oozy	2		200	46°10′57.2″ N 24°35′09.6″ E	Dam of twigs
A3V55	R	45	Arboreal layer	Arable	Oozy	2			46°10′58.9″ N 24°35′08.1″ E	
A3V56	R	45	Arboreal layer	Arable	Oozy	2			46°10′54.5″ N 24°35′07.2″ E	
A3V57	R	109	Arboreal layer	Arable	Oozy	1.5		170	46°10′45.1″ N 24°35′10.2″ E	Cattle farm
A3V58	R	57	Arboreal layer	Arable	Oozy	2		120	46°10′42.7″ N 24°35′07.7″ E	Waste, we are approaching human settlements
A3V59	R	110	Arboreal layer	Arable	Oozy	3		120	46°10′38.9″ N 24°35′07.4″ E	
A3V60	R	30	Arboreal layer	Arable, In locality	Oozy	2		110	46°10′35.1″ N 24°35′06.0″ E	
A3V61	R	115	Arboreal layer	Arable, In locality	Oozy	2		150	46°10′28.2″ N 24°35′06.7″ E	Behind the houses, gardens
A3V62	M	25	Arboreal layer	In locality	Oozy/Gravel	1.5		160	46°10′23.1″ N 24°35′05.6″ E	
A3V63	M	5	Grassy layer	In locality	Oozy/Gravel	0.5		100	46°10′11.9″ N 24°34′58.0″ E	
A3V64	R	25	Grassy layer	In locality	Oozy	1.8		150	46°10′09.2″ N 24°34′55.8″ E	
A3V65	R	40	Shrub layer	Arable	Oozy	1		100	46°10′09.2″ N 24°34′55.8″ E	
A3V66	R	27	Grassy layer	In locality	Oozy	1.5		150	46°09′59.7″ N 24°34′48.1″ E	

Table A3. *Cont.*

Station Code	R-Refuge/M-Downstream Refuge Sector	Refuge Depth (cm)	Banks Description	Land Use	Substrate	Banks Height (m)	Fish Pres-ence	Minor Riverbed Width (cm)	GPS Coordinates	Others
A3V67	R	80	Grassy and arboreal layer	In locality	Oozy	2		150	46°10′00.4″ N 24°34′50.2″ E	
A3V68	R	40	Arboreal layer	Arable	Oozy	0.5		150	46°09′52.0″ N 24°34′44.1″ E	
A3V69	R	100	Arboreal layer	Arable	Oozy	1		120	46°09′52.0″ N 24°34′44.1″ E	
A3V70	R	130	Arboreal layer	Arable	Oozy	1		120	46°08′20.8″ N 24°33′02.6″ E	
A3V71	R	55	Arboreal layer	Arable	Oozy	3		120	46°09′39.0″ N 24°34′40.5″ E	
A3V72	R	110	Arboreal layer	Arable	Oozy	5		120	46°08′13.2″ N 24°32′53.4″ E	Fallen trees in the water
A3V73	R	100	Shrub and arboreal layer	Arable	Oozy	3		150	46°09′16.6″ N 24°34′16.8″ E	
A3V74	M	20	Arboreal layer	Arable	Oozy	4		110	46°08′15.2″ N 24°32′55.9″ E	
A3V75	R	100	Arboreal layer	Arable	Oozy	3		110	46°08′18.9″ N 24°32′57.7″ E	
A3V76	R	130	Arboreal layer	Arable	Oozy	2		150	46°08′15.1″ N 24°32′55.8″ E	
A3V77	R	65	Arboreal layer	Arable	Oozy	2		150	46°08′12.8″ N 24°32′52.7″ E	
A3V78	R	70	Arboreal layer	Arable	Oozy	2		120	46°09′08.4″ N 24°34′08.1″ E	
A3V79	M	25	Shrub and arboreal layer	Arable	Oozy	2		100	46°08′41.8″ N 24°33′26.1″ E	
A3V80	R	80	Shrub and arboreal layer	Pasture	Oozy	3		120	46°08′41.8″ N 24°33′26.1″ E	

Table A3. *Cont.*

Station Code	R-Refuge/M-Downstream Refuge Sector	Refuge Depth (cm)	Banks Description	Land Use	Substrate	Banks Height (m)	Fish Presence	Minor Riverbed Width (cm)	GPS Coordinates	Others
A3V81	R	85	Shrub and arboreal layer	Pasture	Oozy	2		150	46°08′14.6″ N 24°32′55.1″ E	
A3V82	M	60	Shrub and arboreal layer	Pasture	Oozy	2		120	46°08′25.3″ N 24°33′09.3″ E	
A3V83	R	70	Shrub and arboreal layer	Pasture	Oozy	2	+	120	46°08′18.1″ N 24°32′59.7″ E	
A3V84	R	95	Shrub and arboreal layer	Pasture	Oozy	3		100	46°08′16.5″ N 24°32′57.5″ E	
A3V85	R	50	Shrub and arboreal layer	Arable	Oozy	2		110	46°08′15.3″ N 24°32′56.0″ E	
A3V86	R	120	Shrub and arboreal layer	Arable	Oozy	2		120	46°08′34.9″ N 24°33′15.8″ E	Wood dam
A3V87	M	30	Shrub and arboreal layer	Arable	Oozy	3		50	46°08′26.0″ N 24°33′06.8″ E	Construction waste (rubble, brick, plastic and paper packaging)
A3V88	R	100	Shrub and arboreal layer	Arable	Oozy	3		120	46°08′16.3″ N 24°32′56.9″ E	
A3 A3V89	R	55	Shrub and arboreal layer	Arable	Oozy	2		120	46°08′09.4″ N 24°32′55.8″ E	
V90	R	60	Shrub and grassy layer	Ir locality	Oozy	2		120	46°08′01.3″ N 24°32′54.4″ E	Plastic waste
A3V91	R	40	Shrub and grassy layer	Ir locality	Oozy	2		100	46°07′53.7″ N 24°32′51.6″ E	Denuded banks
A3V92	M	25	Shrub and grassy layer	Ir locality	Oozy	2		100	46°07′56.0″ N 24°32′50.3″ E	L bank denuded
A3V93	M	25	Shrub and grassy layer	Ir locality	Oozy	2		100	46°07′50.6″ N 24°32′51.8″ E	Denuded banks, trees cut off from the shore, the banks have undergone works...

Table A3. *Cont.*

Station Code	R-Refuge/M-Downstream Refuge Sector	Refuge Depth (cm)	Banks Description	Land Use	Substrate	Banks Height (m)	Fish Presence	Minor Riverbed Width (cm)	GPS Coordinates	Others
A3V94	R	45	Shrub and grassy layer	In locality	Oozy	1		50	46°07′44.1″ N 24°32′46.4″ E	
A3V95	M	30	Shrub and grassy layer	Arable	Oozy	3		50	46°07′20.3″ N 24°32′57.3″ E	
A3V96	R	45	Shrub and arboreal layer	Arable	Oozy	2		50	46°07′20.1″ N 24°32′57.4″ E	
A3V97	R	35	Shrub and arboreal layer	Arable	Oozy	2	+	50	46°07′20.1″ N 24°32′57.4″ E	Plastic waste
A3V98	R	80	Shrub and arboreal layer	Pasture	Oozy	3		100	46°07′19.0″ N 24°32′58.8″ E	Plastic waste
A3V99	R	80	Shrub and arboreal layer	Pasture	Oozy	1		70	46°07′19.0″ N 24°32′58.8″ E	Cutting trees on the shore, in summer almost dry sector (shepherd)
A3V100	M	10	Grassy layer	Pasture	Oozy	0.5		40	46°06′57.4″ N 24°32′58.5″ E	
A3V101	M	5	Grassy layer and shrub	Pasture	Oozy	3		40	46°06′47.0″ N 24°33′29.9″ E	
A3V102	M	5	Shrub and arboreal layer	Pasture	Oozy	4		30	46°06′35.4″ N 24°33′21.0″ E	We approach the spring, the river branches into several streams, low flow, wetland
A3V103	M	5	Shrub and arboreal layer	Pasture	Oozy	1		20	46°06′23.9″ N 24°33′37.6″ E	Near the spring, unidentified refuges, plastic waste, forest road

Table A4. Laslea River Habitat Characteristics.

Station Code	R-Refuge/M-Downstream Refuge Sector	Refuge Depth (cm)	Banks Description	Land Use	Substrate	Banks Height (m)	Fish Presence	Minor Riverbed Width (cm)	GPS Coordinates	Others
A4L0	R	60	Arboreal layer	Pasture and arable	Oozy	5	+	200	46°13′53.1″ N 24°40′29.8″ E	Confluence of Laslea with Tarnava Mare
A4L1	R	45	Arboreal layer	Pasture and arable	Oozy	5		200	46°13′53.2″ N 24°40′29.5″ E	
A4L2	R	45	Arboreal layer	Pasture and arable	Oozy	5		200	46°13′53.2″ N 24°40′29.5″ E	Significant ripisilva vegetation, meadow, hard to reach
A4L3	R	35	Arboreal layer	Pasture and arable	Oozy	3		200	46°13′39.4″ N 24°40′24.3″ E	
A4L4	R	30	Concreted banks	Arable	Oozy	4			46°13′22.7″ N 24°40′13.3″ E	Concreted banks
A4L5	M	10	Concreted banks	Arable	Concreted	4		500	46°13′23.2″ N 24°40′13.4″ E	Banks and bed concreted on a portion of 100 m
A4L6	R	35	Shrub and grassy layer	Arable	Oozy	2			46°13′16.6″ N 24°40′14.9″ E	
A4L7	R	120	Arboreal layer	Arable	Oozy	2		150	46°13′15.3″ N 24°40′15.9″ E	Dam of twigs, plastic waste, rubble, and packaging waste constructions
A4L8	M	25	Shrub and grassy layer	Arable	Oozy	2		150	46°13′15.3″ N 24°40′15.9″ E	Concrete banks on a portion of 10 m
A4L9	R	100	Arboreal layer	Arable	Oozy	2		150	46°13′11.5″ N 24°40′12.2″ E	Meander, plastic waste
A4L10	R	35	Arboreal layer	Pasture	Oozy	1.5		100	46°13′04.1″ N 24°40′07.2″ E	Rubber, plastic waste
A4L11	R	40	Arboreal layer	Pasture	Oozy	1.5		70	46°12′59.7″ N 24°40′05.7″ E	Angus cows farm, denuded banks, bathing cows
A4L12	R	45	Shrub layer	Pasture	Oozy	1		70	46°12′53.2″ N 24°40′03.3″ E	Plastic waste, polystyrene

Table A4. *Cont.*

Station Code	R-Refuge/M-Downstream Refuge Sector	Refuge Depth (cm)	Banks Description	Land Use	Substrate	Banks Height (m)	Fish Pres-ence	Minor Riverbed Width (cm)	GPS Coordinates	Others
A4L13	R	45	Shrub layer	Pasture	Oozy	1		70	46°12′44.4″ N 24°39′57.3″ E	Angus cows farm, denuded banks, bathing cows, passers-by
A4L14	R	35	Grassy layer	In locality	Oozy	2		50	46°12′25.6″ N 24°39′26.4″ E	
A4L15	M	25	Grassy layer	In locality	Oozy	2		60	46°12′25.5″ N 24°39′26.5″ E	
A4L16	R	35	Arboreal layer	In locality	Oozy	2		50	46°12′06.6″ N 24°39′16.9″ E	
A4L17	R	30	Shrub layer	Arable	Oozy	2		70	46°11′59.2″ N 24°39′14.4″ E	At the exit of Laslea, branches and grasses fallen into the water
A4L18	R	50	Arboreal layer	Arable	Oozy	2		60	46°11′56.3″ N 24°39′01.7″ E	Meandering, branches, and grasses fallen into the water
A4L19	R	35	Arboreal layer	Arable	Oozy	2		70	46°11′46.1″ N 24°38′47.5″ E	Plastic waste, polystyrene, nettles
A4L20	R	40	Arboreal layer	Arable	Oozy	3		60	46°11′46.1″ N 24°38′47.5″ E	Sheep farm, riverbed closed with electric fence on a portion of 500 m
A4L21	R	50	Arboreal layer	Arable	Oozy	3		60	46°11′30.5″ N 24°38′33.3″ E	Waste plastics, meanders
A4L22	R	35	Arboreal layer	Arable	Oozy	3		60	46°11′29.1″ N 24°38′30.7″ E	Twigs fallen into the water
A4L23	R	35	Arboreal layer	Arable and Pasture	Oozy	3		60	46°11′28.8″ N 24°38′24.1″ E	Meandering
A4L24	R	55	Arboreal layer	Arable	Oozy	2		60	46°10′47.0″ N 24°37′41.9″ E	
A4L25	R	55	Arboreal layer	Arable	Oozy	1		150	46°10′45.9″ N 24°37′38.9″ E	Sheep farm, low flow, branches fallen into the water

Table A4. *Cont.*

Station Code	R-Refuge/M-Downstream Refuge Sector	Refuge Depth (cm)	Banks Description	Land Use	Substrate	Banks Height (m)	Fish Presence	Minor Riverbed Width (cm)	GPS Coordinates	Others
A4L26	R	30	Arboreal layer	Arable	Oozy	1		50	46°10′35.5″ N 24°37′20.7″ E	
A4L27	R	45	Arboreal layer	Arable	Oozy	1		50	46°10′26.4″ N 24°37′04.7″ E	
A4L28	R	60	Shrub layer	Pasture	Oozy	1		60	46°10′17.2″ N 24°36′54.1″ E	
A4L29	M	25	Arboreal layer	Arable	Oozy	2		60	46°10′17.3″ N 24°36′54.4″ E	Waste from plastic, rubble, wood, textiles
A4L30	M	25	Arboreal layer	Arable	Oozy	3		60	46°09′51.9″ N 24°36′37.8″ E	Entrance to Roandola, hard to reach, a lot of vegetation, mostly nettles, large quantities of plastic waste, polystyrene, textiles, furniture, electronic, manure storage, animal manure...
A4L31	R	45	Shrub layer	Arable	Oozy	4		60	46°09′52.9″ N 24°36′51.8″ E	Household waste
A4L32	R	60	Arboreal layer	Arable and Pasture	Oozy	4		60	46°09′44.1″ N 24°36′19.1″ E	Meandering, household waste, many nettles, lush vegetation
A4L33	M	25	Arboreal layer	Arable and Pasture	Oozy	0.5		100	46°09′12.1″ N 24°35′59.4″ E	Meander, passing animals, sheep farm, municipal waste
A4L34	R	40	Shrub and arboreal layer	Arable and Pasture	Oozy	1		80	46°09′03.6″ N 24°35′55.0″ E	Lush vegetation, hard to reach, confluence with Malancrav
A4L35	R	100	Shrub and arboreal layer	Arable and Pasture	Oozy	3		70	46°08′53.6″ N 24°35′47.8″ E	Lush vegetation, hard to reach, gas pipe passes over the riverbed, municipal waste
A4L36	M	30	Arboreal layer	Arable and Pasture	Oozy	3		70	46°08′53.6″ N 24°35′48.1″ E	Lush vegetation
A4L37	R	35	Grassy and shrub layer	Arable and Pasture	Oozy	3		60	46°08′33.3″ N 24°35′48.0″ E	The left bank is close to the hill, the coast, the steep bank, the rush, and the floodplain

Table A4. *Cont.*

Station Code	R-Refuge/M-Downstream Refuge Sector	Refuge Depth (cm)	Banks Description	Land Use	Substrate	Banks Height (m)	Fish Presence	Minor Riverbed Width (cm)	GPS Coordinates	Others
A4L38	R	60	Grassy and shrub layer	Arable and Pasture	Oozy	2	+	60	46°08′04.4″ N 24°35′50.0″ E	Lush vegetation, floodplain
A4L39	M	25	Arboreal layer	Arable and Pasture	Oozy	2		70	46°07′37.4″ N 24°35′53.0″ E	
A4L40	M	25	Arboreal layer	Arable and Pasture	Oozy	3		70	46°07′37.4″ N 24°35′53.0″ E	Lush vegetation, nettles
A4L41	R	30	Arboreal layer	In locality	Oozy	1		70	46°06′54.8″ N 24°36′01.1″ E	
A4L42	M	20	Shrub layer	In locality	Oozy	2		70	46°06′33.4″ N 24°36′22.6″ E	Plastic waste, branches fallen into the water
A4L43	R	40	Grassy and shrub layer	In locality and Arable	Oozy	1		70	46°06′18.2″ N 24°36′22.6″ E	Reservoir
A4L44	M	20	Shrub layer	Arable	Oozy	1		60	46°06′07.1″ N 24°36′27.9″ E	200 m upstream of the lake, forest road, rush, wetland
A4L45	M	20	Shrub layer	Arable	Oozy	1		60	46°05′56.8″ N 24°36′32.4″ E	Rush, wetland
A4L46	M	25	Shrub layer	Arable and Pasture	Oozy	1		50	46°05′54.1″ N 24°36′32.5″ E	Cow farm
A4L47	M	25	Shrub layer	Arable and Pasture	Oozy	1		50	46°05′51.1″ N 24°36′33.6″ E	Rush, wetland
A4L48	R	20		Arable and Pasture	Oozy	1		50	46°05′46.5″ N 24°36′34.1″ E	
A4L49	R	20	Grassy and shrub layer	Arable and Pasture	Oozy	1		50	46°05′40.8″ N 24°36′34.3″ E	
A4L50	R	15	Shrub and arboreal layer	Arable and Pasture	Oozy	1		30	46°05′39.7″ N 24°36′35.9″ E	Close to the spring
A4L51	R	10	Shrub layer and arboreal	Pasture	Oozy	1		20	46°05′35.4″ N 24°36′36.3″ E	Nearby spring

Table A5. Mălâncrav River Habitat Characteristics.

Station Code	R-Refuge/M-Downstream Refuge Sector	Refuge Depth (cm)	Banks Description	Land Use	Substrate	Banks Height (m)	Fish Presence	Minor Riverbed Width (cm)	GPS Coordinates	Others
A5M0	R	40	Arboreal layer	Arable and pasture	Oozy	2		80	46°11′23.4″ N 24°38′20.6″ E	
A5M1	R	50	Arboreal layer	Arable and pasture	Oozy	2		80	46°10′57.3″ N 24°38′30.8″ E	
A5M2	R	40	Arboreal layer	Arable and pasture	Oozy	2		80	46°10′16.4″ N 24°38′40.7″ E	Bridge, lush vegetation, ripisilva
A5M3	R	90	Arboreal layer	Arable and pasture	Oozy	2		60	46°10′07.1″ N 24°38′39.0″ E	
A5M4	R	30	Arboreal layer	Arable and pasture	Oozy	5		200	46°09′50.9″ N 24°38′36.2″ E	Ripisilva 50 m high
A5M5	R	110	Arboreal layer	Arable and pasture	Oozy	5		250	6°09′09.7″ N 24°38′27.0″ E	Cow farm, plastic waste
A5M6	M	20	Arboreal layer	Arable and pasture	Oozy	4		250	46°08′57.3″ N 24°38′32.0″ E	Bank close to the hill, steep, logging, plastic waste
A5M7	R	50	Arboreal layer	Arable	Oozy	2		70	46°08′20.9″ N 24°38′19.7″ E	Lush vegetation, floodplain
A5M8	R	60	Arboreal layer	Arable	Oozy	4		70	46°08′20.7″ N 24°38′19.7″ E	Bank close to the hill, steep, logging, plastic waste
A5M9	R	110	Arboreal layer	Arable	Oozy	4		150	46°07′53.3″ N 24°38′22.9″ E	Meandering, lush vegetation, floodplain
A5M10	R	80	Arboreal layer	Arable	Rocky	3		100	46°07′33.5″ N 24°38′40.6″ E	Pass, construction waste
A5M11	R	75	Arboreal layer	Arable	Oozy	0.5		100	46°07′33.5″ N 24°38′40.6″ E	
A5M12	M	40	Shrub layer	Arable	Oozy	2		80	46°07′06.0″ N 24°38′54.4″ E	
A5M13	M	35	Shrub layer	Arable	Oozy	2		70	46°06′23.9″ N 24°39′02.0″ E	Municipal waste

Table A5. *Cont.*

Station Code	R-Refuge/M-Downstream Refuge Sector	Refuge Depth (cm)	Banks Description	Land Use	Substrate	Banks Height (m)	Fish Presence	Minor Riverbed Width (cm)	GPS Coordinates	Others
A5M14	R	75	Arboreal layer	Arable	Oozy	0.5		100	46°05′53.3″ N 24°39′25.2″ E	Entrance to Malancrav village, foam water, smell of detergents, sheep farm
A5M15	M	25	Shrub layer	In locality	Oozy	2		80	46°05′49.0″ N 24°39′34.7″ E	
A5M16	M	20	Shrub layer	In locality	Oozy	2		80	46°05′41.9″ N 24°39′50.7″ E	
A5M17	M	25	Shrub layer	In locality	Oozy	2		70	46°05′41.4″ N 24°40′06.3″ E	Municipal waste
A5M18	R	110	Shrub layer	Arable and pasture	Oozy	2		70	46°05′42.0″ N 24°40′27.6″ E	At the exit of Malancrav, the pig farm, Roma community, large quantities of waste, meandered
A5M19	R	40	Shrub layer	Arable and pasture	Oozy	1.5		60	46°05′43.8″ N 24°40′40.0″ E	Sheep farm
A5M20	R	30	Shrub layer	Arable and pasture	Oozy	1		60	46°05′42.7″ N 24°40′53.1″ E	
A5M21	M	20	Shrub layer	Arable and pasture	Oozy	1		40	46°05′40.6″ N 24°40′59.0″ E	
A5M22	M	20	Shrub layer	Arable and pasture	Oozy	1		30	46°05′36.8″ N 24°41′08.1″ E	Nearby spring

Table A6. Floreşti/Felţa River Habitat Characteristics.

Station Code	R-Refuge/M-Downstream Refuge Sector	Refuge Depth (cm)	Banks Description	Land Use	Substrate	Banks Height (m)	Fish Presence	Minor Riverbed Width (cm)	GPS Coordinates	Others
A6F0	R	60	Shrub layer	Pasture	Oozy	3		100	46°12′03.9″ N 24°39′16.1″ E	
A6F1	R	60	Shrub layer	Pasture	Oozy	5		100	46°12′02.5″ N 24°39′20.3″ E	Municipal waste, at the exit of Laslea
A6F2	R	60	Arboreal layer	Pasture	Oozy	5		100	46°11′58.1″ N 24°39′23.2″ E	
A6F3	R	95	Arboreal layer	Pasture	Oozy	0.5		70	46°11′42.9″ N 24°39′31.7″ E	Cow farm, hop plantation
A6F4	R	40	Arboreal layer	Pasture	Oozy	1		70	46°11′28.5″ N 24°39′39.6″ E	
A6F5	R	100	Shrub layer and arboreal	Arable	Oozy	1		70	46°11′03.9″ N 24°39′43.3″ E	Municipal waste, hop plantation, lush vegetation, burdock
A6F6	R	45	Shrub layer and arboreal	Arable	Oozy	1		70	46°10′41.1″ N 24°39′55.8″ E	
A6F7	R	100	Arboreal layer	Arable	Oozy	0.5		100	46°10′23.7″ N 24°40′04.4″ E	Meandering, lush vegetation
A6F8	R	35	Arboreal layer	Arable	Oozy	0.5		80	46°10′22.7″ N 24°40′03.7″ E	
A6F9	R	30	Arboreal layer	Arable	Oozy	0.5		80	46°10′05.7″ N 24°40′02.7″ E	
A6F10	R	45	Arboreal layer	Arable	Oozy	0.5		60	46°10′03.0″ N 24°40′00.4″ E	100 m upstream chalet, bed closed with electric wire
A6F11	M	25	Arboreal layer	Arable and Pasture	Oozy	0.5		70	46°09′52.6″ N 24°39′53.7″ E	Plastic waste, twigs in the riverbed, cattle
A6F12	R	60	Arboreal layer	Arable and Pasture	Oozy	0.5		70	46°09′52.6″ N 24°39′53.7″ E	
A6F13	R	30	Arboreal layer	Arable and Pasture	Oozy	0.5		70	46°09′40.8″ N 24°39′45.5″ E	Meandering

Table A6. *Cont.*

Station Code	R-Refuge/M-Downstream Refuge Sector	Refuge Depth (cm)	Banks Description	Land Use	Substrate	Banks Height (m)	Fish Presence	Minor Riverbed Width (cm)	GPS Coordinates	Others
A6F14	M	25	Arboreal layer	Arable and Pasture	Oozy	0.5		70	46°09′27.1″ N 24°39′37.8″ E	
A6F15	M	20	Arboreal layer	Pasture	Oozy	3		60	46°09′19.5″ N 24°39′36.8″ E	Farm cattle, twigs in the riverbed, passing cows
A6F16	R	60	Arboreal layer	Pasture	Oozy	0.5		60	46°09′01.3″ N 24°39′31.1″ E	
A6F17	R	25	Arboreal layer	Pasture	Oozy	1.5		80	46°08′47.4″ N 24°39′30.6″ E	Lush vegetation
A6F18	M	25	Arboreal layer	In locality	Oozy				46°08′36.0″ N 24°39′36.2″ E	
A6F19	M	25	Arboreal layer	In locality	Oozy	1		50	46°08′23.9″ N 24°39′35.3″ E	Cemetery
A6F20	R	65	Shrub layer	In locality	Oozy	0.5		50	46°08′21.7″ N 24°39′38.0″ E	
A6F21	R	40	Shrub layer	Arable	Clayey	0.5		70	46°08′18.6″ N 24°39′41.0″ E	At the exit of the village, construction waste, branches in the riverbed
A6F22	R	45	Arboreal layer	Arable and Pasture	Oozy	0.5		60	46°08′15.9″ N 24°39′43.7″ E	Cattle farm
A6F23	M	20	Arboreal layer	Pasture	Oozy	2		50	46°08′15.8″ N 24°39′43.9″ E	Step from twigs and trunk, meander, farm cows
A6F24	M	15	Arboreal layer	Pasture	Oozy	4		60	46°08′06.0″ N 24°39′50.0″ E	
A6F25	M	15	Shrub and arboreal layer	Pasture	Oozy	5		50	46°08′00.8″ N 24°39′51.0″ E	Steep banks, lush vegetation
A6F26	M	15	Shrub layer	Pasture	Oozy	4		35	46°07′55.2″ N 24°39′49.9″ E	

Table A6. *Cont.*

Station Code	R-Refuge/M-Downstream Refuge Sector	Refuge Depth (cm)	Banks Description	Land Use	Substrate	Banks Height (m)	Fish Presence	Minor Riverbed Width (cm)	GPS Coordinates	Others
A6F27	M	15	Shrub and arboreal layer	Fasture	Rocky	4		35	46°07′50.5″ N 24°39′50.1″ E	Plastic waste
A6F27	M	15	Arboreal layer	Forest	Rocky	6		20	46°07′46.6″ N 24°39′49.8″ E	Logging
A6F28	M	15	Arboreal layer	Forest	Rocky	5		25	46°07′41.6″ N 24°39′51.8″ E	Logging
A6F29	M	15	Arboreal layer	Fasture	Oozy	3		25	46°07′35.1″ N 24°39′52.2″ E	
A6F30	M	20	Arboreal layer	Fasture	Oozy	3		25	46°07′31.8″ N 24°39′52.4″ E	
A6F31	M	15	Arboreal layer	Fasture	Oozy	2		20	46°07′26.4″ N 24°39′52.9″ E	Cattle farm
A6F32	M	15	Arboreal layer	Fasture	Oozy	2		15	46°07′20.2″ N 24°39′53.4″ E	Nearby spring

B.

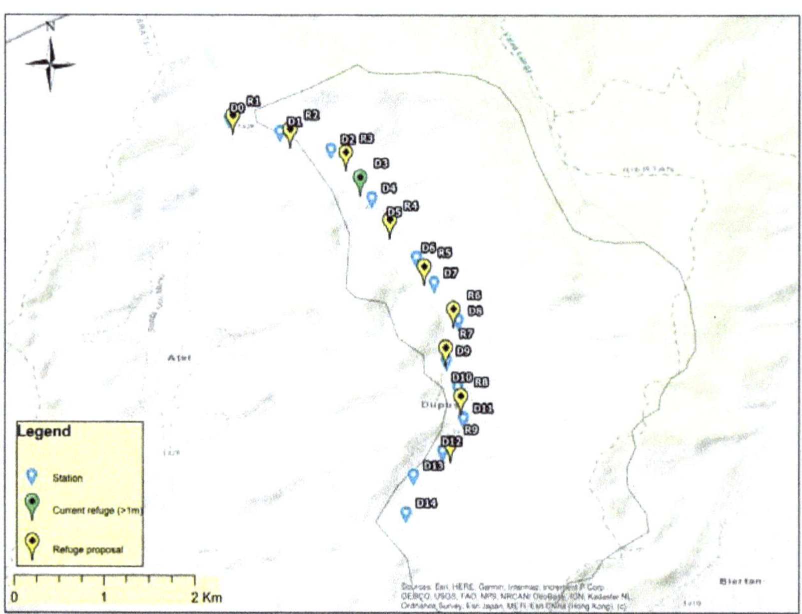

Scheme 1. Dupuş River—present refuge habitats and proposals for new refuge to be created.

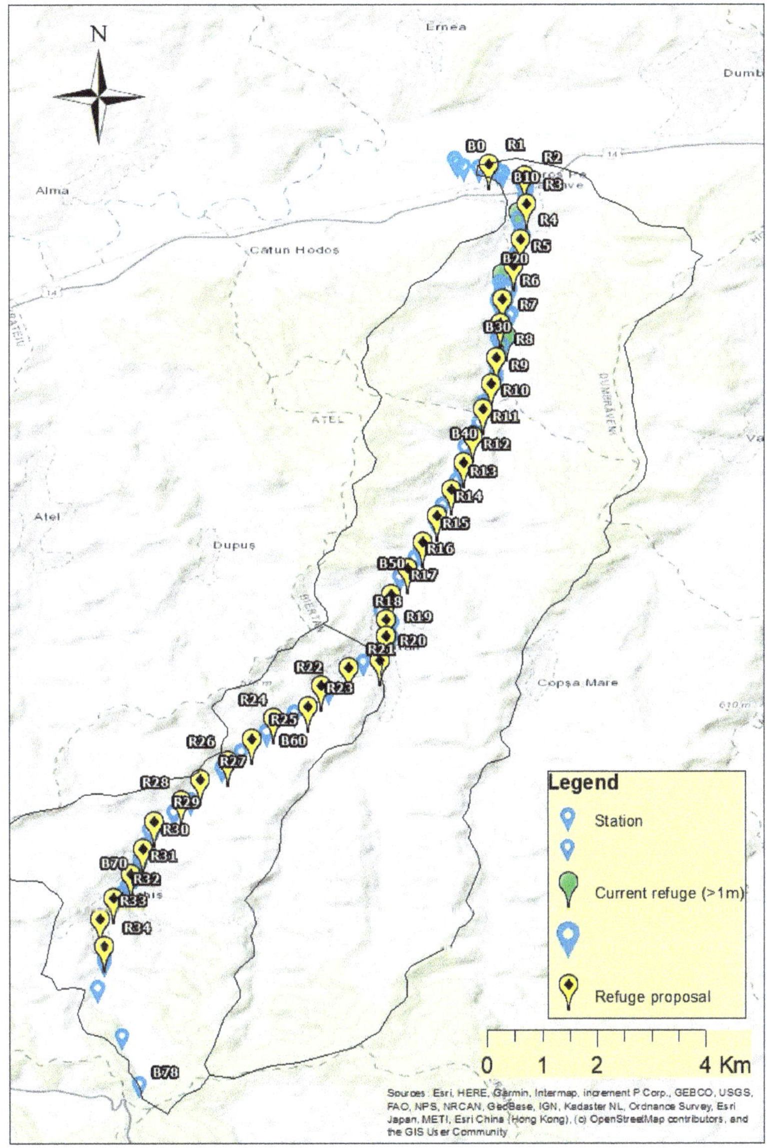

Scheme 2. Biertan River—refuge habitats and proposals for new refuge to be created.

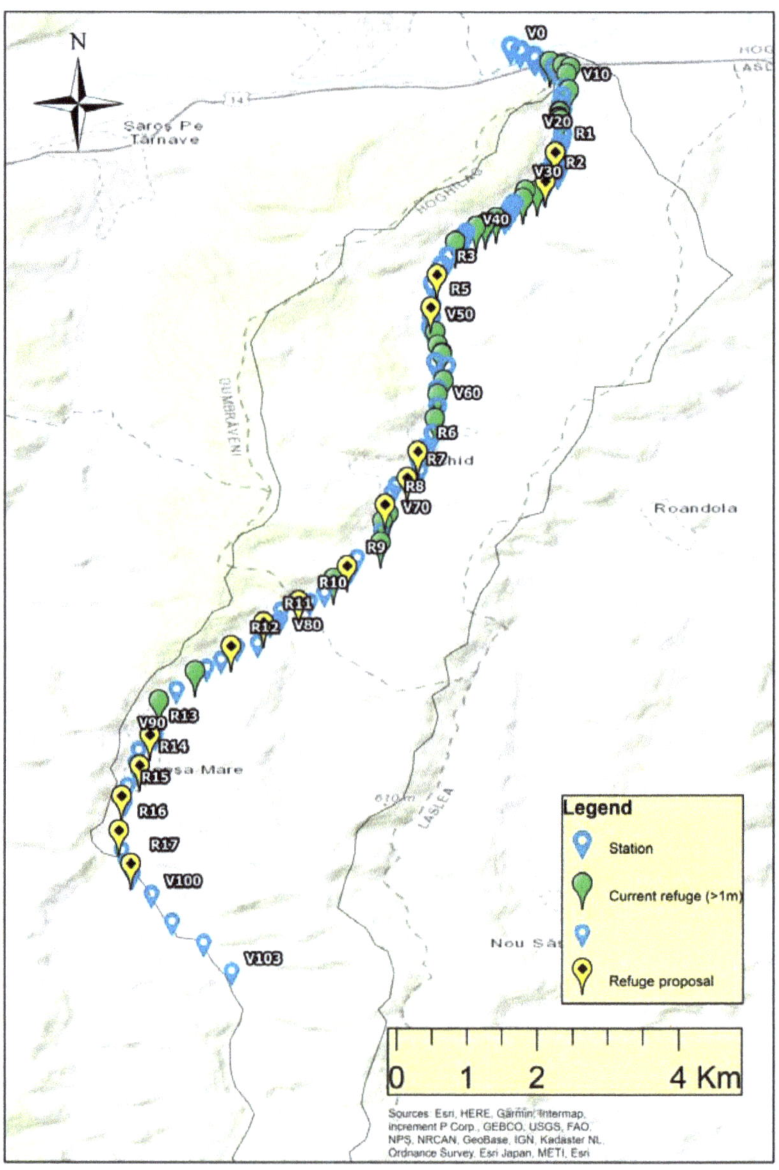

Scheme 3. Valchid River—refuge habitats and proposals for new refuge to be created.

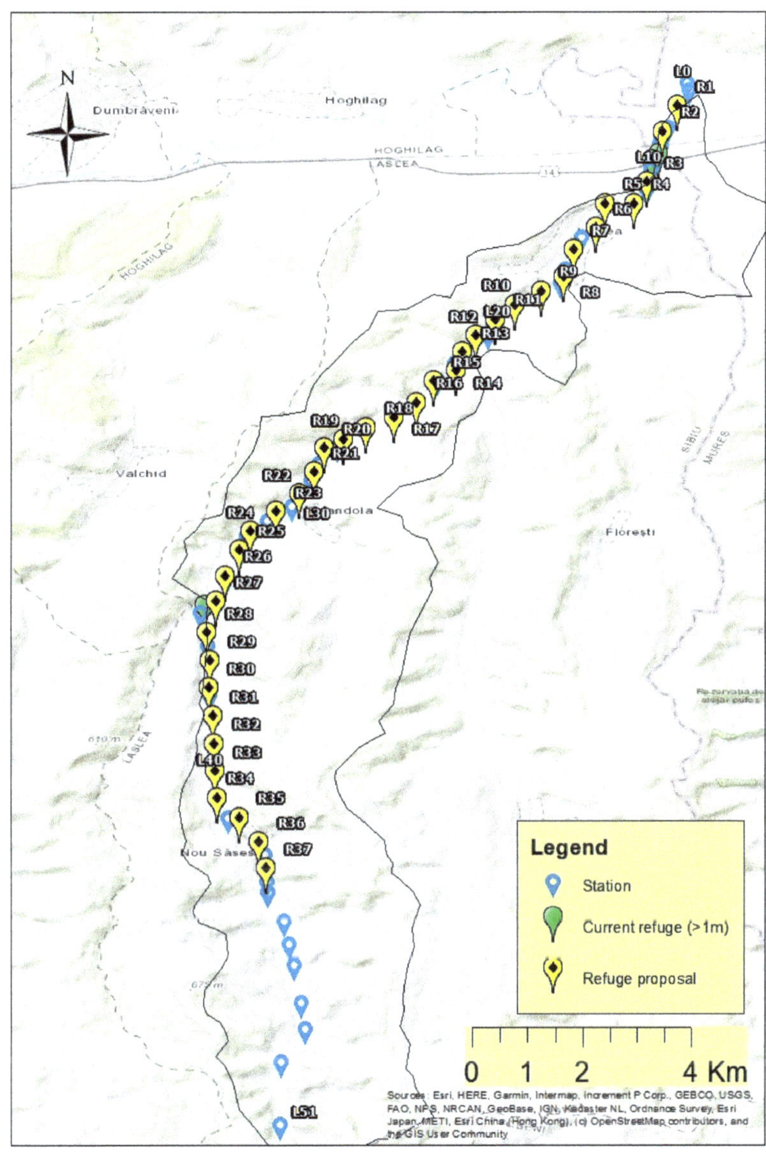

Scheme 4. Laslea River—refuge habitats and proposals for new refuge to be created.

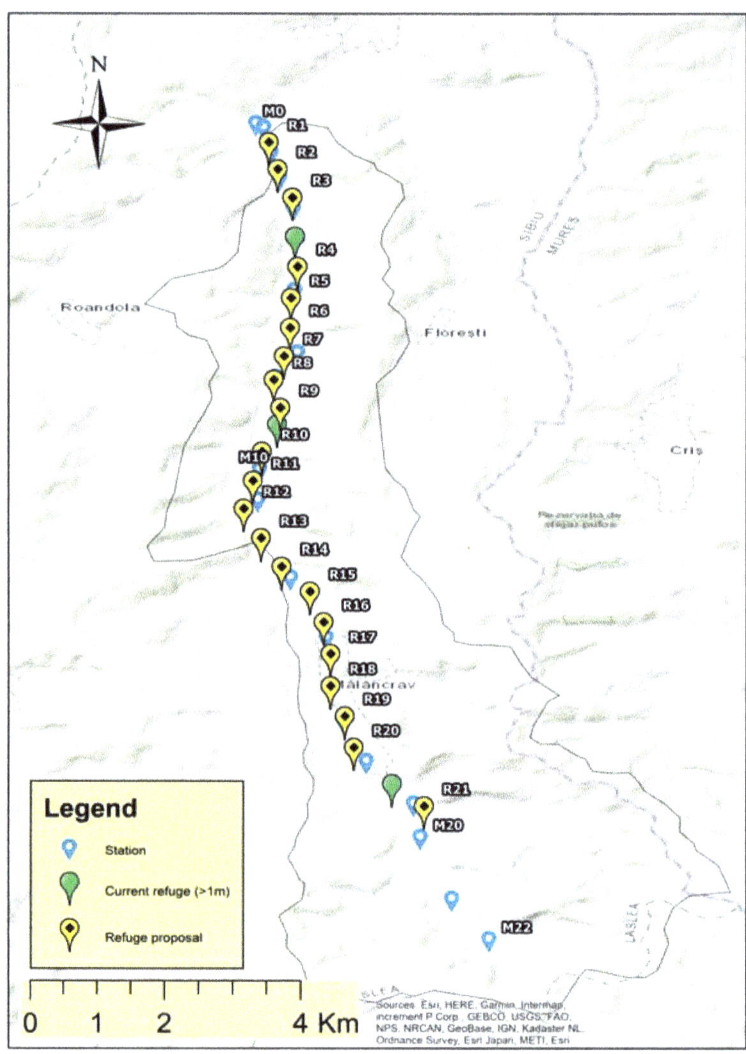

Scheme 5. Mălâncrav River—present refuge habitats and proposals for new refuge to be created.

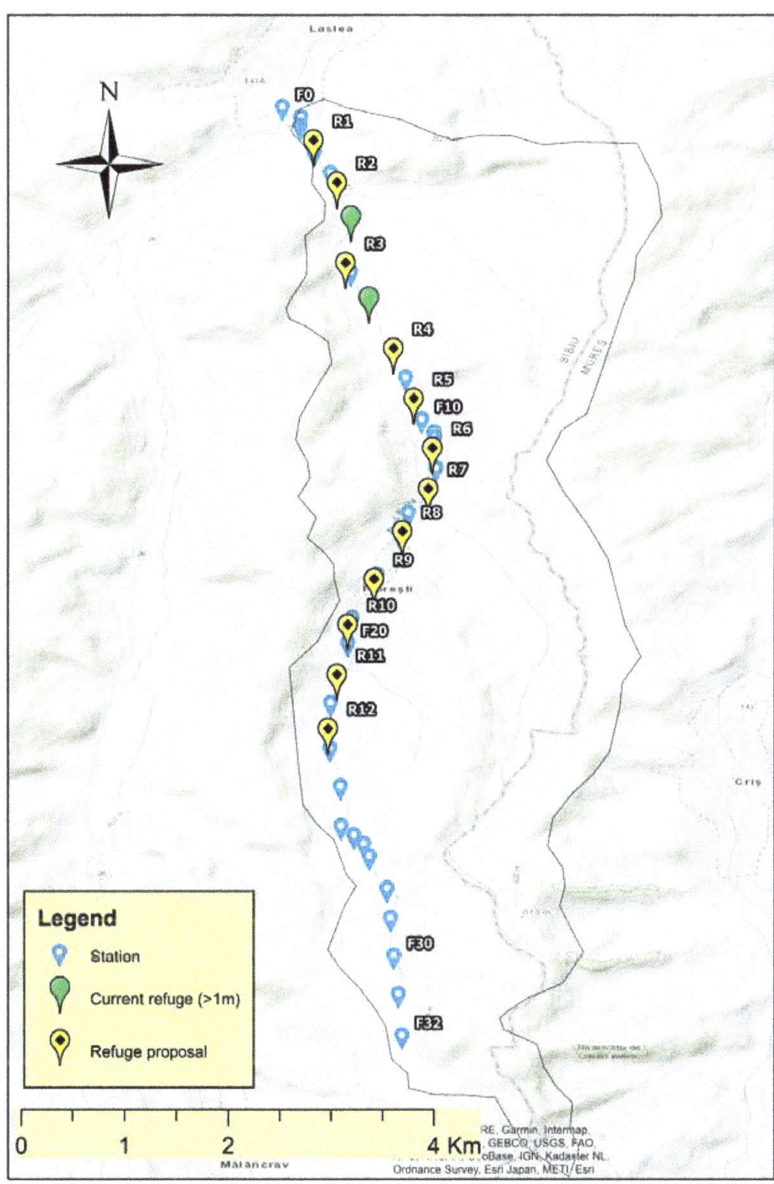

Scheme 6. Felţa River—refuge habitats and proposals for new refuge to be created.

References

1. Maruyama, S.; Ikoma, M.; Genda, H.; Hirose, K.; Yokohama, T.; Santosh, M. The naked planet earth: Most essential pre-requisite for the origin and evolution of life. *Geosci. Front.* **2013**, *4*, 141–165. [CrossRef]
2. Goncharuk, W.; Goncharuk, V.V. Water is everywhere. It holds everything a key to understanding the universe. D. I. Mendeleev's law is the prototype of the universe constitution. *J. Water Chem. Technol.* **2019**, *41*, 341–346. [CrossRef]
3. May, R.M. Biological diversity: Differences between land and sea. *Phil. Trans. R. Soc. Lond.* **1994**, *343*, 105–111.
4. Benton, M.J. Biodiversity on land and in the sea. *Geol. J.* **2001**, *36*, 211–230. [CrossRef]
5. Sandu, C.; Bloesch, J.; Coman, A. Water pollution in the Mureş Catchment and its impact on the aquatic communities (Romania). *Transylv. Rev. Syst. Ecol. Res.* **2008**, *6*, 97–108.

6. Mora, C.; Tittensor, D.P.; Adl, S.; Simpson, A.G.B.; Worm, B. How many species are there on Earth and in the ocean? *PLoS Biol.* **2011**, *9*, 1001127. [CrossRef] [PubMed]
7. Richard, K.; Geerat, G.; Vermeij, J.; Wainwright, P.C. Biodiversity in water and on land. *Curr. Biol.* **2012**, *22*, R900–R903.
8. Barinova, S.S. Empirical model of the functioning of aquatic ecosystems. *Int. J. Oceanogr. Aquac.* **2017**, *1*, 1–9. [CrossRef]
9. Barinova, S. On the classification of water quality from an ecological point of view. *Int. J. Environ. Sci. Nat. Resour.* **2017**, *2*, 1–8. [CrossRef]
10. Sender, J.; Maślanko, W.; Różańska-Boczula, M.; Cianfaglione, K. A new multi-criteria method for the ecological assessment of lakes: A case study from the transboundary biosphere reserve 'West Polesie' (Poland). *J. Limnol.* **2017**, 76. [CrossRef]
11. Cianfaglione, K. Plant landscape and models of french atlantic estuarine systems. Extended summary of the doctoral thesis. *Transylv. Rev. Syst. Ecol. Res.* **2021**, *23*, 15–36.
12. Biggs, J.; von Fumetti, J.S.; Kelly-Quinn, M. The importance of small water bodies for biodiversity and ecosystem services: Implications for policy makers. *Hydrobiologia* **2017**, *793*, 3–39. [CrossRef]
13. Pekárik, L.; Čejka, T.; Čiamporovà-Zatovičovà, Z.; Darolovà, A.; Illèsovà, D.; Illyová, M.; Pastuchovà, Z.; Gatial, E.; Čiampor, F. Multidisciplinary evaluation of the function and importance of the small water reservoirs: The biodiversity aspect. *Transylv. Rev. Syst. Ecol. Res.* **2009**, *8*, 105–112.
14. Agenţia Europeană de Mediu (AEM). Available online: https://www.eea.europa.eu/ro/themes/climate/about-climate-change (accessed on 23 July 2021).
15. Treut, L.; Somerville, R.; Cubasch, U.; Ding, Y.; Mauritzen, C.; Mokssit, A.; Peterson, T.; Prather, M.; Qin, D.; Manning, M.; et al. Historical overview of climate change science. *Earth* **2007**, *43*, 93–127.
16. Heaviside, C. Understanding the impacts of climate change on health to better manage adaptation action. *Atmosphere* **2019**, *10*, 119. [CrossRef]
17. Murray, V.; Waite, T.D. Climate change and human health—The links to the UN landmark agreement on disaster risk reduction. *Atmosphere* **2018**, *9*, 231. [CrossRef]
18. United Nations, Security Council. Available online: https://www.un.org/press/en/2021/sc14445.doc.htm (accessed on 23 July 2021).
19. The Intergovernmental Panel on Climate Change (IPCC-53 BIS). Available online: https://www.ipcc.ch/ (accessed on 23 July 2021).
20. Mann, M.E.; Bradley, R.S.; Hughes, M.K. Global-scale temperature patterns and climate forcing over the past six centuries. *Nature* **1998**, *392*, 779–787. [CrossRef]
21. Krassilov, V.; Barinova, S. Sea level—Geomagnetic polarity correlation as consequence of rotation geodynamics. *Earth Sci.* **2013**, *2*, 1–8. [CrossRef]
22. Bates, B.C.; Kundzewicz, Z.W.; Wu, S.; Palutikof, J.P. Climate change and water. In *Technical Paper of the Intergovernmental Panel on Climate Change*; IPCC Secretariat: Geneva, Switzerland, 2008.
23. United Nations Secretary-General Meetings Coverage and Press Releases. Available online: https://www.un.org/press/en/2021/sgsm20847.doc.htm (accessed on 28 August 2021).
24. Fenoglio, S.; Bo, T.; Cucco, M.; Mercalli, L.; Malacarne, G. Effects of global climate change on freshwater biota: A review with special emphasis on the Italian situation. *Ital. J. Zool.* **2010**, *77*, 374–383. [CrossRef]
25. Svenning, J.C.; Kerr, J.; Rahbek, C. Predicting future shifts in species diversity. *Ecography* **2009**, *32*, 3–4. [CrossRef]
26. Kühn, I.; Sykes, M.T.; Berry, P.M.; Thuiller, W.; Piper, J.M.; Nigmann, U. MACIS: Minimization of and adaptation to climate change impacts on biodiversity. *GAIA—Ecol. Per. Sci. Soc.* **2009**, *17*, 393–395. [CrossRef]
27. Quero, J.C. Changes in the Euro-Atlantic fish species composition resulting from fishing and ocean warming. *Ital. J. Zool.* **1998**, *65*, 493–499. [CrossRef]
28. Munday, P.L.; Jones, G.P.; Pratchett, M.S.; Williams, A.J. Climate change and the future for coral reef fishes. *Fish Fish.* **1998**, *9*, 261–285. [CrossRef]
29. Gomi, T.; Nagasaka, M.; Fukuda, T.; Hagihara, H. Shifting of the life cycle and life-history traits of the fall webworm in relation to climate change. *Entomol. Exp. Appl.* **1998**, *125*, 179–184. [CrossRef]
30. Balletto, E.; Barbero, F.; Casacci, L.P.; Cerrato, C.; Patricelli, D.; Bonelli, S. *L'Impatto dei Cambiamenti Climatici Sulle Farfalle Italiane*; XVIII Convegno Gruppo Per l'Ecologia di Base G. Gadio: Alessandria, Italy, 2008; pp. 9–11.
31. Pounds, J.A.; Fogden, P.L.; Campbell, J.H. Biological response to climate change on a tropical mountain. *Nature* **1999**, *398*, 611–615. [CrossRef]
32. Arthington, A.H.; Balcombe, S.R.; Wilson, G.A.; Thoms, M.C.; Marschall, J.C. Spatial and temporal variation in fish assemblage structure in isolated waterholes during the 2001 dry season of an arid-zone river, Cooper Creek, Australia. *Mar. Freshw. Res.* **2005**, *56*, 25–35. [CrossRef]
33. Zubcov, N.; Zubcov, E.; Schlenk, D. The dynamics of metals in fish from Nistru and Prut rivers (Moldova). *Transylv. Rev. Syst. Ecol. Res.* **2008**, *6*, 51–58.
34. Vassilev, M.; Botev, I. Assessment of the ecological status of some Bulgarian rivers from the Aegean Sea Basin based on both environmental and fish parameters. *Transylv. Rev. Syst. Ecol. Res.* **2008**, *6*, 71–80.
35. Momeu, L.; Battes, K.; Battes, K.; Stoica, I.; Avram, A.; Cîmpean, M.; Pricpe, F.; Ureche, D. Algae, macroinvertebrate and fish communities from the Arieş River catchment area (Transylvania, Romania). *Transylv. Rev. Syst. Ecol. Res.* **2009**, *7*, 149–180.

36. Trichokva, T.; Stefanov, T.; Vassilev, M.; Zivcov, M. Fish species diversity in the rivers of the north-west Bulgaria. *Transylv. Rev. Syst. Ecol. Res.* **2009**, *8*, 161–168.
37. Jeeva, V.; Kumar, S.; Verma, D.; Rumana, S. River fragmentation and connectivity problems in Gange River of upper Himalayas: The effect on the fish communities (India). *Transylv. Rev. Syst. Ecol. Res.* **2011**, *12*, 75–90.
38. Florea, L.; Strătilă, S.D.; Costache, M. The assessment of community interest fish species from protected area ROSCI0229. *Transylv. Rev. Syst. Ecol. Res.* **2014**, *16*, 73–96. [CrossRef]
39. Sosai, A.S. Illegal fishing in southern Mannar Island coastal area (Sri Lanka). *Transylv. Rev. Syst. Ecol. Res.* **2015**, *17*, 95–108.
40. Khoshnood, Z.; Khoshnood, R. Effect of industrial wastewater on fish in Karoon River. *Transylv. Rev. Syst. Ecol. Res.* **2015**, *17*, 109–120. [CrossRef]
41. Rumana, H.S.; Jeeva, V.; Kumar, S. Impact of the low head dam/barrage on fisheries—A case study of Giri River of Yamuna Basin (India). *Transylv. Rev. Syst. Ecol. Res.* **2015**, *17*, 119–138. [CrossRef]
42. Del Monte-Luna, P.; Lluch-Belda, D.; Arreguín-Sánchez, F.; Lluch-Cota, S.; Villalobos-Ortiz, H. Approaching the potential of world marine fish. *Transylv. Rev. Syst. Ecol. Res.* **2015**, *18*, 45–56.
43. Taiwo, I.O.; Olopade, O.A.; Bamidele, N.A. Heavy metal concentration in eight fish species from Epe Lagoon (Nigeria). *Transylv. Rev. Syst. Ecol. Res.* **2019**, *21*, 69–82. [CrossRef]
44. Kar, D. Wetlands and their fish diversity in Assam (India). *Transylv. Rev. Syst. Ecol. Res.* **2019**, *21*, 47–94. [CrossRef]
45. Rios, J.M. Predation by nonnative Rainbow Trout, Oncorhynchus mykiss (Walbaum, 1792), on the native biota from freshwater environment of the Central Andes (Argentina). *Transylv. Rev. Syst. Ecol. Res.* **2019**, *23*, 67–72.
46. Jenkins, M. Prospects for biodiversity. *Science* **2003**, *302*, 1175–1177. [CrossRef]
47. Antonescu, B. In Verbis, 11 August 2021. Available online: https://www.digi24.ro/stiri/actualitate/romania-in-raportul-cod-rosu-pentru-umanitate-expert-meteo-vom-avea-valuri-de-caldura-precipitatii-intense-si-seceta-1629497 (accessed on 28 August 2021).
48. Intergovernmental Panel on Climate Change (IPCC). Climate Change 2014, Synthesis Report. Available online: https://www.ipcc.ch/report/ar5/syr/ (accessed on 8 August 2021).
49. Lupo, A.R.; Kononova, N.K.; Semenova, I.G.; Lebedeva, M.G. A comparison of the characteristic drought during the late 20th and early 21st centuries over Eastern Europe, Western Russia and Central North America. *Atmosphere* **2021**, *12*, 1033. [CrossRef]
50. Kjellström, E.; Nikulin, G.; Hanson, U.; Strandberg, G.; Ullerstig, A. 21st century changes in the European climate: Uncertainties derived from an ensemble of regional climate model simulations. *Tellus A* **2010**, *63*, 24–40. [CrossRef]
51. Wong, W.K.; Beldring, S. Climate change effects on spatiotemporal patterns of hydroclimatological summer droughts in Norway. *J. Hydrometeorol.* **2011**, *12*, 1205–1220. [CrossRef]
52. Cianfaglione, K.; Chelli, S.; Campetella, G.; Wellstein, C.; Cerivellini, M.; Ballelli, S.; Lucarini, D.; Canullo, R.; Jentsch, A. European grasslands gradient and the resilience to extreme climate events: The SIGNAL project in Italy. In *Climate Gradients and Biodiversity in Mountains of Italy*; Pedrotti, F., Ed.; Springer: Berlin/Heidelberg, Germany, 2008; p. 175.
53. Brönnimann, S. Early twentieth-century warming. *Nature* **2009**, *2*, 735–736. [CrossRef]
54. Levitus, S.; Antonov, J.I.; Wang, J.; Delworth, T.L.; Dixon, K.W.; Broccoli, A.J. Anthropogenic warming of Earth's climate system. *Science* **2001**, *292*, 267–270. [CrossRef] [PubMed]
55. Lloyd-Hughes, B. The impracticality of a universal drought definition. *Theor. Appl. Climatol.* **2014**, *117*, 607–611. [CrossRef]
56. IPCC 2012. Summary for policymakers. In *Managing the Risks of Extreme Events and Disasters to Advance Climate Change Adaptation*; Field, C.B., Barros, V.R., Stocker, D., Qin, D.J., Dokken, K.L., Ebi, M.D., Mastrandrea, K.J., Mach, G.-K., Plattner, S.K., Allen, M.T., et al., Eds.; Cambridge University Press: Cambridge, UK; New York, NY, USA, 2012; pp. 3–21.
57. Stahle, D.W. Anthropogenic megadrought, human-driven climate warming worsens an otherwise moderate drought. *Science* **2020**, *368*, 238–239. [CrossRef]
58. Wilhite, D.A.; Pulwarty, R.S. (Eds.) Drought as hazard: Understanding the natural and social context. In *Drought and Water Crises*; CRC Press: Boca Raton, FL, USA, 2017.
59. Gerald, A.M.; Washington, W.M.; Arblaster, J.M.; Hu, A.; Teng, H.; Tebaldi, C.; Sanderson, B.N.; Lamarque, J.-F.; Conley, A.J.; Strand, W.G.; et al. Climate system response to external forcings and climate change projections in CCSM4. *J. Clim.* **2007**, *25*, 3661–3683.
60. Bănăduc, D.; Joy, M.; Olosutean, H.; Afanasyev, S.; Curtean-Bănăduc, A. Natural and anthropogenic driving forces as key elements in the Lower Danube Basin-South-Eastern Carpathians-North-Western Black Sea coast area lakes, a broken stepping stones for fish in a climatic change scenario? *Environ. Sci. Eur.* **2020**, *32*, 73. [CrossRef]
61. Singh, V.; Barinova, S. Cladocera from the sediment of high Arctic lake in Svalbard. *Transylv. Rev. Syst. Ecol. Res.* **2021**, *23*, 17–24.
62. Cianfaglione, K. On the potential of *Quercus pubescens* willd and other species of *Quercus* in the camerino syncline (central Italy). In *Warm-Temperate Deciduous Forests around the Northern Hemisphere*; Box, E., Fujiwara, K., Eds.; Springer: Cham, Switzerland, 2015.
63. Wellstein, C.; Cianfaglione, K. Impact of extreme drought and warming on survival and growth characteristics of different provenences of juvenile *Quercus pubescens* willd. *Folia Geobot.* **2014**, *49*, 31–47. [CrossRef]
64. Arthington, A. The challenge of providing environmental flow rules to sustain river ecosystems. *Ecol. Appl.* **2006**, *16*, 1311–1318. [CrossRef]
65. Bunn, S.E.; Arthington, A.H. Basic principles and ecological consequences of altered flow regimes for aquatic biodiversity. *Environ. Manag.* **2002**, *30*, 492–507. [CrossRef]

66. Kuriqi, A.; Pinheiro, A.N.; Sordo-Ward, A.; Garrote, L. Influence of hydrological based environmental flow methods on flow alteration and energy production in a run-of-river hydropower plant. *J. Clean. Prod.* **2019**, *232*, 1028–1042. [CrossRef]
67. Kuriqi, A.; Pinheiro, A.N.; Sordo-Ward, A.; Bejarano, M.D.; Garrote, L. Ecological impacts of run-of-river hydropower plants—Current status and future prospects on the brink of energy transition. *Renew. Sustain. Energy Rev.* **2021**, *142*, 110833. [CrossRef]
68. Popa, G.-O.; Curtean-Bănăduc, A.; Bănăduc, D.; Florescu, I.E.; Burcea, A.; Dudu, A.; Georgescu, S.E.; Costache, M. Molecular markers reveal reduced genetic diversity in Romanian populations of Brown Trout, *Salmo trutta* L., 1758 (Salmonidae). *Acta Zool. Bulg.* **2016**, *68*, 399–406.
69. Marić, S.; Stanković, D.; Wazenbök, J.; Šanda, R.; Erös, T.; Takács, A.; Specziár, A.; Sekulić, A.; Sekulić, N.; Bănăduc, D.; et al. Phylogeography and population genetics of the European mudminnow (*Umbra krameri*) with a time-calibrated phylogeny for the family Umbridae. *Hydrobiologia* **2017**, *792*, 151–168. [CrossRef]
70. Curtean-Bănăduc, A.; Olosutean, H.; Bănăduc, D. Influence of environmental variables on the structure and diversity of ephemeropteran communities: A case study of the Timiş River, Romania. *Acta Zool. Bulg.* **2016**, *68*, 215–224.
71. Curtean-Bănăduc, A.; Didenko, A.; Guti, G.; Bănăduc, D. *Telestes souffia* (Risso, 1827) species conservation at the eastern limit of range—Vişeu River basin, Romania. *Appl. Ecol. Environ. Res.* **2018**, *16*, 291–303. [CrossRef]
72. Curtean-Bănăduc, A.; Marić, S.; Gabor, G.; Didenko, A.; Rey Planellas, S.; Bănăduc, D. *Hucho hucho* (Linnaeus, 1758): Last natural viable population in the Eastern Carpathians—Conservation elements. *Turk. J. Zool.* **2019**, *43*, 215–223. [CrossRef]
73. Burcea, A.; Boeraş, I.; Mihuţ, C.-M.; Bănăduc, D.; Matei, C.; Curtean-Bănăduc, A. Adding the Mureş River Basin (Transylvania, Romania) to the list of hotspots with high contamination with pharmaceuticals. *Sustainability* **2020**, *12*, 10197. [CrossRef]
74. Curtean-Bănăduc, A.; Burcea, A.; Mihuţ, C.-M.; Berg, V.; Lyche, J.L.; Bănăduc, D. Bioaccumulation of persistent organic pollutants in the gonads of Barbus barbus (Linnaeus, 1758). *Ecotoxicol. Environ. Saf.* **2020**, *201*, 10. [CrossRef]
75. Bănăduc, D.; Curtean-Bănăduc, A.; Cianfaglione, K.; Akeroyd, J.R.; Cioca, L.-I. Proposed environmental risk management elements in a Carpathian valley basin, within the Roşia Montană European historical mining area. *Int. J. Environ. Res. Public Health* **2021**, *18*, 4565. [CrossRef]
76. Bartholy, J.; Pongracz, R. Regional analysis of extreme temperature and precipitation indices for the Carpathian Basin from 1946 to 2001. *Glob. Planet. Chang.* **2007**, *57*, 83–95. [CrossRef]
77. Bartholy, J.; Pongracz, R.; Torma, C.; Pieczka, I.; Kards, I.; Hunyady, A. Analysis of regional climate change modeling experiments for the Carpathian Basin. *Int. J. Glob. Warm.* **2009**, *1*, 238–252. [CrossRef]
78. Krüzselyi, I.; Bartholy, J.; Horányi, A.; Piecza, I.; Pongrácz, R.; Szabó, P.; Szépszó, G.; Torma, C.S. The future climate characteristics of the Carpathian Basin based on a regional climate model mini-ensemble. *Adv. Sci. Contrib. Appl. Meteorol. Climatol.* **2011**, *29*, 69–73. [CrossRef]
79. Bartholy, J.; Pongracz, R. Analysis of precipitation conditions for the Carpathian Basin based on extreme indices in the 20th century and climate simulations for 2050 and 2100. *Phys. Chem. Earth Parts A/B/C* **2010**, *35*, 43–52. [CrossRef]
80. Rakonczai, J. Effects and Consequences of Global Climate Change in the Carpathian Basin. In *Climate Change—Geophysical Foundations and Ecological Effects*; Intechopen: London, UK, 2011; pp. 297–321.
81. Mezösi, G.; Blanka, V.; Ladányi, Z.; Bata, T.; Udrea, P.; Frank, A.; Burghard, C. Expected mid- and long-term changes in drought hazard for the south-eastern Carpathian Basin. *Carpathian J. Earth Environ. Sci.* **2016**, *11*, 355–366.
82. Sábitz, J.; Pongrácz, R.; Bartholy, J. Estimated changes of drought tendency in the Carpathian Basin. *Hung. Geogr. Bull.* **2014**, *63*, 4. [CrossRef]
83. Spinoni, J.; Antofie, T.; Barbosa, P.; Bihari, Z.; Lakatos, M.; Szalai, S.; Szentimrey, T.; Vogt, J. An overview of drought events in the Carpathian Region in 1961–2010. *Adv. Sci. Res.* **2013**, *10*, 21–32. [CrossRef]
84. Antofie, T.; Naumann, G.; Spinoni, J.; Vogt, J. Estimating the water needed to end the drought or reduce the drought severity in the Carpathian region. *Hydrol. Earth Syst. Sci.* **2015**, *19*, 177–193. [CrossRef]
85. Gheorghe, C. Action Plan for Water Scarcity and Drought in Romania. Water Resources Management Directorate. Available online: https://www.icpdr.org/main/sites/default/files/08_CONSTANTIN_Climate_Change_Romania_Action_Plan.pdf (accessed on 8 August 2021).
86. Larned, S.T.; Datry, T.; Arscott, D.B.; Tockner, K. Emerging concepts in temporary-river ecology. *Freshw. Biol.* **2010**, *55*, 717–738. [CrossRef]
87. Krupa, E.G.; Barinova, S.S.; Romanova, S.M. Ecological mapping in assessing the impact of environmental factors on the aquatic ecosystem of the Arys River Basin, South Kazakhstan. *Diversity* **2019**, *11*, 239. [CrossRef]
88. Tsarenko, P.M.; Bilous, O.P.; Kryvosheia-Zakharova, O.M.; Lilitska, H.H.; Barinova, S. Diversity of algae and cyanobacteria and bioindication characteristics of the alpine lake nesamovyte (Eastern Carpathians, Ukraine) from 100 years ago to the present. *Diversity* **2021**, *13*, 256. [CrossRef]
89. Popa, G.-O.; Dudu, A.; Bănăduc, D.; Curtean-Bănăduc, A.; Burcea, A.; Ureche, D.; Nechifor, R.; Georgescu, S.E.; Costache, M. Genetic analysis of populations of brown trout (*Salmo trutta* L.) from the Romanian Carpathians. *Aquat. Living Resour.* **2019**, *32*, 10. [CrossRef]
90. Curtean-Bănăduc, A.; Bănăduc, D. The Transylvanian Water Tower through history. *Danub. News* **2015**, *17*, 1–4.
91. Curtean-Bănăduc, A.; Burcea, A.; Mihuţ, C.-M.; Bănăduc, D. The benthic trophic corner stone compartment in POPs transfer from abiotic environment to higher trophic levels—Trichoptera and Ephemeroptera pre-alert indicator role. *Water* **2021**, *13*, 1778. [CrossRef]

92. Costea, G.; Push, M.T.; Bănăduc, D.; Cosmoiu, D.; Curtean-Bănăduc, A. A review of hydropower plants in Romania: Distribution, current knowledge, and their effects on fish in headwater streams. *Renew. Sustain. Energy Rev.* **2021**, *54*, 111003. [CrossRef]
93. Curtean-Bănăduc, A.; Bănăduc, D.; Bucşa, C. *Watersheds Management (Transylvania/Romania): Implications, Risks, Solutions, Strategies to Enhance Environmental Security in Transition Countries, NATO Science for Peace and Security Series Series C-Environmental Security*; Springer: Geneva, Switzerland, 2007; pp. 225–238. ISSN 1971-4668.
94. Mac, I. *Subcarpaţii Transilvăneni Dintre Mureş şi Olt*; Academic Press: Bucureşti, Romania, 1972; p. 156.
95. Oancea, D.; Velcea, V. *Geografia României, III, Carpaţii şi Depresiunea Transilvaniei*; Academic Press: Bucharest, Romania, 1987; p. 647.
96. Posea, G.R. *Geomorfologia României*; Fund România de Mâine: Bucharest, Romania, 2002; p. 444.
97. Dobros, V. Aspects concerning the liquid flow in Târnava Mică River hydrographic basin (Transylvania, Romania). *Transylv. Rev. Syst. Ecol. Res.* **2005**, *2*, 1–4.
98. Dobros, V. Aspects concerning the superficial liquid flow in the Târnava Mare River middle hydrographic basin (Transylvania, Romania). *Transylv. Rev. Syst. Ecol. Res.* **2007**, *4*, 1–6.
99. *Raport Annual 2016 Administraţia Naţională de Meteorologie, Bucureşti, România*; Meteo Romania: Bucureşti, Romania, 2016; pp. 1–154.
100. *Raport Annual 2017 Administraţia Naţională de Meteorologie, Bucureşti, România*; Meteo Romania: Bucureşti, Romania, 2017; pp. 1–80.
101. *Raport Annual 2018 Administraţia Naţională de Meteorologie, Bucureşti, România*; Meteo Romania: Bucureşti, Romania, 2018; pp. 1–181.
102. *Raport Annual 2019 Administraţia Naţională de Meteorologie, Bucureşti, România*; Meteo Romania: Bucureşti, Romania, 2019; pp. 1–220.
103. *Raport Annual 2020 Administraţia Naţională de Meteorologie, Bucureşti, România*; Meteo Romania: Bucureşti, Romania, 2020; pp. 1–170.
104. Drăgulescu, C. The Hydrophilous and Hygrophilous Flora and Vegetation of Târnave rivers. *Transylv. Rev. Syst. Ecol. Res.* **2005**, *2*, 13–30.
105. Mountford, O.J.; Akeroyd, J.R. A biodiversity assessment of the saxon villages region of Transylvania. *Transylv. Rev. Syst. Ecol. Res.* **2005**, *2*, 31–42.
106. Cupșa, D. Preliminary note on the aquatic Oligochaeta from the Târnave rivers. *Transylv. Rev. Syst. Ecol. Res.* **2005**, *2*, 43–50.
107. Sîrbu, I. The freshwater mollusks from the Târnava rivers basin (Romania). *Transylv. Rev. Syst. Ecol. Res.* **2005**, *2*, 51–61.
108. Curtean-Bănăduc, A. Study regarding Târnava Mare and Târnava Mică rivers (Transylvania, Romania) stonefly (Insecta. Plecoptera) larvae communities. *Transylv. Rev. Syst. Ecol. Res.* **2005**, *2*, 75–84.
109. Curtean-Bănăduc, A. Târnava Mare River (Romania) ecological assessment, based on the benthic macroinvertebrates communities. *Transylv. Rev. Syst. Ecol. Res.* **2005**, *2*, 109–122.
110. Robert, S.; Curtean-Bănăduc, A. Aspects concerning Târnava Mare and Târnava Mică rivers (Transylvania, Romania) caddis fly (Insecta. Trichoptera) larvae communities. *Transylv. Rev. Syst. Ecol. Res.* **2005**, *2*, 89–98.
111. Momeu, L.; Péterfi, L.Ş. The structure of Diatom communities inhabiting the Târnava Mică and Târnava Mare rivers. *Transylv. Rev. Syst. Ecol. Res.* **2005**, *2*, 5–12.
112. Bănăduc, D. Fish associations—Habitats quality relation in the Târnave rivers (Transylvania, Romania) ecological assessment. *Transylv. Rev. Syst. Ecol. Res.* **2005**, *2*, 137–144.
113. Curtean-Bănăduc, A.; Bănăduc, D. Benthic macro-invertebrate and fish communities of some southern Târnava Mare River tributaries (Transylvania, Romania). *Transylv. Rev. Syst. Ecol. Res.* **2007**, *4*, 135–148.
114. Tian, L.; Yuan, S.; Quiring, S.M. Evaluation of six indices for monitoring agricultural drought in the south-central United States. *Agric. For. Meteorol.* **2018**, *249*, 107–119. [CrossRef]
115. Gray, C.; Mueller, V. Drought and population mobility in rural Ethiopia. *World Dev.* **2012**, *40*, 134–145. [CrossRef]
116. Grolle, J. Historical case studies of famines and migrations in the West African Sahel and their possible relevance now and in the future. *Popul. Environ.* **2015**, *37*, 181–206. [CrossRef]
117. Sweet, S.K.; Wolfe, D.W.; DeGaetano, A.; Benner, R. Anatomy of the 2016 drought in the Northeastern United States: Implications for agriculture and water resources in humid climates. *Agric. For. Meteorol.* **2017**, *247*, 571–581. [CrossRef]
118. Schneider-Binder, E. The habitats along the upper danube in germany and changes to them induced by human impacts. In *Human Impact on Danube Watershed Biodiversity in the XXI Century. Geobotany Studies (Basics, Methods and Case Studies)*; Bănăduc, D., Curtean-Bănăduc, A., Pedrotti, F., Cianfaglione, K., Akeroyd, J., Eds.; Springer: Cham, Switzerland, 2020; pp. 27–48.
119. Cianfaglione, K.; Pedrotti, F. Italy in the Danube geography: Territory, landscape, environment, vegetation, fauna, culture, Human management and outlooks for the future. In *Human Impact on Danube Watershed Biodiversity in the XXI Century. Geobotany Studies (Basics, Methods and Case Studies)*; Bănăduc, D., Curtean-Bănăduc, A., Pedrotti, F., Cianfaglione, K., Akeroyd, J., Eds.; Springer: Cham, Switzerland, 2020; pp. 87–118.
120. Adámek, Z.; Jurajdová, Z.; Janáč, M.; Zahrádková, S.; Němejcová, D.; Jurajda, P. The response of fish assemblages to human impacts along the lower stretch of the Rivers Morava and Dyje (Danube River Basin, Czech Republic). In *Human Impact on Danube Watershed Biodiversity in the XXI Century. Geobotany Studies (Basics, Methods and Case Studies)*; Bănăduc, D., Curtean-Bănăduc, A., Pedrotti, F., Cianfaglione, K., Akeroyd, J., Eds.; Springer: Cham, Switzerland, 2020; pp. 135–150.

121. Ćaleta, M.; Mustafić, P.; Zanella, D.; Buj, I.; Marčić, Z.; Mrakovčić, M. Human impact on the Dobra River (Croatia). In *Human Impact on Danube Watershed Biodiversity in the XXI Century. Geobotany Studies (Basics, Methods and Case Studies)*; Bănăduc, D., Curtean-Bănăduc, A., Pedrotti, F., Cianfaglione, K., Akeroyd, J., Eds.; Springer: Cham, Switzerland, 2020; pp. 151–168.
122. Dekić, R.; Ivanc, A.; Ćetković, D.; Lolić, A. Anthropogenic impact and environmental quality of different tributaries of the River Vrbas (Bosnia and Hertzegovina). In *Human Impact on Danube Watershed Biodiversity in the XXI Century. Geobotany Studies (Basics, Methods and Case Studies)*; Bănăduc, D., Curtean-Bănăduc, A., Pedrotti, F., Cianfaglione, K., Akeroyd, J., Eds.; Springer: Cham, Switzerland, 2020; pp. 169–214.
123. Guti, G. Assessment of long-term changes in the Szigetköz Floodplain of the Danube River. In *Human Impact on Danube Watershed Biodiversity in the XXI Century. Geobotany Studies (Basics, Methods and Case Studies)*; Bănăduc, D., Curtean-Bănăduc, A., Pedrotti, F., Cianfaglione, K., Akeroyd, J., Eds.; Springer: Cham, Switzerland, 2020; pp. 215–240.
124. Đikanović, V.; Nikčević, M.; Mićković, B.; Hegediš, A.; Mrdak, D.; Pešić, V. Anthropogenic pressures on watercourses of the Danube River Basin in Montenegro. In *Human Impact on Danube Watershed Biodiversity in the XXI Century. Geobotany Studies (Basics, Methods and Case Studies)*; Bănăduc, D., Curtean-Bănăduc, A., Pedrotti, F., Cianfaglione, K., Akeroyd, J., Eds.; Springer: Cham, Switzerland, 2020; pp. 241–256.
125. Lenhardt, M.; Snederevac-Lalić, M.; Hegediš, A.; Skorić, S.; Cvijanović, G.; Višnjić-Jeftić, Ž.; Djikanović, V.; Jovičić, K.; Jaćimović, M.; Jarić, I. Human impacts on fish fauna in the Danube River in Serbia: Current status and ecological implications. In *Human Impact on Danube Watershed Biodiversity in the XXI Century. Geobotany Studies (Basics, Methods and Case Studies)*; Bănăduc, D., Curtean-Bănăduc, A., Pedrotti, F., Cianfaglione, K., Akeroyd, J., Eds.; Springer: Cham, Switzerland, 2020; pp. 257–280.
126. Mišiková Elexová, E.; Makovinská, J. Assessment of the aquatic ecosystem in the slovak stretch of the Danube River. In *Human Impact on Danube Watershed Biodiversity in the XXI Century. Geobotany Studies (Basics, Methods and Case Studies)*; Bănăduc, D., Curtean-Bănăduc, A., Pedrotti, F., Cianfaglione, K., Akeroyd, J., Eds.; Springer: Cham, Switzerland, 2020; pp. 281–300.
127. Maślanko, W.; Ferencz, B.; Dawidek, J. State and changes of natural environment in polish part of the Danube River Basin Poland. In *Human Impact on Danube Watershed Biodiversity in the XXI Century. Geobotany Studies (Basics, Methods and Case Studies)*; Bănăduc, D., Curtean-Bănăduc, A., Pedrotti, F., Cianfaglione, K., Akeroyd, J., Eds.; Springer: Cham, Switzerland, 2020; pp. 301–326.
128. Afanasyev, S.; Lyashenko, A.; Iarochevitch, A.; Lietytska, O.; Zorina-Sakharova, K.; Marushevska, O. Pressures and impacts on ecological status of Surface Water Bodies in Ukrainian part of the Danube River Basin. In *Human Impact on Danube Watershed Biodiversity in the XXI Century. Geobotany Studies (Basics, Methods and Case Studies)*; Bănăduc, D., Curtean-Bănăduc, A., Pedrotti, F., Cianfaglione, K., Akeroyd, J., Eds.; Springer: Cham, Switzerland, 2020; pp. 327–359.
129. Bakiu, R. Drina River (Sava's Tributary of Danube River) and Human Impact in Albania. In *Human Impact on Danube Watershed Biodiversity in the XXI Century. Geobotany Studies (Basics, Methods and Case Studies)*; Bănăduc, D., Curtean-Bănăduc, A., Pedrotti, F., Cianfaglione, K., Akeroyd, J., Eds.; Springer: Cham, Switzerland, 2020; pp. 359–380.
130. Kostov, V.; Slavevska-Stamenkovic, V.; Ristovska, M.; Stojov, V.; Marić, S. Characteristics of the Danube Drainage area in the Republic of Macedonia. In *Human Impact on Danube Watershed Biodiversity in the XXI Century. Geobotany Studies (Basics, Methods and Case Studies)*; Bănăduc, D., Curtean-Bănăduc, A., Pedrotti, F., Cianfaglione, K., Akeroyd, J., Eds.; Springer: Cham, Switzerland, 2020; pp. 381–392.
131. Snigirova, A.; Bogatova, Y.; Barinova, S. Assessment of river-sea interaction in the Danube Nearshore Area (Ukraine) by Bioindicators and Statistical Mapping. *Land* **2021**, *10*, 310. [CrossRef]
132. Kenderov, L.; Trichkova, T. Long-term changes in the ecological conditions of the Iskar River (Danube River Basin, Bulgaria). In *Human Impact on Danube Watershed Biodiversity in the XXI Century. Geobotany Studies (Basics, Methods and Case Studies)*; Bănăduc, D., Curtean-Bănăduc, A., Pedrotti, F., Cianfaglione, K., Akeroyd, J., Eds.; Springer: Cham, Switzerland, 2020; pp. 393–424.
133. Curtean-Bănăduc, A.; Bănăduc, D. Human impact effects on Târnava Mare Basin aquatic biodiversity (Transylvania, Romania). In *Human Impact on Danube Watershed Biodiversity in the XXI Century. Geobotany Studies (Basics, Methods and Case Studies)*; Bănăduc, D., Curtean-Bănăduc, A., Pedrotti, F., Cianfaglione, K., Akeroyd, J., Eds.; Springer: Cham, Switzerland, 2020; pp. 425–437.
134. Bănărescu, P.M.; Bănăduc, D. Habitats Directive (92/43/EEC) fish species (Osteichthyes) on the Romanian territory. *Acta Ichtiologica Romanica* **2007**, *2*, 43–78.
135. McEwan, A.J.; Joy, M.K. Responses of three PIT-tagged native fish species to floods in a small, upland stream in New Zealand. *N. Z. J. Mar. Freshw. Res.* **2013**, *47*, 225–234. [CrossRef]
136. Bănăduc, D.; Oprean, L.; Bogdan, A. Fish species of community interest management issues in Natura 2000 site Sighişoara-Târnava Mare (Transylvania, Romania). *Rev. Econ.* **2011**, *3*, 23–27.
137. Smederevac-Lalic, M.M.; Kalauzi, A.J.; Regner, S.B.; Lenhardt, M.B.; Naunovic, Z.Z.; Hegedis, A.W. Prediction of fish catch in the Danube River based on long-term variability environmental parameters and catch statistics. *Sci. Total. Environ.* **2009**, *609*, 664–671. [CrossRef] [PubMed]
138. Romanenko, D.; Afanasyev, S.; Vasenko, A.G. Review and status of fisheries and aquaculture in the Dnipro region in relation to biodiversity conservation. *Water Qual. Res. J. Can.* **2005**, *40*, 42–53.
139. Didenko, A.V. Gobiids of the Dneprodzerzhinsk Reservoir (Dnieper River, Ukraine): Distribution and habitat preferences. *Acta Ichthyologica Piscatoria* **2013**, *43*, 257–266. [CrossRef]
140. Askeyev, A.; Askeyev, O.; Yanybaev, N.; Askeyev, I.; Monakhov, M.; Marić, S.; Hulsman, K. River fish assemblages along an elevation gradient in the eastern extremity of Europe. *Environ. Biol. Fishes* **2017**, *100*, 585–596. [CrossRef]

141. Grapci-Kotori, L.; Vavalidis, T.; Zogaris, D.; Šanda, R.; Vukic, J.; Geci, D.; Ibrahimi, H.; Bilalli, A.; Zogaris, S. Fish distribution patterns in the White Drin (Drini I Bardhë) river, Kosovo. *Knowl. Manag. Aquat. Ecosyst.* **2020**, *421*, 29. [CrossRef]
142. Nikolsky, G.V. Theory of fish population dynamics as the biological background for rational exploitation and management of fishery resources. *J. Appl. Ecol.* **1970**, *7*, 660–662.
143. Erdoğan, Z.; Torcu Koç, H.; Özdemir, F. Reproductive biolgy of the Tigris Scraper, Capoeta umbla (Heckel, 1843) population living in Solhan Creek of Murat River (Bingöl, Turkey). *Transylv. Rev. Syst. Ecol. Res.* **2021**, *23*, 39–50.
144. Hodges, S.W.; Magoulick, D.D. Refuge habitats for fishes during seasonal drying in an intermittent stream: Movement, survival and abundance of three minnow species. *Aquat. Sci.* **2011**, *73*, 513–522. [CrossRef]
145. Oliva-Paterna, F.J.; Miñnano, P.A.; Torralva, M. Habitat quality affects the condition of *Barbus sclateri* in Mediterranean semi-arid streams. *Environ. Biol. Fishes* **2003**, *67*, 13–22. [CrossRef]
146. Seo-Yeon, P.; Chanyang, S.J.-H.L.; Jong-Suk, K. Ecological drought monitoring through fish habitat-based flow assessment in the Gam river basin of Korea. *Ecol. Indic.* **2020**, *109*, 105830.
147. Pulliam, H.R. On the relationship between niche and distribution. *Ecol. Lett.* **2000**, *3*, 349–361. [CrossRef]
148. Bain, M.B.; Haifeng, J. A habitat model for fish communities in large streams and small rivers. *Int. J. Ecol.* **2012**, *2012*, 962071. [CrossRef]
149. Hirzel, A.H.; Le Lay, G. Habitat suitability modeling and niche theory. *J. Appl. Ecol.* **2008**, *45*, 1372–1381. [CrossRef]
150. Palmer, M.; Bernhardt, E.; Chornesky, E.; Collins, S.; Dobson, A.; Duke, C.; Gold, B.; Jacobson, R.; Kingsland, S.; Kranz, R.; et al. Ecology for a crowded planet. *Science* **2004**, *304*, 1251–1252. [CrossRef] [PubMed]
151. Costea, M. Characteristics of the relief from the Central-Eastern part of the Târnavelor Plateau, with reference to present modeling and the associate geomorphologic risk (Transylvania, Romania). *Transylv. Rev. Ecol. Res.* **2007**, *4*, 7–22.
152. Stăncioiu, P.T.; Lazăr, G.; Tudoran, G.M.; Candrea Bozga, Ş.B.; Predoiu, G.; Şofletea, N. *Habitate Forestiere de Interes Comunitar Incluse în Proiectul Life 05/NAT/RO/000176: Habitate Prioritare Alpine, Subalpine şi Forestiere din România, Măsuri de Gospodărire*; Universităţii Transilvania din Braşov: Braşov, România, 2008; pp. 1–184. (In Romanian)
153. Bănăduc, D.; Voicu, R.; Curtean-Bănăduc, A. Sediments as factor in the fate of the threatened endemic fish species *Romanichthys valsanicola* Dumitrescu, Bănărescu and Stoica, (Vâlsan River basin, Danube Basin). *Transylv. Rev. Syst. Ecol. Res.* **2021**, *22*, 15–30.
154. Kareiva, P.; Watts, S.; McDonald, R.; Boucher, T. Domesticated nature: Shaping landscapes and ecosystems for human welfare. *Science* **2007**, *316*, 1866–1869. [CrossRef] [PubMed]
155. Bond, N.R.; Lake, P.S. Ecological restoration and large-scale ecological disturbance: The effects of drought on the response by fish to a habitat restoration experiment, restoration ecology. *J. Soc. Ecol. Rest.* **2005**, *13*, 39–48. [CrossRef]
156. Magoulick, D.D.; Kobza, R.M. The role of refugia for fishes during drought: A review and synthesis. *Freshw. Biol.* **2003**, *48*, 1186–1198. [CrossRef]
157. Magoulick, D.D.; Dekar, M.P.; Hodges, S.W.; Scott, M.K.; Rabalais, M.R.; Bare, C.M. Hydrologic variation influences stream fish assemblage dynamics through flow regime and drought. *Sci. Rep.* **2021**, *11*, 10704. [CrossRef] [PubMed]
158. Barbosa, F.; Găldean, N. Ecological taxonomy: A basic tool for biodiversity conservation. *Trends Ecol. Evol.* **1997**, *12*, 359–360. [CrossRef]
159. Marshall, J.C.; Lobegeiger, J.S.; Starkey, A. Risks to fish populations in Dryland Rivers from the combined threats of drought and instream barriers. *Front. Environ. Sci.* **2021**, *9*, 325. [CrossRef]
160. Hopper, G.W.; Gido, B.; Pennock, C.A.; Hedden, S.C.; Frenette, B.D.; Barts, N.; Hedden, C.K.; Lindsey Bruckerhoff, A. Nowhere to swim: Interspecific responses of prairie stream fishes in isolated pools during severe drought. *Aquat. Sci.* **2020**, *82*, 42. [CrossRef]

MDPI

Article

Forecasting of Drought: A Case Study of Water-Stressed Region of Pakistan

Prem Kumar [1], Syed Feroz Shah [1,*], Mohammad Aslam Uqaili [2], Laveet Kumar [3] and Raja Fawad Zafar [4]

1 Department of Basic Sciences and Related Studies, Mehran University of Engineering and Technology, Jamshoro 76090, Pakistan; prem.kumar@faculty.muet.edu.pk
2 Department of Electrical Engineering, Mehran University of Engineering and Technology, Jamshoro 76090, Pakistan; aslam.uqaili@faculty.muet.edu.pk
3 Department of Mechanical Engineering, Mehran University of Engineering and Technology, Jamshoro 76090, Pakistan; laveet.kumar@faculty.muet.edu.pk
4 Department of Mathematics, Sukkur Institute of Business Administration University, Sukkur 65200, Pakistan; fawad.zafar@iba-suk.edu.pk
* Correspondence: feroz.shah@faculty.muet.edu.pk

Abstract: Demand for water resources has increased dramatically due to the global increase in consumption of water, which has resulted in water depletion. Additionally, global climate change has further resulted as an impediment to human survival. Moreover, Pakistan is among the countries that have already crossed the water scarcity line, experiencing drought in the water-stressed Thar desert. Drought mitigation actions can be effectively achieved by forecasting techniques. This research describes the application of a linear stochastic model, i.e., Autoregressive Integrated Moving Average (ARIMA), to predict the drought pattern. The Standardized Precipitation Evapotranspiration Index (SPEI) is calculated to develop ARIMA models to forecast drought in a hyper-arid environment. In this study, drought forecast is demonstrated by results achieved from ARIMA models for various time periods. Result shows that the values of p, d, and q (non-seasonal model parameter) and P, D, and Q (seasonal model parameter) for the same SPEI period in the proposed models are analogous where "p" is the order of autoregressive lags, q is the order of moving average lags and d is the order of integration. Additionally, these parameters show the strong likeness for Moving Average (M.A) and Autoregressive (A.R) parameter values. From the various developed models for the Thar region, it has been concluded that the model (0,1,0)(1,0,2) is the best ARIMA model at 24 SPEI and could be considered as a generalized model. In the (0,1,0) model, the A.R term is 0, the difference/order of integration is 1 and the moving average is 0, and in the model (1,0,2) whose A.R has the 1st lag, the difference/order of integration is 0 and the moving average has 2 lags. Larger values for R^2 greater than 0.9 and smaller values of Mean Error (ME), Mean Absolute Error (MAE), Mean Percentile Error (MPE), Mean Absolute Percentile Error (MAPE), and Mean Absolute Square Error (MASE) provide the acceptance of the generalized model. Consequently, this research suggests that drought forecasting can be effectively fulfilled by using ARIMA models, which can be assist policy planners of water resources to place safeguards keeping in view the future severity of the drought.

Keywords: forecasting; ARIMA; Standardized Precipitation Evapotranspiration Index (SPEI); mitigation; drought

Citation: Kumar, P.; Shah, S.F.; Uqaili, M.A.; Kumar, L.; Zafar, R.F. Forecasting of Drought: A Case Study of Water-Stressed Region of Pakistan. *Atmosphere* **2021**, *12*, 1248. https://doi.org/10.3390/atmos12101248

Academic Editors: Andrzej Walega and Agnieszka Ziernicka-Wojtaszek

Received: 11 August 2021
Accepted: 22 September 2021
Published: 26 September 2021

1. Introduction

With the increasing human consumption rate of water around the world, and especially in highly arid regions that have experienced water depletion, the demand for water resources has increased dramatically [1]. Furthermore, changing patterns of global climate have further resulted as an impediment to human survival [2]. Thus, the human population and other species on Earth are subject to more and more droughts. In arid as well as semi-arid regions, droughts are common and repeating. Drastic change in the projection of floods and droughts has been seen in the 21st century compared to the 20th century. Mainly, the effect of prolonged droughts on natural ecosystems has highly deteriorated regional agriculture, water resources and the environment [3–5]. In such complex situations, the unavailability of proper evaluation of drought may result in wrong decisions and actions by policy makers and monitors [6,7]. For this reason, scientists divide drought conditions into four main categories, namely meteorology [8], agriculture [9], hydrology and socioeconomics [10]. In order to describe these drought categories, drought detection and monitoring indicators have different natural variables over the required time period, such as Precipitation (PCPN), soil moisture, Potential Evapotranspiration (PET), vegetation condition, and ground water along with surface water. In simple words, a drought index measures the actual features and their correlated effects [11]. The prominent attributes or qualities that must be focused on in the index are the time span, intensity, magnitude and spatial extent of the drought; nevertheless, incorporating all these features into one index is highly an arduous task. For this, numerous indices have been put forward by researchers. Some of them are, the Palme Drought Severity Index (PDSI) [12], Surface Water Supply Index (SWSI) [13], Palfai Aridity Index (PAI) [14], Standardized Precipitation Index (SPI) [15,16], and Standardized Precipitation Evapotranspiration Index (SPEI). The examination through any of above indices is subject to the essence of the index, local surroundings, accessibility of facts and rationality [17]. The advantages and disadvantages of drought indices are based upon the clarity and provisional flexibilities of their administration Standardized Precipitation Index (SPI) has been endorsed as a standard drought index by the world meteorological organization that is likely due to its simple calculations, as precipitation data is itself enough to direct and test without the demand of statistical barriers. Furthermore, the SPI is also capable of showing high performance in finding and computing drought potency [18,19]. Despite of all this ease, the SPI still has some issues regarding water balance. This led to the development of SPEI and eliminated these problems in the SPI release. SPEI focuses on PDSI sensitivity to PET mutations and the multifaceted compound of SPI [20]. Both SPEI and SPI have almost the same evaluation procedure. In this way, the use of climate water balance by SPEI is the differentiation between the two indices [21,22]. For this purpose, now days, SPEI has been widely used as a relevant index to observe the drought in various regions worldwide [23].

At present, the crucial challenge in the research of droughts is in the development of reasonable techniques and methods to forecast the start and end points of droughts. Thus, the significance of drought prediction advances from the reduction of drought effects [24]. It has been attempted multiple times to utilize statistical models in meteorological drought forecasting, depending upon time series methods such as ARIMA models, exponential smoothing and neural networks [25]. ARIMA is a classic model for statistical evaluation that makes use of time series data to foresee subsequent tendencies in meteorological variables such as annual and monthly temperature and precipitation can also be estimated through ARIMA models [26]. In a meteorological time series, the ARIMA model approach can exceed multiple new models such as exponential smoothing and neural network. Thus, due to its statistical properties together with effective methodology in establishing the model, ARIMA has relative advantage to the other models [27].

A lot of research has focused on contemporary drought predictions. It uses a new imitation hybrid wavelet-Bayesian regression model to develop a meteorological time series for extended lead-time drought prediction [28], which is a combined model using wavelet and fuzzy logic [29]. A group of authors also discussed the application of a wavelike fuzzy logic model based on meteorological variables to PDSI measurement [30]. Mishra also provides a drought prediction model that uses hybrid, single-neural and synthetic randomized Artificial Neural Network (ANN) models to predict droughts based on SPI. All these efforts are fruitful and sufficient in predicting time-series droughts [31]. The Standardized Precipitation Index (SPI) is currently widely utilized in both scholarly and operational applications around the world. The SPI at short time scales is lower limited, referring to non-normally distributed for arid climates or those with a distinct dry season when zero values are typical. The SPI is always greater than a particular number in certain instances, and thus fails to signal the onset of a drought. The non-normality rates appear to be closely associated to local precipitation climates. Herein, the authors present the Standardized Precipitation Evapotranspiration Index (SPEI) as an enhanced drought indicator that is particularly well suited for research of the effect of global warming on drought severity. The SPEI evaluates the effect of reference evapotranspiration on drought severity, similar to the Palmer Drought Severity Index (PDSI), but its multi-scalar nature allows identification of distinct drought types and drought consequences on various systems [19].

Thus, the SPEI, like the standardized precipitation index, has the sensitivity of the PDSI in measuring evapotranspiration demand (induced by variations and trends in climatic variables other than precipitation), is easier to compute, and is multi-scalar (SPI). Detailed discussions of the SPEI's theory, computations, and comparisons to other widely used drought indicators including the PDSI and SPI are given by [32]. The SPEI is calculated in the same way as the SPI is calculated. However, rather than using precipitation as an input, the SPEI employs "climatic water balance," which is the difference between precipitation and reference Potential evapotranspiration. At various time frames, the climatic water balance is estimated (i.e., over three month, six months, nine months, twelve months, and 24 months). Despite the fact that the SPEI was just recently developed, it has already been employed in a number of research looking at drought variability.

The first step in the SPI calculation is to determine the probability density function (PDF), which describes the long-term observed precipitation. It also allows the SPI to be computed at any location and at any number of time scales, depending upon the impacts of interest to the user. Ratios of drought on the basis of analysis of stations across Colorado are given in Table 1 [33].

Table 1. Ratio of drought in Colorado [33].

Drought Condition	Ratio in Percentage
Mild	24%
Moderate	9.2%
Severe	4.4%
Extreme	2.3%

With the help of normal distribution of SPI above percentage are expected. The cumulative probability is then applied to the inverse normal (Gaussian) function, yielding the SPI. This approach is a transformation of equivalence. The equiprobability transformation's key feature is that the probability of being less than a particular value of the produced cumulative probability should be the same as the probability of being less than the normal distribution's equivalent value [33]. Based on the time series of drought monitoring findings of the Vegetation Temperature Condition Index (VTCI), Autoregressive Integrated Moving Average (ARIMA) models were developed. From the erecting stage to the maturity stage of winter wheat (early March to late May in each year at a ten-day interval) of the years 2000 to 2009, about 90 VTCI pictures produced from Advanced Very High-Resolution Radiometer (AVHRR) data were selected to create the ARIMA models. The ARIMA models' category drought predictions findings in April 2009 are more severe in the northeast of the Plain, which accord well with the monitoring results. The AR(1) models have smaller absolute errors than the SARIMA models, both in terms of frequency distributions and statistic findings. SARIMA models, on the other hand, are better at detecting changes in the drought situation than AR(1) models. These findings suggest that ARIMA models can better predict the type and extent of droughts, and that they can be used to predict droughts in the plain [34]. Time series forecasting has been extensively used and has emerged as a key method for drought forecasting. The Autoregressive Integrated Moving Average (ARIMA) model is one of the most extensively used time series models. The ARIMA model's wide application owes to its flexibility and methodical search (identification, estimation, and diagnostic check) for an acceptable model at each stage. The ARIMA model offers various advantages over other approaches, such as moving average, exponential smoothing, and neural networks, including predicting and more information on time-related changes. Hydrologic time series have also been analyzed and modelled using ARIMA models [35]. From the literature review and retrospective analysis, it has been found that the SPEI was not utilized as a drought index previously. Therefore, the main focus of this research investigation is to establish the stochastic model ARIMA to lay and predict the SPEI series on discrete time series. In addition, it also provides a subjective method to deal with climate-related parameters for 14 years of drought, i.e., (2005–2019).

2. Research Methodology

2.1. Area of the Study

In two districts of Sindh's Tharparkar desert, the resistance of rural households to food insecurity was studied. Drought has been the most dangerous risk for the study region because of its severe consequences on food, income, health, people's adaptability, and livestock survival. A resilience index was used by scholars to measure the inhabitants' resilience to the severe dry circumstances. Income and food access, agricultural assets, non-agricultural assets, access to basic services, social safety nets, sensitivity, adaptive capability, climate change, agricultural methods and technology, and enabling institutional environment were used to generate the household resilience index.

The Tharparkar region was vulnerable since it relied heavily on natural resources for its livelihood. Any community's access to water resources supported their survival even in the most adverse conditions. Because Nagarparkar was close to publicly accessible water, it was more robust than Islamkot, which had no water nearby [36].

Climate change is now a reality, exacerbating the misery of people who live in arid ecosystems. Rainfall has decreased, temperatures have risen, and the frequency of extreme events has increased in the semi-arid desert of district Tharparkar. People of Tharparkar have been coping with drought and aridity of the terrain for thousands of years by employing traditional wisdom. However, global shifts in weather patterns, as well as worsening social and economic situations, have forced the people of this desert region into a precarious position. From the perspective of changing climatic patterns, this research analyzes the link between climate-induced natural disasters, notably drought, and food insecurity and water scarcity.

Drought in the district has shifted from its typical pattern of little or no rainfall to increased but unpredictable rainfall, posing a greater threat to people's livelihoods and, as a result, a multiplier effect on water and food poverty. In the absence of social protection and basic essentials for existence during a drought, women are particularly vulnerable. Women, for example, have traditionally carried the burden of managing water resources, resulting in increased workload during droughts and water scarcity [37]. The research for this paper has been conducted on the basis of data collected from Tharparkar region of Sindh province in Pakistan. The Figures 1 and 2 shows the directorial regions of the district of Tharparkar, Sindh in overall map of South Asia and Pakistan, respectively.

Figure 2 shows the location of Mithi weather station. The main localities of the region of Thar are predominately plain deserts together with some mountain peaks greater than 3000 m. The main cities of the Tharparkar where this study has been conducted are Mithi, Diplo, Islamkot, Nagarparkar, and Dhali. Mainly the climate of all these areas is extremely hot in daytime, but at night the temperature of the region decreases. Tharparkar is listed among top hottest areas of Pakistan with extreme temperature in summer ranges between 35 to 45 °C, while in winter it ranges between 9 to 28 °C. Additionally, the rain gauge station within Tharparkar is located at Mithi that has recorded 277 mm of rainfall annually, but this differs significantly yearly [38].

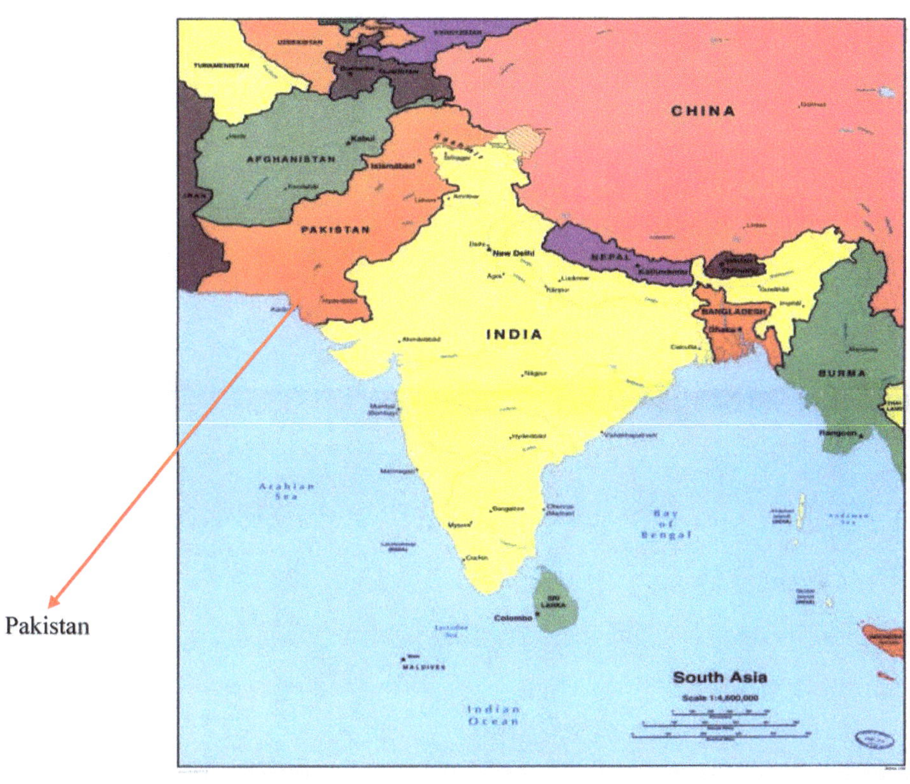

(a)

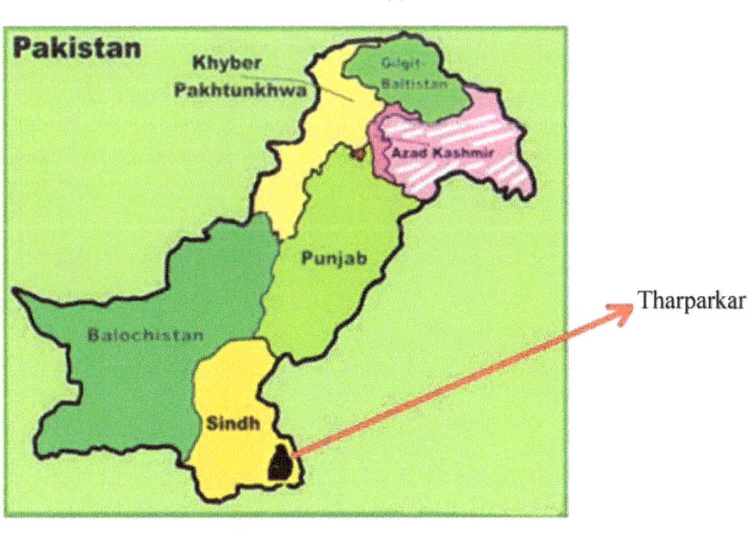

(b)

Figure 1. (**a**) Location of Pakistan in South Asia; (**b**) location of Tharparkar district in Pakistan. Both are highlighted with an arrow.

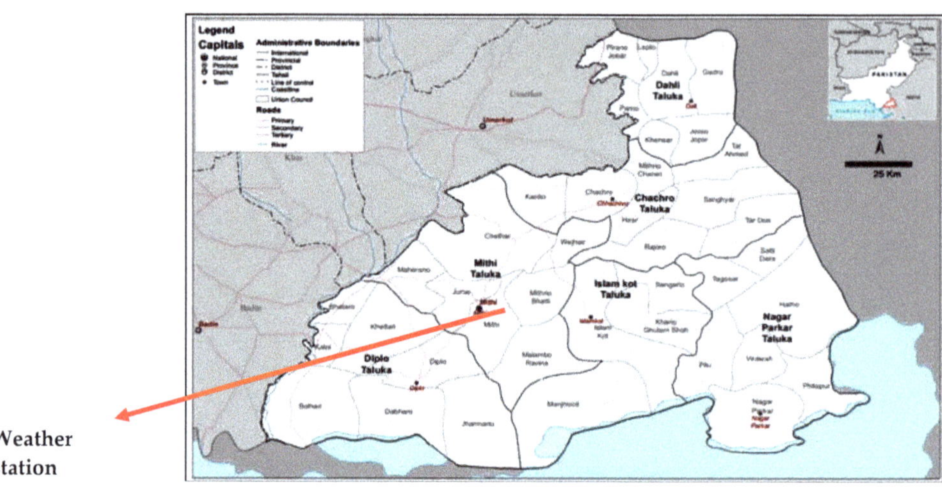

Figure 2. Location of Mithi weather station (Meteorological Station); the data is collected for this station that was established in 2004.

2.2. Data Sources and Preparation

This study has been carried out on the basis of data collected from Pakistan Meteorological Department (PMD) Karachi. The research includes precipitation records and varying air temperatures of Mithi Meteorological station, Tharparkar. The data also include some facts obtained from close localities of Mithi such as Islamkot and Diplo by utilizing linear regression methods. With the help of nonparametric tests, the parameters of the acquired data values were thoroughly checked. In this research, data has been obtained from the Mithi weather station for 2005–2019 period, whose altitude is 42 m (138 feet), longitude is 69.800430 m and latitude is 24.740065 m [38].

The future climate data will be generated through ARIMA model using R-programming language. From the collected data, a long-term dataset for SPEI series of 3, 6, 9, 12- and 24-month time scales will be established. Time-series plots provide long-term and concrete information regarding drought situations at the regional scale with a 0.5-degree geographic resolution and monthly time resolution. It also has a multi-scale nature that gives timescales for the SPEI between 1 and 8 months. The potential evapotranspiration (PET) equation to calculate the SPEI indices is given as:

$$\text{PET} = \frac{0.408\Delta(R_n - G) + \gamma\left(\frac{900}{T+273}\right)U_2(e_s - e_a)}{\Delta + (1 + 0.34U_2)} \tag{1}$$

where Δ is the slope of vapor pressure curve (kPa °C^{-1}), R_n the surface net radian (MJ.m^{-2} day^{-1}), G the soil heat flux density (MJ.m^{-2} day^{-1}), γ the psychometric constant (kPa °C^{-1}), T the mean daily air temperature (°C^{-1}), U_2 the wind speed (m.s^{-1}), e_s the saturated vapor pressure and e_a the actual vapor pressure [39]. Table 2 illustrates the standard precipitation evaporation and transpiration (SPEI) value of the dry/wet classification value [33,40,41].

2.3. Autoregressive and Moving Average Model (ARMA)

From the statistical analysis of time series, ARMA model gives a wretched representation of a static stochastic technique as far as 2 polynomials, the 1st for the A.R and the 2nd for the M.A. The A.R part contains reverting the variable all alone slacked values. The M.A part includes modeling the error term as a linear combination of error terms arising contemporaneously and at changed time in earlier. The model is typically mentioned to as the ARMA (p, q) display where p is the order of the A.R part and q is the order of the M.A [42].

$$Z_t = \omega + \eta_t + \sum_{i=1}^{p} \beta_i Z_{t-i} + \sum_{i=1}^{q} \alpha_i \eta_{t-i} \quad (2)$$

where ω is mean of the series, α_i is the parameter of the moving average model, and β_i is the parameter of the autoregressive model, whereas η_t, η_{t-i}, and Z_{t-i} are error terms known as white noise and Z_t is the time series.

Table 2. SPEI scale [33,40,41].

Conditions		SPEI
Dry	Extremely	$\text{SPEI} \leq -2$
	Severely	$-2 < \text{SPEI} \leq -1.5$
	Moderately	$-1.5 < \text{SPEI} \leq -1$
Normal	Near	$-1 < \text{SPEI} \leq 1$
Wet	Moderately	$1 < \text{SPEI} \leq 1.5$
	Severely	$1.5 < \text{SPEI} \leq 2$
	Extremely	$\text{SPEI} \geq 2$

2.4. *Autoregressive Integrated Moving Average Model (ARIMA)*

In econometrics and statistics and specifically in time series analysis, the ARIMA model is speculation of ARMA model. Together the formulations are fixed to time series data either to all the more likely comprehend the data or to forecast future outcomes in row. ARIMA models are related sometimes where data shows non-stationarity proofs, where an underlying differencing step can be connected at least multiple times to wipe out the non-stationarity [43].

2.4.1. Non-Seasonal Model

In general the non-seasonal ARIMA model is A.R having order p and M.A of order q and operate on the time series differences z_t; thus ARIMA family formulation is categorized by three parameter (p, d, q) that can either have 0 or positive integral values.

Generally non-Seasonal ARIMA model is written as:

$$\beta(C)\nabla^d z_t - \alpha(C)a_t \quad (3)$$

where $\beta(C)$ and $\alpha(C)$ are polynomials of order p and q, ∇ shows the order of difference [44].

2.4.2. Seasonal Model

General multiplicative seasonal ARIMA model, which is known as SARIMA model defined as ARIMA $(p,d,q)(P,D,Q)_s$ where (p,d,q) the non-seasonal part of model is and $(P,D,Q)_s$ is the seasonal part of the model is given as:

$$\beta_p(C)\Phi_p(C^s)\nabla^d \nabla_s^D z_t = \alpha_q(C)\Theta_Q(C^s)a_t \quad (4)$$

where p is the order of non-seasonal auto regression, d is the number of regular differencing, q is the order of non-seasonal MA, P is the order of seasonal auto regression, D is the number of seasonal differencing, Q is the order of seasonal MA, s is the length of season, Φ_p is the seasonal AR parameter of order P, and Θ_Q is the seasonal MA parameter of order Q [45].

2.5. *Model Identification*

Model identification comprises recognizing the possible ARIMA model that depicts the nature of time series. In order to ascertain the order of model, the Autocorrelation Function (ACF) and Partial Autocorrelation Function (PACF) were utilized for assistance. The information obtained through the utility of ACF and PACF was also helpful in advocating various types of new models that could be established. The selection of the ultimate model is performed by employing the penalty function statistics through the Akaike Information Criterion (AIC) and Schwarz Bayesian Criterion (SBC). These criteria assist in ranking the models, where the models which have the least value of criterion are considered the best. AIC is an estimator of prediction error and thereby relative quality of statistical models for a given set of data. Given a collection of models for the data, AIC estimates the quality of each model, relative to each of the other models. Thus, AIC provides a means for model selection.

$$\mathrm{AIC} = 2k - 2\ln(\hat{L})\ \hat{L} \tag{5}$$

Let k be the number of estimated parameters in the model. Let $\hat{L}$ be the maximum value of the likelihood function for the model.

In statistics, SBC is a criterion for model selection among a finite set of models; the model with the lowest BIC is preferred. It is based, in part, on the likelihood function and it is closely related to the AIC.

$$\mathrm{SBC} = -2\ln(\hat{L}) + k\ln(n) \tag{6}$$

Here k represents number of parameters in the model, (p + q + P + Q); whereas, L depicts likelihood function of ARIMA model. Additionally, n shows number of observations [46,47].

2.6. Parameter Estimation

After the reorganization of suitable model, the assessment of model parameters is attained with the help of the procedure and methods proposed by Box and Jenkins, the evaluation of model values for A.R and M.A parts were made possible. To ascertain the statistical significance of A.R and M.A parameters, they were tested whether they are important or not. The related parameters such as standard error of estimates and their linked t-values are also determined [48].

2.7. Diagnostic Checking

The final step in the development of model is diagnosing the ARIMA model. It is one of the significant steps of model development it point towards the appropriateness of model that inspects the assumptions of model unquestionably. The acceptability or appropriateness of model guarantees that the time series is in the time with model assumptions and that the prophecy of values is well founded. To examine the correlation of residuals with error terms, various diagnostic statistics and plots of residuals have been inspected to make it sure that whether these residuals correspond with error terms or not.

$$\mathrm{MAE} = \frac{1}{N}\sum_{i=1}^{N}|(X_m)_i - (X_s)_i| \tag{7}$$

$$\mathrm{RMSE} = \sqrt{\frac{1}{N}\sum_{i=1}^{N}[(X_m)_i - (X_s)_i]^2} \tag{8}$$

where N is the number of forecasting events, X_m the observed SPEI and X_s the predicted SPEI [49].

$$\mathrm{MAE} = \frac{1}{N}\sum_{i=1}^{N}|x_i - \hat{x}_i| \tag{9}$$

$$\mathrm{MPE} = \frac{100\%}{N}\sum_{i=1}^{N}\left(\frac{x_i - \hat{x}_i}{x_i}\right) \tag{10}$$

$$\text{MAPE} = \frac{100\%}{N}\sum_{i=1}^{N}\left|\frac{x_i - \hat{x}_i}{x_i}\right| \tag{11}$$

$$\text{MASE} = \frac{1}{N}\sum_{i=1}^{N}|x_i - \hat{x}_i|^2 \tag{12}$$

where $\frac{1}{N}\sum_{i=1}^{N}$ is test set, x_i predicted value and $\hat{x}_i$ is actual value. (N is the number of total data points).

In this research, all of the SPEI forecasting ARIMA models have been developed using the forecasting packages already available in R-programming language. Packages used are t series, forecast, SPEI and Uurca [50].

3. Results and Discussion

3.1. Climate Descriptive Analysis

For better understanding of the nature of drought in the area of research, a detailed examination of climate parameters was employed to achieve more accurate results. In Figure 3, annual means of high drought-pertinent parameters have been shown. In the region of the Tharparkar, the monthly highest temperature is recorded mostly in June, July and August. In this way, the lowest mean temperature occurs in December, January and February. In general, Tharparkar is warm area in the province of Sindh. However, due to the variability in the altitudes of various localities of the area, there is a small rise in monthly mean temperature throughout the region which is due to lake of rain fall and changing climatic conditions. Together with changing mean temperatures, there has also been seen dissimilarity in the monthly mean precipitation over the year. Zero mean precipitation is reported in the months of June, July, August and September. With the changes of temperature and precipitation, the region also faces acute water crisis throughout the year except in the times of monsoons. The compensation of this lack of water availability requires to be fulfilled by utilizing alternative water resources or through sound water management and rationalization methods. Such methods are mostly common across globe.

Through the examination of drought-linked climatic parameters for the region of Tharparkar, remarkable decrease in the level of precipitation has been proved. This continuous decline in the rate of precipitation has resulted in the significant rise of temperature. Thus, the prevailing circumstances may drive a slight increase in frequency and magnitude of drought patterns. Actually, the prime rationale concerned with changing drought circumstances is that it will deplete actual water resources. In this way, lack of water availability may further deteriorate the conditions considered fit for human survival.

3.2. Drought Frequency Variations

The SPEI time series illustrates at variable time scales that covers the period from 2005 to 2019 for Tharparkar region shown in Figure 2. The outcome of this research manifests that the area of Tharparkar will face more and more drought in future. Figure 3. shows a clear idea that there is a continuous rise in the conditions of drought. In the starting years, the drought tendency proceeds towards the natural limits of close normal and reasonably. Wet scales with few years showing irregular non-typical values. These abnormal values are related to the shortage of rainfall, while on the contrary, in previous decade, the beginning of drought conditions with extreme dry patterns started to occur. All the localities of the region of Tharparkar depict time evolution, with slight deviations among others. This situation is the result of changing climatic patterns and their effect. The other parts of the world, e.g., Egypt [51], Turkey [52], Portugal [53], and China [54] also show such circumstances due to climatic change impact.

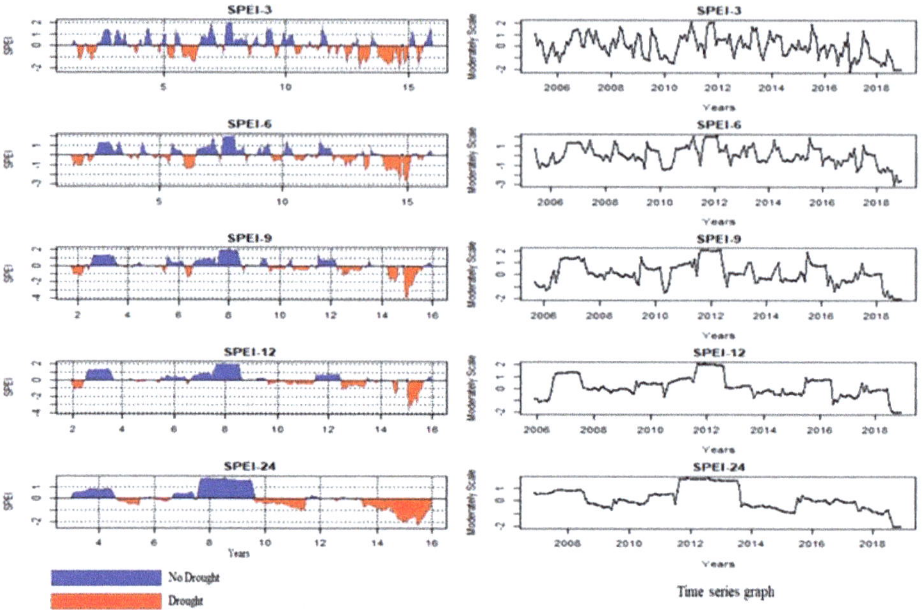

Figure 3. Time series plots of SPEI- 3, 6, 9, 12 and 24 are in right panel, and the left panel shows drought condition plots of SPEI- 3, 6, 9, 12 and 24.

The Figure 3 has two panels, in which the right side panel shows time series graphs of the given data, in which we can see the pattern of drought yearly; in the left side panel the color red shows the drought and color blue means no drought. The right panel shows time series graphs of SPEI-3, 6,9,12 and 24, showing the highest drought in the years on the x-axis, which is different in each scale of SPEI, but most of the SPEI shows in 2009 to 2014. Moreover, the time series and annual cycle of the precipitation and PET are shown in Figures 4–6, respectively. Our data source is Karachi Pakistan Meteorological Department (PMD) (https://www.pmd.gov.pk/en/, accessed on 21 September 2021) and the trend of drought is clearly shown, with the highest drought during the period of 2009 to 2014 and the lowest trend in the period of 2017 to 2019; the same period is also verified by the SPEI time series graph in Figure 3.

Additionally, as our selected region is desert area, there is no other source of water such as a canal or river and there is no source of water without rain for agriculture and other uses. Therefore, severe drought occurs, which can be seen from the annual precipitation graph (Figures 4–6) of the Mithi weather station.

Unit Root Test

In the literature there are various tests but, in this research, we have selected the three most important and different tests on SPEI, given below as:

1. Augmented Dickey–Fuller test
2. Phillips–Perron unit root test
3. KPSS test for level stationarity.

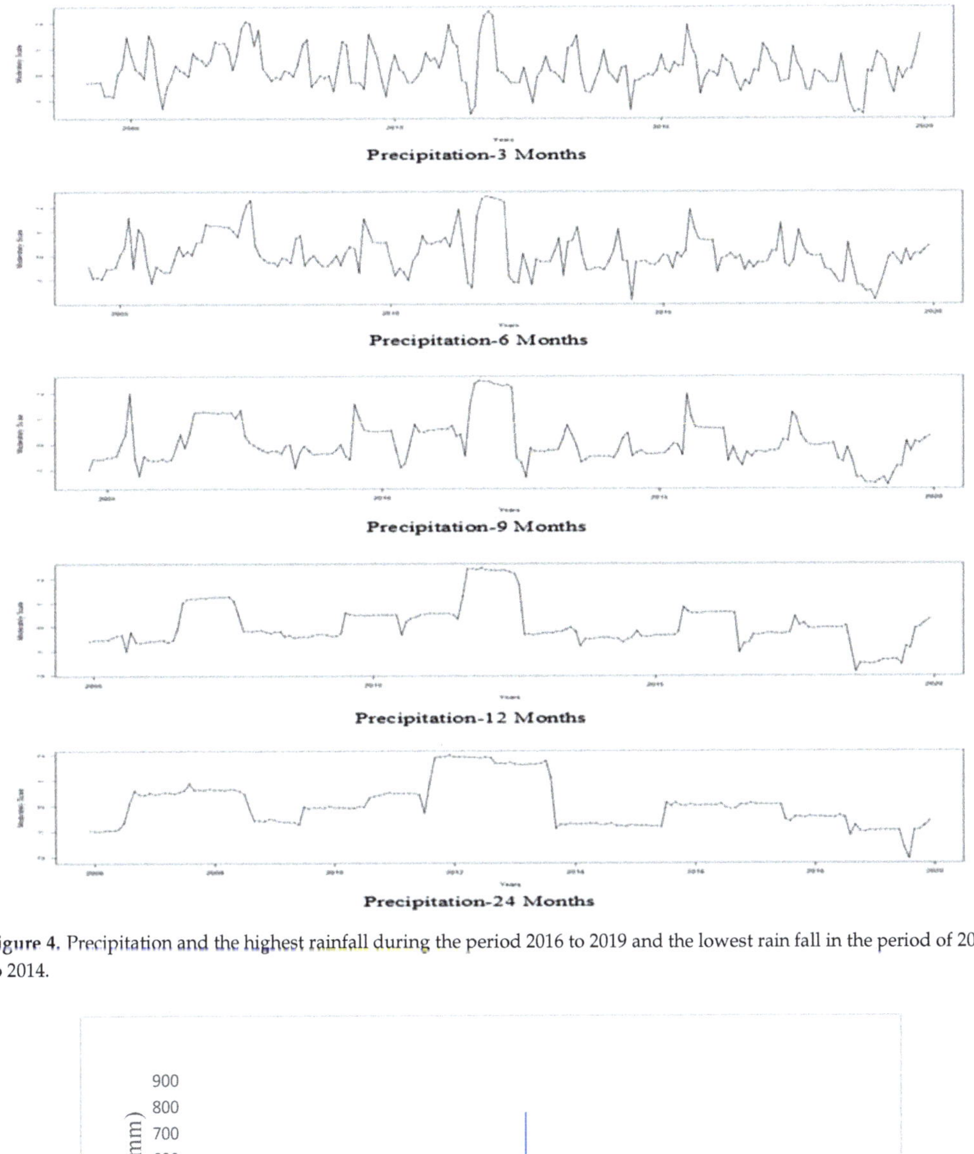

Precipitation-3 Months

Precipitation-6 Months

Precipitation-9 Months

Precipitation-12 Months

Precipitation-24 Months

Figure 4. Precipitation and the highest rainfall during the period 2016 to 2019 and the lowest rain fall in the period of 2009 to 2014.

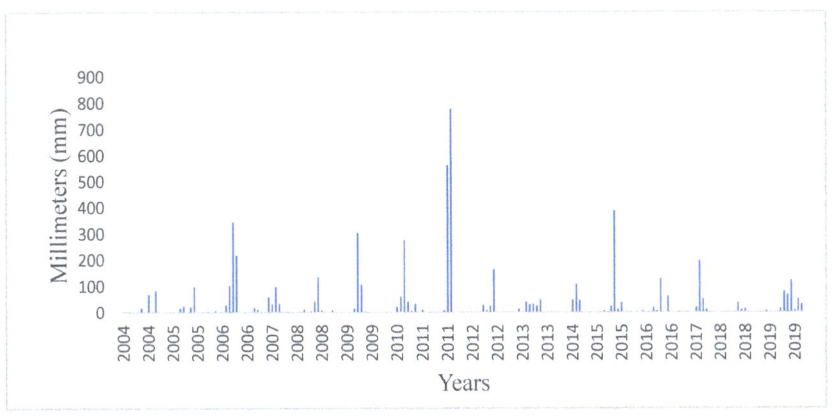

Figure 5. Annual cycle of the precipitation the input data starts from 2004 to 2019.

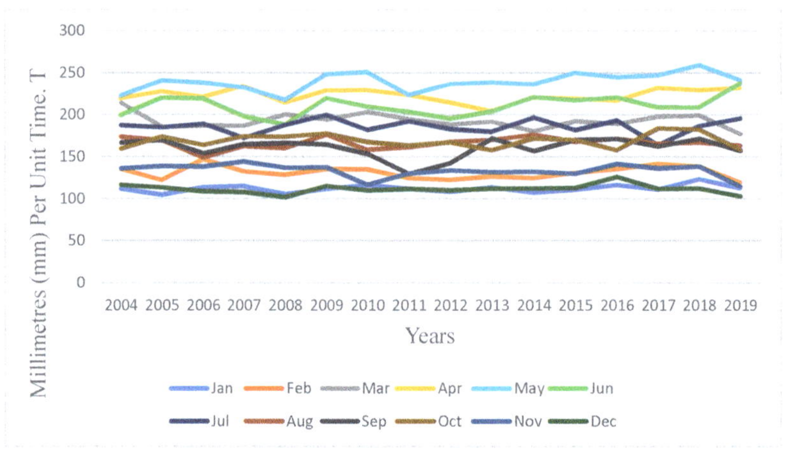

Figure 6. Annual cycle of the PET from 2004 to 2019 (Monthly variation).

Table 3 shows that the SPEI- 3, 6, 9, 12 and 24.have the 1st difference unit root test whereas the SPEI 24 has the 2nd difference unit root test.

Table 3. Augmented Dickey–Fuller test.

SPEI	Decision	*p*-Value of Unit Root at Level	Unit Root at 1st Difference	Unit Root at 2nd Difference
SPEI-3	No	0.01	N/A	N/A
SPEI-6	No	0.01	N/A	N/A
SPEI-9	No	0.01134	N/A	N/A
SPEI-12	No	0.03514	N/A	N/A
SPEI-24	Yes	0.5635	0.01	N/A

Table 4 shows that the SPEI- 3, 6, 9, 12 and 24 have the 1st difference unit root test whereas the SPEI 24 has the 2nd difference unit root test.

Table 4. Phillips–Perron unit root test.

SPEI	Decision	*p*-Value of Unit Root at Level	Unit Root at 1st Difference	Unit Root at 2nd Difference
SPEI-3	No	0.01	N/A	N/A
SPEI-6	No	0.01	N/A	N/A
SPEI-9	No	0.01	N/A	N/A
SPEI-12	No	0.09258	N/A	N/A
SPEI-24	Yes	0.5561	0.01	N/A

Table 5 shows that the all SPEI- 3, 6, 9, 12 and 24 has the 1st difference unit root test.

Table 5. KPSS test for level stationarity.

SPEI	Decision	*p*-Value of Unit Root at Level	Unit Root at 1st Difference
SPEI-3	No	0.1	N/A
SPEI-6	No	0.1	N/A
SPEI-9	No	0.1	N/A
SPEI-12	No	0.1	N/A
SPEI-24	No	0.0469	N/A

3.3. Estimation of Model Parameters

In Tables 6–13, the model parameters are standard error, p-value and related significance value at a significance level less than 0.05 for Tharparkar region. In comparison with the parametric values it has been observed to be very small. The above proposition bears exclusion of the model parametric values of SPEI at the three-month time scale. Furthermore, at the significance level of less than 0.05, almost all ARIMA model parameters are significant. Therefore, these parameters ought to be incorporated in the model. Other models also showed identical results. Table 6 describes the generalized ARIMA seasonal and non-seasonal models for Tharparkar.

Table 6. Seasonal ARIMA and non-seasonal ARIMA models.

Time Scales	Seasonal ARIMA Models	Non-Seasonal ARIMA Models
SPEI-3	ARIMA (1,1,3)(0,0,0)	ARIMA (1,1,3)
SPEI-6	ARIMA (1,1,1)(0,0,2)	ARIMA (1,1,1)
SPEI-9	ARIMA (1,1,1)(1,0,0)	ARIMA (0,1,0)
SPEI-12	ARIMA (0,1,0)(1,0,2)	ARIMA (0,1,0)
SPEI-24	ARIMA (0,1,0)(1,0,2)	ARIMA (0,1,0)

Table 7. SPEI-3 ARIMA (1,1,3)(0,0,0) for Seasonal.

SPEI−3	Coefficient	Standard Error	t-Value	p-Value
AR_{t-1}	−0.0392	0.2060	−0.1902913	0.3912019
MA_{t-1}	−0.2654	0.1910	−1.389529	0.15169
MA_{t-2}	−0.2065	0.0991	−2.083754	0.04607481
MA_{t-3}	−0.3757	0.1038	−3.619461	0.0006903546

Table 8. SPEI-3 ARIMA (1,1,3) for non-seasonal.

SPEI−3	Coefficient	Standard Error	t-Value	p-Value
AR_{t-1}	−0.0392	0.2060	−0.1902913	0.3912019
MA_{t-1}	−0.2654	0.1910	−1.389529	0.15169
MA_{t-2}	−0.2065	0.0991	−2.083754	0.04607481
MA_{t-3}	−0.3757	0.1038	−3.619461	0.0006903546

Table 9. SPEI-6 ARIMA (1,1,1)(0,0,2) for seasonal.

SPEI−6	Coefficient	Standard Error	t-Value	p-Value
AR_{t-1}	0.6700	0.0895	7.486034	8.854317×10^{-12}
MA_{t-1}	−0.9151	0.0528	−17.33144	1.045039×10^{-39}
MA_{t-2}	−0.2601	0.1105	−2.353846	0.02562977
MA_{t-3}	−0.2932	0.1369	−2.141709	0.04085877

Table 10. SPEI-6 ARIMA (1,1,1) for non-seasonal.

SPEI−6	Coefficient	Standard Error	t-Value	p-Value
AR_{t-1}	0.6939	0.0812	8.545567	1.644041×10^{-14}
MA_{t-1}	−0.9456	0.0428	−22.09346	1.130022×10^{-52}

Table 11. SPEI-9 ARIMA (1,1,1)(1,0,0) for seasonal.

SPEI−9	Coefficient	Standard Error	*t*-Value	*p*-Value
AR_{t-1}	0.8070	0.0717	11.25523	4.54369×10^{-22}
MA_{t-1}	−0.9619	0.0412	−23.34709	6.873699×10^{-56}
MA_{t-2}	−0.1550	0.0825	−1.878788	0.06870885

Table 12. SPEI-12 ARIMA (0,1,0)(1,0,2) for Seasonal.

SPEI−12	Coefficient	Standard Error	*t*-Value	*p*-Value
AR_{t-1}	0.3903	0.4020	0.9708955	0.2483222
MA_{t-1}	−0.9433	0.4071	−2.317121	0.02786761
MA_{t-2}	0.1335	0.2550	0.5235294	0.3471406

Table 13. SPEI-24 ARIMA (0,1,0)(1,0,2) for seasonal.

SPEI−24	Coefficient	Standard Error	*t*-Value	*p*-Value
AR_{t-1}	0.1253	0.1356	0.9240413	0.2596008
MA_{t-1}	−0.0125	0.1671	−0.07480551	0.3972693
MA_{t-2}	−0.7308	0.1208	−6.049669	2.116665×10^{-8}

In the case of (1,1,3)(0,0,0) model estimation, the series is stationary and has no time dependence so the best prediction for this kind of series is the average of the series. In (1,1,3) model, A.R term is 1, difference/order of integration is 1 and moving average is 3, and in the model (0,0,0) whose A.R has 0 lag, difference/order of integration is 0 and moving average has also 0 lag.

In the case of (1,1,3) model estimation, the series is stationary, and prediction for this kind of series is the average of the series, whose A.R. term is 1 lag, difference/order of integration is 1 and moving average is 3 lags. This is the best ARIMA model at SPEI-3.

The (1,1,1)(0,0,2) model estimation, the series is stationary, and prediction for this kind of series is the average of the series, model (1,1,1) whose A.R. term is 1 lag, difference/order of integration is 1 and moving average is 1 lag, and the (0,0,2) model who's A.R has 0, difference/order of integration is 0 and moving average has 2 lags.

In the case of (1,1,1) model estimation, the series is stationary, and prediction for this kind of series is the average of the series, whose A.R. term is 1 lag, difference/order of integration is 1 and moving average is 1 lag. This is the best ARIMA model at SPEI-6.

For the (1,1,1)(1,0,0) model estimation, the series is stationary, and prediction for this kind of series is the average of the series, model (1,1,1) whose A.R. term is 1 lag, difference/order of integration is 1 and moving average is 1 lag, and the (1,0,0) model who's A.R has 1 lag, difference/order of integration is 0 and moving average has 0.

For the (0,1,0)(1,0,2) model estimation, the series is stationary and prediction for this kind of series is the average of the series, model (0,1,0) whose A.R. term is 0, difference/order of integration is 1 and moving average is 0, and the (1,0,2) model who's A.R has 1 lag, difference/order of integration is 0 and moving average has 2 lags.

3.4. *Diagnostic Checking of Residuals*

In order to test the authenticity of the model, diagnostic examination was carried out after the assessment of model parameters. Figure 7 depicts the ACF and PACF of the residuals at various time scales. All the values of the ACF and PACF are found within the limit of 0.01 range for all lags. Thus, no significant association is found between residuals in Figure 8 and the normally distributed histograms of residuals for the SPEI at varying time scales have been represented. This result for the shaped model is sufficient for the

SPEI time series data and residuals to error terms. The greatest accuracy predicting models linked to every examined SPEI time scale with accuracy fit measures (ME, RMSE, MAE, MPE, MAPE, MASE and Theil's U) are shown in Table 14. In general, substantial results have been obtained for drought predicting with the help of ARIMA models in Tharparkar. In short explanation, the ARIMA models that have longer time scales shows profound ability of forecast and fit exactly with drought prediction in upcoming times.

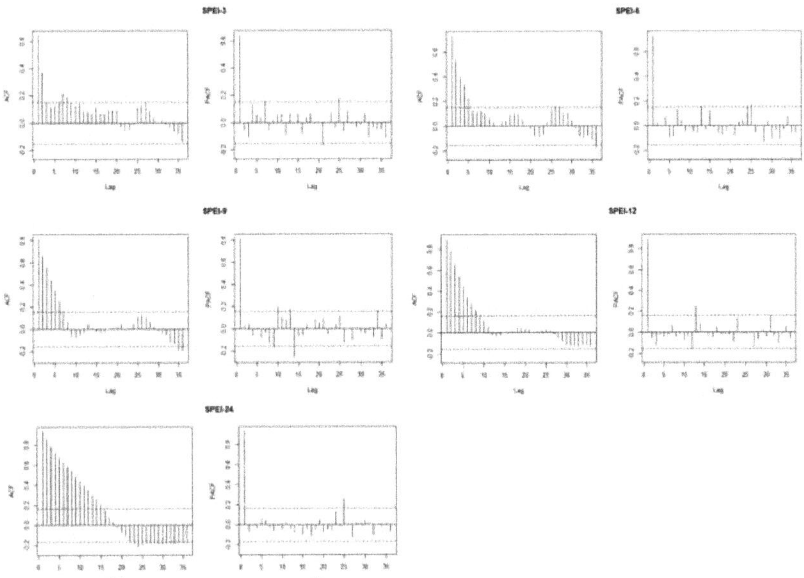

Figure 7. ACF and PACF of SPEI- 3, 6, 9, 12 and 24 from this we can identified the data is stationary.

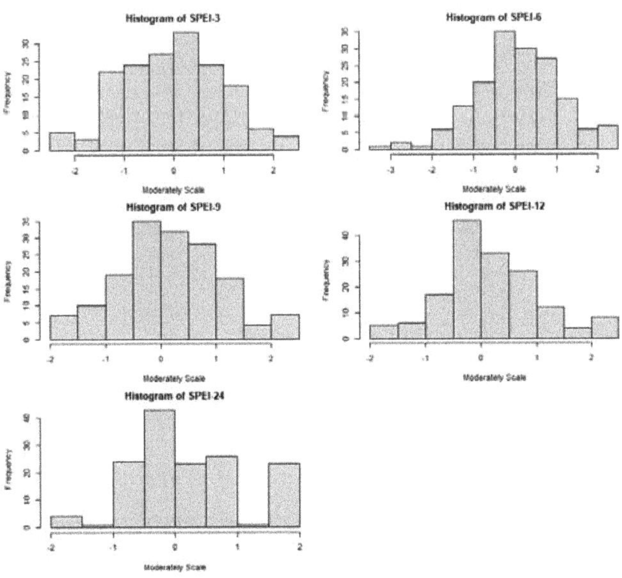

Figure 8. Residuals of SPEI- 3, 6, 9, 12 and 24 are normally distributed.

Table 14. Errors of SPEI-3, 6, 9, 12 and 24.

Accuracy	ME	RMSE	MAE	MPE	MAPE	MASE	Theil's U
SPEI-3 Test set	−0.2274	0.8534	0.7261	120.104	120.104	0.6972	0.9804
SPEI-6 Test set	−0.7831	0.9648	0.8858	117.805	117.805	0.8577	1.048
SPEI-9 Test set	−0.4988	0.8747	0.7941	100.101	100.101	0.7973	1.0798
SPEI-12 Test set	−0.0784	1.009	0.9462	91.841	91.841	0.9632	1.067
SPEI-24 Test set	1.4086	1.4139	1.4086	206.6870	206.687	1.9954	1.067

Almost familiar results have been shown across world and put into the SPI that forms the core of SPEI, i.e., China [54] and India [35]. These research studies have shown that the time series of SPEI for Mithi, Tharparkar has the same nature as of Figure 3. In addition to this, the time series of SPEI-12 and -24 has a similar trend, and likewise for the time series of SPEI-3 and -6, respectively. The identical order seems to be for the time series depending on 12 and 24 months however, the related 3- and 6-month scale do not show this result.

From the Figure 5 for PACF and ACF it is clearly shown that the selected data is stationary.

3.5. Model Forecasting

In fact, forecasting is one of the prime factors in decision making. It bears significant importance for the process about decision making and future planning. It assists in predicting the uncertain future by utilizing the behavior of past and ongoing experiments and observations. The forecasting that is performed using the ARIMA models lays out a sound basis for meteorological phenomena. The forecasting of drought is carried out by selecting the city of Tharparkar and then the data of that location has been utilized to foresee the data series of the SPEI at various time scales, from the 2005 to 2019 period to assess the agreement of data, where the examined and detected SPEI were plotted for its evaluation. Through the prism of comparison within predicted and observed data in Figures 9 and 10, high authenticity of forecasted data is observed. No doubt, with the increase in number of SPEI time series, the forecasting ability of model will be improved. This enhancement in the ability of model is due to rising number of SPEI time series that filters the final values, resulting in the decline of sudden shifts in the curve of SPEI.

The comparison of A.R and M.A coefficients suggests that the ARIMA models of 24-month time scale for Tharparkar is quite accurate. ARIMA models of 3-month time scale are similar in the surrounding regions of Tharparkar. The ARIMA models of 24-month time scale also showed the very accurate results. In Tables 7–13, the estimation of like parameters of developed ARIMA model has been shown. From the outcome in Table 14, the value of p, d, q, P, D, and Q received from the models shaped for Tharparkar are almost alike at the same time scales. Hence, the ARIMA model (0,1,0), (1,0,2) at 24-SPEI could be summarized for the whole region of Tharparkar. In addition to this, the ARIMA model (1,1,3)(0,0,0) at 3-SPEI is also applicable to the neighboring cities of Tharparkar as they are very close to it. In Tables 15–19, the point forecasted values of SPEI-3, 6, 9, 12 and 24 for five years model has been shown.

For the (0,1,0)(1,0,2) model estimation, the series is stationary and prediction for this kind of series is the average of the series, model (0,1,0) whose A.R. term is 0, difference/order of integration is 1 and moving average is 0, and the (1,0,2) model who's A.R has 1 lag, difference/order of integration is 0 and moving average has 2 lags.

3.6. Comparison with Previous Study

The comparative results of present and a previous study has been shown in Table 20. It is investigated that the present standard error and t-value of different SPEI (3, 6 and 24) have significant coherence with previous study [39].

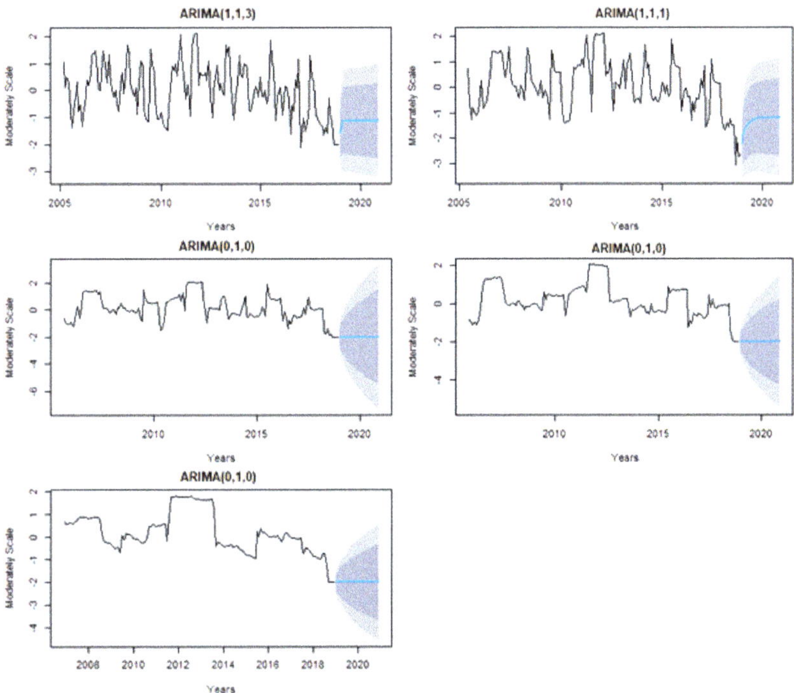

Figure 9. The long-term projection of SPEI for 5 year.

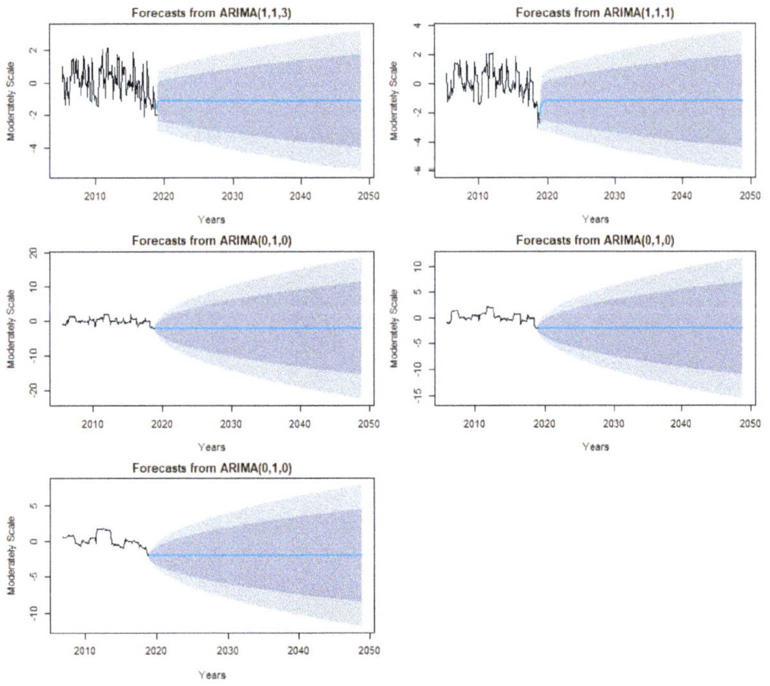

Figure 10. The long-term projection of SPEI for 30 years.

Table 15. Point forecasted value of SPEI-3 for five years.

	Point Forecast											
	January	**February**	**March**	**April**	**May**	**June**	**July**	**August**	**September**	**October**	**November**	**December**
2019	−1.05594	−1.00396	−0.97157	−0.9729	−0.9729	−0.9729	−0.9729	−0.9729	−0.9729	−0.9729	−0.9729	−0.9729
2020	−0.9729	−0.9729	−0.9729	−0.9729	−0.9729	−0.9729	−0.9729	−0.9729	−0.9729	−0.9729	−0.9729	−0.9729
2021	−0.9729	−0.9729	−0.9729	−0.9729	−0.9729	−0.9729	−0.9729	−0.9729	−0.9729	−0.9729	−0.9729	−0.9729
2022	−0.9729	−0.9729	−0.9729	−0.9729	−0.9729	−0.9729	−0.9729	−0.9729	−0.9729	−0.9729	−0.9729	−0.9729
2023	−0.9729	−0.9729	−0.9729	−0.9729	−0.9729	−0.9729	−0.9729	−0.9729	−0.9729	−0.9729	−0.9729	−0.9729

Table 16. Point forecasted value of SPEI-6 for five years.

	Point Forecast											
	January	**February**	**March**	**April**	**May**	**June**	**July**	**August**	**September**	**October**	**November**	**December**
2019	−2.19456	−1.88657	−1.67284	−1.52453	−1.42161	−1.35019	−1.30063	−1.26624	−1.24237	−1.22581	−1.21432	−1.20635
2020	−1.20081	−1.19697	−1.19246	−1.19118	−1.19029	−1.18967	−1.18924	−1.18894	−1.18874	−1.18874	−1.18859	−1.18849
2021	−1.18842	−1.18838	−1.18834	−1.18832	−1.1883	−1.18829	−1.18828	−1.18828	−1.18828	−1.18827	−1.18827	−1.18827
2022	−1.18827	−1.18827	−1.18827	−1.18827	−1.18827	−1.18827	−1.18827	−1.18827	−1.18827	−1.18827	−1.18827	−1.18827
2023	−1.18827	−1.18827	−1.18827	−1.18827	−1.18827	−1.18827	−1.18827	−1.18827	−1.18827	−1.18827	−1.18827	−1.18827

Table 17. Point forecasted value of SPEI-9 for five years.

	Point Forecast											
	January	**February**	**March**	**April**	**May**	**June**	**July**	**August**	**September**	**October**	**November**	**December**
2019	−2	−2	−2	−2	−2	−2	−2	−2	−2	−2	−2	−2
2020	−2	−2	−2	−2	−2	−2	−2	−2	−2	−2	−2	−2
2021	−2	−2	−2	−2	−2	−2	−2	−2	−2	−2	−2	−2
2022	−2	−2	−2	−2	−2	−2	−2	−2	−2	−2	−2	−2
2023	−2	−2	−2	−2	−2	−2	−2	−2	−2	−2	−2	−2

Table 18. Point forecasted value of SPEI-12 for five years.

	Point Forecast											
	January	**February**	**March**	**April**	**May**	**June**	**July**	**August**	**September**	**October**	**November**	**December**
2019	−2	−2	−2	−2	−2	−2	−2	−2	−2	−2	−2	−2
2020	−2	−2	−2	−2	−2	−2	−2	−2	−2	−2	−2	−2
2021	−2	−2	−2	−2	−2	−2	−2	−2	−2	−2	−2	−2
2022	−2	−2	−2	−2	−2	−2	−2	−2	−2	−2	−2	−2
2023	−2	−2	−2	−2	−2	−2	−2	−2	−2	−2	−2	−2

Table 19. Point forecasted value of SPEI-24 for five years.

	Point Forecast											
	January	**February**	**March**	**April**	**May**	**June**	**July**	**August**	**September**	**October**	**November**	**December**
2019	−1.90309	−1.90309	−1.90309	−1.90309	−1.90309	−1.90309	−1.90309	−1.90309	−1.90309	−1.90309	−1.90309	−1.90309
2020	−1.90309	−1.90309	−1.90309	−1.90309	−1.90309	−1.90309	−1.90309	−1.90309	−1.90309	−1.90309	−1.90309	−1.90309
2021	−1.90309	−1.90309	−1.90309	−1.90309	−1.90309	−1.90309	−1.90309	−1.90309	−1.90309	−1.90309	−1.90309	−1.90309
2022	−1.90309	−1.90309	−1.90309	−1.90309	−1.90309	−1.90309	−1.90309	−1.90309	−1.90309	−1.90309	−1.90309	−1.90309
2023	−1.90309	−1.90309	−1.90309	−1.90309	−1.90309	−1.90309	−1.90309	−1.90309	−1.90309	−1.90309	−1.90309	−1.90309

Table 20. A comparison of the present and previous study [39].

SPEI	Standard Error of Previous Study	*t*-Value of Previous Study	Standard Error of Present Study	*t*-Value of Present Study
3 AR	0.105	5.84	0.2060	−0.1902913
3 MA	0.117	−3.27	0.1910	−1.38925
	0.104	−3.46	0.0991	−2.083754
	0.094	2.43	0.1038	−3.619461
6 AR	0.054	7.54	0.0895	7.486034
6 MA	0.048	−15.49	0.0528	−17.3314
	0.054	−12.01	0.1105	−2.353846
	0.052	−9.99	0.1369	−2.141709
	0.043	−11.30		
	0.033	−16.74		
24 AR	0.036	6.98	0.1356	0.9240413
24 MA	0.048	−7.47	0.1671	−0.07480551
	0.034	−14.54	0.1208	−6.049669

4. Conclusions

This research proves SPEI as a unique and powerful multi-scalar drought index for the examination of drought event variations in the region of Tharparkar, Sindh. The prime objective of this research work is categorized into two parts. The first part deals with the evaluation of climatic parameters and drought frequency on the basis of SPEI. This research concluded that the water crisis are a result of overlapping of two unfavorable factors—a decrease in precipitation amounts and an increase of temperature in the region of Tharparkar. Therefore, due to the deficiency of water, the likelihood of drought events has been greatly increased such situation is ultimately generating immense threat to available water resources. In addition to this, the variation in SPEI also shows unusual course of drought (extremely dry) since preceding decade. However, from these results, it is also evident that the situation of hyper-arid regions could be more alarming and eye-opening. It should be noted that the forecast of drought events is one of the most troublesome issues faced be meteorologists.

In this way, the second objective of this research was related to the development and test of ARIMA models for the forecast of drought by utilizing SPEI with 3, 6, 9, 12, and 24-month time scales. The identification of the ARIMA models was conducted on the basis of AIC and SBC values. The basic point for researchers is the credibility of forecasted values. Because the implementation of drought alleviation policies depends upon these forecasted values. In this way, a series of diagnostic checking tests were conducted after the inspection of the parameters of said models. The ARIMA model (0,1,0)(1,0,2) at 24-SPEI could be selected from other possible models for the region of Tharparkar. Additionally, the ARIMA model (1,1,3)(0,0,0) at 3-SPEI, the ARIMA model (1,1,1)(0,0,2) at 6-SPEI, the ARIMA model (1,1,1)(1,0,0) at 9-SPEI and the ARIMA model (0,1,0)(1,0,2) at 12-SPEI can be generalized for Tharparkar region. This is because other localities are very close to the Mithi region. It was also observed that the result obtained through the ARIMA model at the 24-SPEI time scale was the best forecasting model, that follows the lower values of ME, RMSE, MAE, MPE, MAPE and MASE. The ARIMA model at SPEI 3-time scale was found to be the worst model for the prediction of drought for the region of Tharparkar. The best ARIMA models represent profound accuracy in foretelling the droughts, as these can perform a very significant role for planners and water resources managers in measures for such regions as well as in view of drought.

In fact, the connectivity between climate change shown in droughts and the present water resources in Tharparkar is the need of the hour. Thus, in the Tharparkar region, it is very important to overcome the forecasted drought conditions and this should be considered as a significant future study. Additionally, unfortunately the Tharparkar region in Pakistan has only one meteorological station located at Mithi, which is the limitation of our study. Therefore, this study can be extended using different models and a larger set of data in future.

Author Contributions: Conceptualization, P.K. and S.F.S.; Data curation, P.K. and L.K.; Formal analysis, S.F.S., M.A.U., L.K. and R.F.Z.; Methodology, P.K. and R.F.Z.; Project administration, S.F.S. and M.A.U.; Software, P.K., L.K. and R.F.Z. All authors have read and agreed to the published version of the manuscript.

Funding: This research was supported from the Higher Education Commission (HEC), Islamabad, Pakistan under the National Research Program for Universities (NRPU 6872).

Institutional Review Board Statement: Not applicable.

Informed Consent Statement: Not applicable.

Data Availability Statement: Data sharing not applicable.

Acknowledgments: Authors acknowledge the support of the Higher Education Commission (HEC), Islamabad, Pakistan and the Mehran University of Engineering and Technology, Jamshoro, Pakistan. This research was supported from the Higher Education Commission (HEC), Islamabad, Pakistan under the National Research Program for Universities (NRPU-6872).

Conflicts of Interest: The authors declare that they have no known competing financial interest or personal relationships that could have appeared to influence the work reported in this paper.

Nomenclature

ACF	Autocorrelation Function
AIC	Akaike information criterion
ANN	Artificial Neural Network
A.R	Autoregressive
ARIMA	Autoregressive Integrated moving Average
M.A	Moving Average
MAE	Mean Absolute Error
MAPE	Mean Absolute Percentile Error
MASE	Mean Absolute Square Error
ME	Mean Error
MPE	Mean Percentile Error
PACF	Partial Autocorrelation Function
PAI	Palfai aridity index
PCPN	precipitation
PDSI	Palme drought severity index
PET	potential evapotranspiration
RMSE	Root Mean Square Error
SBC	Schwarz's Bayesian criterion
SPEI	Standardized Precipitation Evapotranspiration Index
SPI	Standardized precipitation index
SWSI	Surface water supply index
Symbols	
Δ	slope of vapor pressure
∇^d	order of difference
R_n	surface net radian
G	soil heat flux density
γ	psychometric constant
T	mean daily air temperature

U_2	the wind speed
e_s	saturated vapor pressure
e_a	actual vapor pressure
ω	Mean of the series
η_t	error terms
η_{t-i}	MA error terms
α_i	Parameter of Moving average model
β_i	Parameter of Autoregressive model
Z_t	Time series
Z_{t-i}	AR error terms
$\beta(C)$	Polynomial
$\alpha(C)$	Polynomial
Φ_p	seasonal AR parameter of order P
Θ_Q	seasonal MA parameter of order Q
X_m	observed SPEI
X_s	predicted SPEI
x_i	predicted value
$\hat{x}_i$	actual value
d	number of regular difference for Non-Seasonal Model
p	Autoregressive Parameter for Non-Seasonal Model
q	Moving Average Parameter for Non-Seasonal Model
D	number of regular difference for Seasonal Model
P	Autoregressive Parameter for Seasonal Model
Q	Moving Average Parameter for Seasonal Model

References

1. Boretti, A.; Rosa, L. Reassessing the projections of the World Water Development Report. *NPJ Clean Water* **2019**, *2*, 1–6. [CrossRef]
2. Diffenbaugh, N.S.; Pal, J.S.; Trapp, R.J.; Giorgi, F. Fine-scale processes regulate the response of extreme events to global climate change. *Proc. Natl. Acad. Sci. USA* **2005**, *102*, 15774–15778. [CrossRef] [PubMed]
3. Sheffield, J.; Wood, E. Projected changes in drought occurrence under future global warming from multi-model, multi-scenario, IPCC AR4 simulations. *Clim. Dyn.* **2008**, *31*, 79–105. [CrossRef]
4. Berbel, J.; Esteban, E. Droughts as a catalyst for water policy change. Analysis of Spain, Australia (MDB), and California. *Glob. Environ. Chang.* **2019**, *58*, 101969. [CrossRef]
5. Zandalinas, S.I.; Mittler, R.; Balfagón, D.; Arbona, V.; Gómez-Cadenas, A. Plant adaptations to the combination of drought and high temperatures. *Physiol. Plant.* **2018**, *162*, 2–12. [CrossRef]
6. Madani, K.; AghaKouchak, A.; Mirchi, A. Iran's Socio-economic Drought: Challenges of a Water-Bankrupt Nation. *Iran. Stud.* **2016**, *49*, 997–1016. [CrossRef]
7. Marengo, J.A.; Alves, L.M.; Alvala, R.C.; A Cunha, A.P.M.; Brito, S.; Moraes, O. Climatic characteristics of the 2010-2016 drought in the semiarid Northeast Brazil region. *Anais Acad. Bras. Cienc.* **2018**, *90*, 1973–1985. [CrossRef]
8. Montaseri, M.; Amirataee, B. Comprehensive stochastic assessment of meteorological drought indices. *Int. J. Clim.* **2017**, *37*, 998–1013. [CrossRef]
9. Belayneh, A.; Adamowski, J. Drought forecasting using new machine learning methods. *J. Water Land Dev.* **2013**, *18*, 3–12. [CrossRef]
10. Jesus, E.T.d.; Amorim, J.d.S.; Junqueira, R.; Viola, M.R.; Mello, C.R. *Meteorological and Hydrological Drought from 1987 to 2017 in Doce River Basin, Southeastern Brazil*; RBRH: Porto Alegre, Brazil, 2020; Volume 25.
11. Vangelis, H.; Tigkas, D.; Tsakiris, G. The effect of PET method on Reconnaissance Drought Index (RDI) calculation. *J. Arid. Environ.* **2013**, *88*, 130–140. [CrossRef]
12. Li, J.; Wang, Z.; Wu, X.; Xu, C.-Y.; Guo, S.; Chen, X. Toward Monitoring Short-Term Droughts Using a Novel Daily Scale, Standardized Antecedent Precipitation Evapotranspiration Index. *J. Hydrometeorol.* **2020**, *21*, 891–908. [CrossRef]
13. Kwon, H.-J.; Kim, S.-J. Assessment of distributed hydrological drought based on hydrological unit map using SWSI drought index in South Korea. *KSCE J. Civ. Eng.* **2010**, *14*, 923–929. [CrossRef]
14. Bucur, D.; Corduneanu, F.; Vintu, V.; Balan, I.; Crenganis, L. Impact of drought on water resources in North Eastern Romania. Case study—The prut river. *Environ. Eng. Manag. J.* **2016**, *15*, 1213–1222. [CrossRef]
15. Kalisa, W.; Zhang, J.; Igbawua, T.; Ujoh, F.; Ebohon, O.J.; Namugize, J.N.; Yao, F. Spatio-temporal analysis of drought and return periods over the East African region using Standardized Precipitation Index from 1920 to 2016. *Agric. Water Manag.* **2020**, *237*, 106195. [CrossRef]
16. Bong, C.H.J.; Richard, J. Drought and climate change assessment using Standardized Precipitation Index (SPI) for Sarawak River Basin. *J. Water Clim. Chang.* **2020**, *11*, 956–965. [CrossRef]

17. Hoffmann, D.; Gallant, A.J.E.; Arblaster, J.M. Uncertainties in Drought from Index and Data Selection. *J. Geophys. Res. Atmos.* **2020**, *125*, 031946. [CrossRef]
18. Anshuka, A.; Buzacott, A.J.V.; Vervoort, R.W.; van Ogtrop, F.F. Developing drought index–based forecasts for tropical climates using wavelet neural network: An application in Fiji. *Theor. Appl. Clim.* **2021**, *143*, 557–569. [CrossRef]
19. Vicente-Serrano, S.M.; Beguería, S.; López-Moreno, J.I. A multiscalar drought index sensitive to global warming: The standardized precipitation evapotranspiration index. *J. Clim.* **2010**, *23*, 1696–1718. [CrossRef]
20. Wainwright, H.M.; Steefel, C.; Trutner, S.D.; Henderson, A.N.; Nikolopoulos, E.I.; Wilmer, C.F.; Chadwick, K.D.; Falco, N.; Schaettle, K.B.; Brown, J.B.; et al. Satellite-derived foresummer drought sensitivity of plant productivity in Rocky Mountain headwater catch-ments: Spatial heterogeneity and geological-geomorphological control. *Environ. Res. Lett.* **2020**, *15*, 084018. [CrossRef]
21. Zarei, A.R.; Shabani, A.; Moghimi, M.M. Accuracy Assessment of the SPEI, RDI and SPI Drought Indices in Regions of Iran with Different Climate Conditions. *Pure Appl. Geophys. PAGEOPH* **2021**, *178*, 1387–1403. [CrossRef]
22. Li, L.; She, D.; Zheng, H.; Lin, P.; Yang, Z.-L. Elucidating Diverse Drought Characteristics from Two Meteorological Drought Indices (SPI and SPEI) in China. *J. Hydrometeorol.* **2020**, *21*, 1513–1530. [CrossRef]
23. Bai, X.; Shen, W.; Wu, X.; Wang, P. Applicability of long-term satellite-based precipitation products for drought indices considering global warming. *J. Environ. Manag.* **2020**, *255*, 109846. [CrossRef]
24. Hao, Z.; Singh, V.P.; Xia, Y. Seasonal Drought Prediction: Advances, Challenges, and Future Prospects. *Rev. Geophys.* **2018**, *56*, 108–141. [CrossRef]
25. Rhee, J.; Im, J. Meteorological drought forecasting for ungauged areas based on machine learning: Using long-range climate forecast and remote sensing data. *Agric. For. Meteorol.* **2017**, *237–238*, 105–122. [CrossRef]
26. Papacharalampous, G.; Tyralis, H.; Koutsoyiannis, D. Predictability of monthly temperature and precipitation using automatic time series forecasting methods. *Acta Geophys.* **2018**, *66*, 807–831. [CrossRef]
27. Chen, P.; Niu, A.; Liu, D.; Jiang, W.; Ma, B. Time Series Forecasting of Temperatures using SARIMA: An Example from Nanjing. In *IOP Conference Series: Materials Science and Engineering*; IOP Publishing: England, UK, 2018.
28. Hao, Z.; Hao, F.; Singh, V.P.; Ouyang, W.; Cheng, H. An integrated package for drought monitoring, prediction and analysis to aid drought modeling and assessment. *Environ. Model. Softw.* **2017**, *91*, 199–209. [CrossRef]
29. Ozger, M. Significant wave height forecasting using wavelet fuzzy logic approach. *Ocean Eng.* **2010**, *37*, 1443–1451. [CrossRef]
30. Hassanzadeh, Y.; Ghazvinian, M.; Abdi, A.; Baharvand, S.; Jozaghi, A. Prediction of short and long-term droughts using artificial neural networks and hydro-meteorological variables. *arXiv* **2020**, arXiv:2006.02581.
31. Ozger, M.; Mishra, A.K.; Singh, V.P. Estimating Palmer Drought Severity Index using a wavelet fuzzy logic model based on mete-orological variables. *Int. J. Climatol.* **2011**, *31*, 2021–2032. [CrossRef]
32. Akinremi, O.O.; McGinn, S.M.; Barr, A. Evaluation of the Palmer Drought Index on the Canadian Prairies. *J. Clim.* **1996**, *9*, 897–905. [CrossRef]
33. McKee, T.B.; Doesken, N.J.; Kleist, J. The relationship of drought frequency and duration to time scales. In Proceedings of the 8th Conference on Applied Climatology, Boston, MA, USA, 17–22 January 1993.
34. Tian, M.; Wang, P.; Khan, J. Drought forecasting with vegetation temperature condition index using ARIMA models in the Guan-zhong Plain. *Remote Sens.* **2016**, *8*, 690. [CrossRef]
35. Han, P.; Wang, P.; Tian, M.; Zhang, S.; Liu, J.; Zhu, D. Application of the ARIMA Models in Drought Forecasting Using the Standardized Precipitation Index. In Proceedings of the International Conference on Computer and Computing Technologies in Agriculture, Beijing, China, 18–20 September 2013.
36. Shahzad, L.; Yasin, A. Analyzing Resilience and Food Insecurity of Drought Prone Communities of Tharparkar Desert. Available online: https://www.researchsquare.com/article/rs-472894/v1 (accessed on 21 September 2021).
37. Memon, M.H.; Aamir, N.; Ahmed, N. Climate Change and Drought: Impact of Food Insecurity on Gender Based Vulnerability in District Tharparkar. *Pak. Dev. Rev.* **2018**, *57*, 307–321. [CrossRef]
38. Pakistan Meteorological Department. PMD. Karachi. 1947. Available online: https://www.pmd.gov.pk/en/ (accessed on 21 September 2021).
39. Mossad, A.; Alazba, A.A. Drought Forecasting Using Stochastic Models in a Hyper-Arid Climate. *Atmosphere* **2015**, *6*, 410–430. [CrossRef]
40. Dikshit, A.; Pradhan, B.; Huete, A. An improved SPEI drought forecasting approach using the long short-term memory neural network. *J. Environ. Manag.* **2021**, *283*, 111979. [CrossRef]
41. Agnew, C.T. *Using the SPI to Identify Drought*; University College London: London, UK, 2000.
42. Valipour, M.; Banihabib, M.E.; Behbahani, S.M.R. Parameters estimate of autoregressive moving average and autoregressive in-tegrated moving average models and compare their ability for inflow forecasting. *J. Math. Stat.* **2012**, *8*, 330–338.
43. Yuan, C.; Liu, S.; Fang, Z. Comparison of China's primary energy consumption forecasting by using ARIMA (the autoregressive integrated moving average) model and GM (1, 1) model. *Energy* **2016**, *100*, 384–390. [CrossRef]
44. Peiris, M.S.; Singh, N. Predictors for Seasonal and Nonseasonal Fractionally Integrated ARIMA Models. *Biom. J.* **1996**, *38*, 741–752. [CrossRef]
45. Chang, Y.-W.; Liao, M.-Y. A Seasonal ARIMA Model of Tourism Forecasting: The Case of Taiwan. *Asia Pac. J. Tour. Res.* **2010**, *15*, 215–221. [CrossRef]

46. Akaike, H. A new look at the statistical model identification. *IEEE Trans. Autom. Control.* **1974**, *19*, 716–723. [CrossRef]
47. Mishra, A.K.; Singh, V.P. A review of drought concepts. *J. Hydrol.* **2010**, *391*, 202–216. [CrossRef]
48. Makridakis, S.; Hibon, M. ARMA Models and the Box-Jenkins Methodology. *J. Forecast.* **1997**, *16*, 147–163. [CrossRef]
49. Afrifa-Yamoah, E.; Mueller, U.; Taylor, S.M.; Fisher, A.J. Missing data imputation of high-resolution temporal climate time series data. *Meteorol. Appl.* **2019**, *27*, 1873. [CrossRef]
50. RC Team. R: A Language and Environment for Statistical Computing. R Foundation for Statistical Computing. Available online: http://www.R-project.org/ (accessed on 15 July 2021).
51. Mossad, A.; Mehawed, H.S.; El-Araby, A. Seasonal drought dynamics in El-Beheira governorate, Egypt. *Am. J. Environ. Sci.* **2014**, *10*, 140–147. [CrossRef]
52. Sönmez, F.K.; Komuscu, A.U.; Erkan, A.; Turgu, E. An analysis of spatial and temporal dimension of drought vulnerability in Turkey using the standardized precipita-tion index. *Nat. Hazards* **2005**, *35*, 243–264. [CrossRef]
53. Paulo, A.; Rosa, R.D.; Pereira, L.S. Climate trends and behaviour of drought indices based on precipitation and evapotranspiration in Portugal. *Nat. Hazards Earth Syst. Sci.* **2012**, *12*, 1481–1491. [CrossRef]
54. Yu, M.; Li, Q.; Hayes, M.J.; Svoboda, M.; Heim, R.R., Jr. Are droughts becoming more frequent or severe in China based on the standardized precipitation evapotranspiration in-dex: 1951–2010? *Int. J. Climatol.* **2014**, *34*, 545–558. [CrossRef]

MDPI

Article

Drought Trends in the Polish Carpathian Mts. in the Years 1991–2020

Anita Bokwa *, Mariusz Klimek, Paweł Krzaklewski and Wojciech Kukułka

Faculty of Geography and Geology, Institute of Geography and Spatial Management, Jagiellonian University, 7 Gronostajowa St., 30-387 Kraków, Poland; mariusz.klimek@uj.edu.pl (M.K.); pawel.krzaklewski@uj.edu.pl (P.K.); wojciech.kukulka@uj.edu.pl (W.K.)
* Correspondence: anita.bokwa@uj.edu.pl; Tel.: +48-12-664-53-27

Abstract: Mountains are highly sensitive to the effects of climate change, including extreme short- and long-term weather phenomena. Therefore, in spite of relatively high annual precipitation totals, mountains might become endangered by droughts. The paper presents drought trends in the Polish Carpathians located in Central Europe. Data from the period 1991–2020 from 12 meteorological stations located in various vertical climate zones of the mountains were used to define drought conditions using the following indices: Standardized Precipitation (SPI), Standardized Precipitation Evapotranspiration (SPEI), Relative Precipitation (RPI) and Sielianinov. Additionally, four forest drought indices were used in order to estimate the impact of drought on beech as a typical Carpathian tree species, i.e., the Ellenberg (EQ), Forestry Aridity (FAI), Mayr Tetratherm (MT) and De Martonne Aridity (AI) indices. Statistically significant but weak trends were obtained for the 6-month SPI for four stations (indicating an increase in seasonal to mid-term precipitation), for the 1-month SPEI for three stations, for the 3-month SPEI for four stations, and for MT for all stations (indicating an increase in drought intensity). The analysis of dry month frequency according to particular indices shows that at most of the stations during the last decade of the study period, the frequency of dry months was much higher than in previous decades, especially in the cold half-year. Two zones of the Polish Carpathians are the most prone to drought occurrence: the peak zone due to the shift in climatic vertical zones triggered by the air temperature increase, and the forelands and foothills, together with basins located about 200–400 m a.s.l., where the mean annual air temperature is the highest in all the vertical profile, the annual sums of precipitation are very diversified, and the conditions for beech are already unfavorable.

Keywords: atmospheric drought; forest drought; Carpathian Mts.; beech; vertical climate zones

Citation: Bokwa, A.; Klimek, M.; Krzaklewski, P.; Kukułka, W. Drought Trends in the Polish Carpathian Mts. in the Years 1991–2020. *Atmosphere* **2021**, *12*, 1259. https://doi.org/10.3390/atmos12101259

Academic Editors: Alexander V. Chernokulsky, Andrzej Walega and Agnieszka Ziernicka-Wojtaszek

Received: 28 July 2021
Accepted: 20 September 2021
Published: 27 September 2021

1. Introduction

Drought is a phenomenon that negatively affects many economic sectors. Depending on the duration, effects and intensity, drought can be classified into four types: meteorological, agricultural, hydrological and socio-economic [1]. According to IPCC [2], mountains are among the areas most endangered by climate change, and droughts have been observed and are predicted in various mountain ranges, for example, in the Alps [3]. The pan-European study showed that in the 1990s and 2000s, drought hotspots were identified in the Mediterranean and Carpathian Regions; in the latter, drought severity and duration were highest in Hungary and Slovakia [4]. In the period 1961–2010, the worst droughts occurred in 1990, 2000, and 2003; less intense or prolonged droughts took place in 1964, 1970, 1973/74, 1983, 1987, 1992, and 2007 [5]. The Carpathians are located in Central Europe where the future drought risk is estimated to be relatively high with projected increases in hydrological, agricultural and ecological droughts at mid-century warming levels of 2 °C or above, regardless of greenhouse gas emission scenarios [6]. The Carpathian region includes the Carpathian Mts. and the Pannonian Basin. Studies concerning droughts in the whole region were based on gridded data (e.g., [5,7,8]) and showed that the region's lowlands are much more endangered than the mountains. National-level studies confirmed the results for the lowlands and revealed the complexity of the issue in the mountains. In the lowlands of Hungary, Serbia, Romania and Slovakia, droughts have already caused significant agricultural yield losses in recent decades ([9–11]). Studies for the Carpathian Mts. are more limited and encompass only a few mountain ranges. For the part of the Western Carpathians located in Slovakia, in the period 1951–2007 precipitation exceeded potential evapotranspiration [9,12]. Those results have been confirmed for the upper Hron region [13]. In a study for Romania, with data for the period 1961–2010, it was shown that drought may affect areas with both low and high precipitation averages, and can occur in mountains or lowlands. Large-scale atmospheric circulation is the major drought driver in Romania in winter, while thermodynamic factors (such as air temperature and humidity levels) are the major drivers in summer. The Carpathian mountain chain itself is the second regional factor influencing drought spatial variability, triggering differences between intra-Carpathian and extra-Carpathian regions in wintertime [14]. The Tatra Mts are the highest range in the Carpathians, located at the border between Poland and Slovakia. For the Slovakian part, data from the period 1961–2010 were used to show that the occurrence of drought has a cyclic pattern over an approximately 30-year period. The core areas of the biosphere reserve of the Tatra National Park inhabited by unique species (altitudes over 1500 m a.s.l.) are in the relatively "drought-safe altitudinal zone". Unfortunately, ecosystems at lower altitudes (up to 900 m a.s.l.) could be impacted by drought due to a lower precipitation surplus. The occurrence of drought episodes was influenced by the precipitation shadow of the Tatra Mts range and of the surrounding mountains situated to their north and northwest. Thus the occurrence of drought is more likely in the south and southeastern regions of the mountains than on the windward north or northeastern parts. In addition, another drought-prone area was indicated in the Western Tatra Mts. This area is influenced by the Oravsk'e Beskydy and Oravsk'a Magura ranges located to the northwest [15]. A study concerning the Polish part of the Tatra Mts was compared with a study for the Ukrainian Carpathians for the period 1984–2015, concerning the occurrence of dry months. At least one extremely dry month at each meteorological station was detected, but only in November 2011 was an extreme drought at all the stations observed. That was a month with precipitation less than 5% of the long-term average at specific meteorological stations [16]. In Poland, atmospheric drought is observed most often from April to September [17], and it affects mainly the lowlands where agriculture is concentrated [18,19]. The Carpathians receive much more precipitation than the rest of the country (with the exception of the Sudeten Mts where precipitation is comparable), but long periods without precipitation have been observed more frequently during the last few decades both on the mountain foreland [20] and on the mountain ranges [21,22]. Such a tendency combined with increased runoff and decreased retention (due to human activity) is a huge disadvantage for water resource management, and for the functioning of ecosystems. Additionally, an increase in hydrological drought risk has been observed in the Carpathians over the period 1901–2002 [23]. Agricultural drought has been studied for Poland for the period 1961–2010, but mountain areas were excluded from that research. However, the study showed that foreland areas were relatively little endangered by agricultural drought [24].

In spite of receiving the highest annual precipitation totals, in comparison to the rest of the country, the Polish Carpathians are affected by current climate change too, and drought has to be considered one of the potential new threats to the region. Therefore, this paper is aimed to show whether atmospheric drought risk varies in the vertical climatic profile of the Polish Carpathians and whether the eastern part of the mountain chain is more endangered by drought than the western part due to more continental climatic conditions. This aspect of the present climate has not yet been studied for that part of the Carpathians in contrast to some other parts. The present paper shows variability and trends in atmospheric drought occurrence in the most recent 30-year period. The middle of the 1980s represents a turning point for all the climatic variables in the Carpathian region [25] and shows the beginnings of presently observed climate change. The data from the period 1991–2020 represent the current long-term climatic period characterized by ecosystems entering a new state of dynamic balance. The Polish Carpathians are not an important agricultural region in the country; the main economic sector developed there is tourism and the properties of the natural environment are one of the most important factors in its development. The Carpathians, apart from offering picturesque landscapes, are a European biodiversity hotspot, with rich flora and endemic plants, and including the most extensive primeval forests across the whole of Europe; there are many different bird species and it is home to the largest communities of carnivores and predators such as bears and wolves [5,26]. Beech is the main species of the *Dentario glandulosae-Fagetum* community, a typical element of the Carpathian environment. Therefore, forest drought indices have also been used in the present study to estimate trends in atmospheric conditions favorable for beech. This is one of the indicators of the long-term impact of the drought trend on the natural environment of the Carpathians.

2. Study Area

The Polish Carpathian Mts. are part of the huge Carpathian mountain chain in Central Europe. It is divided into the Southern Carpathians (located in Romania), the Eastern Carpathians (Ukraine, Slovakia, Hungary, Poland), and the Western Carpathians (Poland, Slovakia, Czech Republic, Hungary, Austria). The relief of the Carpathians varies from undulating foothills to typical alpine landscapes in the Tatra, Rodna, Fagaras, and Retezat mountains [27]. The highest peaks can be found in the Tatra Mts, in Slovakia: Gerlach, 2655 m a.s.l.; and in the Fagaras Mts, in Romania: Moldoveanu, 2543 m a.s.l. As much as 88% of the Carpathian area located in Poland belongs to the Western Carpathians [28]. The climatic and hydrological conditions of the Polish Carpathians have been the subject of many studies, for example, [29–35], but rarely in the context of atmospheric drought as it is the region with the highest precipitation on a national scale. Most of the climatic parameters show increasing climate continentality from the west toward the east; for example, the mean annual air temperature range increases by 0.49 °C per degree of longitude on convex landforms, and by 0.35 °C on concave ones [31]. Even more important are changes in climatic conditions with altitude which are visible as vertical climate-vegetation zones [29]. The location of zonal boundaries (i.e., altitudes where a certain mean annual air temperature is found) depends on the scale of a particular mountain range, slopes aspect, the main geomorphological features and the prevailing direction of air mass advection [36]. According to the original vertical zone pattern [29], the study area contains six zones from "cold" with a mean annual air temperature from −4 to −2 °C, to "moderately warm" with a mean temperature from 6 to 8 °C. However, one of the effects of global warming is a shift in zonal boundaries [33] and this will be discussed further in the present study. The vertical climate-vegetation zone that occupies the largest area in the Polish Carpathians is that of deciduous forest, with its dominating *Dentario glandulosae-Fagetum* plant community [37]. European beech (*Fagus sylvatica*) is the main species of that community. The deciduous forest zone is located in the mountain foothills and on medium-height mountain ranges which are areas of intensive anthropopressure due to tourist activity. Therefore, a potential impact of drought on beech forest conditions will be presented later in the paper.

3. Materials and Methods

The data used in the present study come from 12 meteorological stations located in the Polish Carpathians (Figure 1 and Table 1). The stations represent the highest parts of the Carpathians, that is, the peaks of the Tatra Mts (Kasprowy Wierch) and the neighboring basin (Zakopane), the Beskidy ranges which are medium-height mountains (Limanowa, Nowy Sącz, Krynica, Lesko and Komańcza), the foothills (Gaik-Brzezowa, Łazy and Bielsko-Biała), and the foreland (Kraków and Katowice). The stations in Gaik-Brzezowa and Łazy belong to the Institute of Geography and Spatial Management, Jagiellonian University, Kraków, while the others are administered by the Institute of Meteorology and Water Management—National Research Institute. The spatial distribution of the stations allows drought occurrence to be studied in the Polish Carpathians both vertically and from west to east. %clearpage

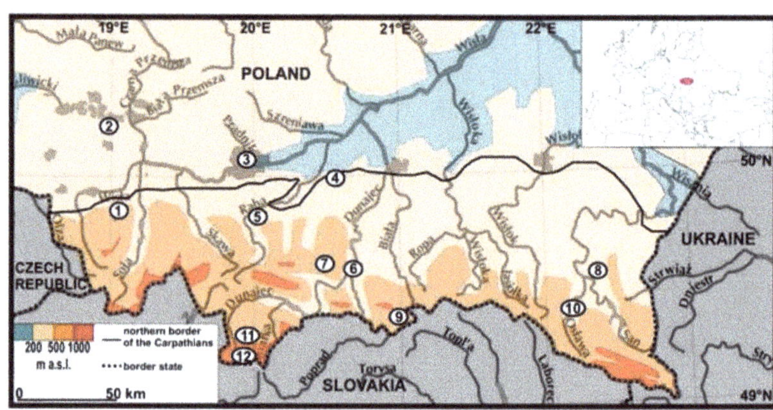

Figure 1. Location of the meteorological stations used in the study numbers as in Table 1.

Table 1. Coordinates and altitude of the meteorological stations used in the study numbers as in Figure 1. TPZ concept is explained in Section 4.1.

No.	Station Name	Latitude (φ)	Longitude (λ)	Altitude m a.s.l	TPZ
1	Bielsko-Biała	49°48′26″ N	19°00′01″ E	396	4
2	Katowice	50°14′26″ N	19°01′58″ E	278	4
3	Kraków	50°04′40″ N	19°47′42″ E	237	4
4	Łazy	49°57′55″ N	20°29′43″ E	245	4
5	Gaik- Brzezowa	49°52′00″ N	20°05′00″ E	303	4
6	Nowy Sącz	49°37′38″ N	20°41′19″ E	292	4
7	Limanowa	49°41′37″ N	20°25′06″ E	515	3
8	Lesko	49°27′59″ N	22°20′30″ E	420	3
9	Krynica-Zdrój	49°24′28″ N	20°57′39″ E	582	2
10	Komańcza	49°20′21″ N	22°03′48″ E	478	2
11	Zakopane	49°17′38″ N	19°57′37″ E	855	2
12	Kasprowy Wierch	49°13′57″ N	19°58′55″ E	1991	1

The data used come from the 30-year period 1991–2020 (while for Gaik-Brzezowa the data cover that of 1991–2019) and consist of mean monthly air temperatures and monthly precipitation totals. The choice of study period is linked to data availability and to the fact that since the 1990s, a significant change in climatic conditions has begun on both global and regional scales [38]. Therefore, the analyses presented show the contemporary situation and trends which are the effect of the present climate drivers. The data were used first to determine the variability of air temperature and precipitation in the study period and to distinguish areas with different air temperature and precipitation patterns.

Then the data were used to calculate indices widely used to determine the occurrence of drought (see also Appendix A):

1. SPI (Standardized Precipitation Index) [39]: it is based on the probability of precipitation which is the only input parameter. SPI was calculated for 1-, 3- and 6- monthly timescales for each station: SPI ≤ -2.0—extreme drought, $-1.99 <$ SPI ≤ -1.50—strong drought, $-1.49 <$ SPI ≤ -1.00—moderate drought, $-0.99 <$ SPI < 0.99 near normal conditions, $1.0 <$ SPI < 1.49 moderately wet, $1.5 <$ SPI < 1.99 very wet.

2. RPI (Relative Precipitation Index) is the ratio of precipitation sum for the given period and the long-term average for the same period expressed in percent [40]. It was calculated for each month and station; the values for particular months were interpreted following the intensity scale: <25%—extremely dry, 25–50%—very dry, 51–75%—dry, 76–125%—normal, 126–150%—wet, 151–200%—very wet, >200%—extremely wet.
3. Selianinov index describes humidity conditions in relation to the needs of the environment; the volume of precipitation is determined together with the potential for its use by plants, depending on the thermal conditions [41]. According to its formula (Appendix A), it was calculated for those months only when mean daily air temperature exceeded 10 °C; then the months were classified as follows: <0.4—extremely dry, 0.4–0.7—very dry, 0.8–1.0—dry, 1.1–1.3—moderately dry, 1.4–1.6—optimal, 1.7–2.0—moderately wet, 2.1–2.5—wet, 2.6–3.0—very wet, >3.0—extremely wet.
4. SPEI (Standardized Precipitation Evapotranspiration Index) is a standardized monthly climatic balance computed as the difference between the cumulative precipitation and the potential evapotranspiration [42]. It was calculated for 1-, 3- and 6-each monthly timescales and for each station. The following drought classes were applied: ≥2—exceptionally wet, 1.6–1.99—extremely wet, 1.3–1.59—severely wet, 0.8–1.29—moderately wet, 0.5–0.79—slightly wet, 0.49 to −0.49—normal, −0.5 to −0.79—slightly dry, −0.8 to −1.29—moderate drought, −1.3 to −1.59—severe drought, −1.6 to −1.99—extreme drought, ≤−2—exceptional drought.

In order to estimate the impact of drought on beech, the following indices of forest drought were used and calculated for each year and station:

1. EQ (Ellenberg index) [43]; the values optimal for beech are <30 while at EQ > 40, beech cannot survive.
2. FAI (Forestry Aridity Index) [44]; the values optimal for beech are at FAI < 4.75.
3. MT (Mayr Tetratherm Index) [45]; the values optimal for beech are 13–18 °C [46]
4. AI (De Martonne Aridy Index) [47]; the values optimal for beech are 35–40 [46].

The SPI index was calculated with SPI Generator software [48], SPEI was calculated with the Package 'SPEI' software (https://cran.r-project.org/web/packages/SPEI/SPEI.pdf, accessed on 20 August 2021) and other indices were calculated with MS Excel, with the formulas listed in Appendix A and provided in the publications mentioned above.

In the calculation of SPI and RPI, only precipitation data are taken into consideration, while the Selianinov index and SPEI include also air temperature data. The forest drought indices are not only based on data for air temperature and precipitation but are calculated at various temporal resolutions, which allows different aspects of the phenomenon to be observed.

The series of air temperature, precipitation and index values were tested using regression analysis; linear trends were determined together with their equations, R^2 and p level with Statistica software (https://www.statsoft.pl/, accessed on 1 August 2021). Then the data series were additionally tested for the presence of trends by the Mann–Kendall test [49–51] using XLSTAT software (https://www.xlstat.com/en/, accessed on 5 August 2021). That software was also used to calculate standard errors for mean values. For each data series, the variability coefficient was calculated with MS Excel following the formula:

$$Vc = (\sigma/m) \times 100 \quad (1)$$

where:
Vc—variability coefficient (in %)
σ—standard deviation
m—mean value

The values of the coefficient calculated were interpreted in the following way:

<25%—low variability, 25–45%—mean variability, 46–100%—high variability, >100%—very high variability.

The drought occurrence estimation was based on a comparison of the SPI, SPEI, RPI and Selianinov index outcomes, that is, the number of dry months defined as SPI ≤ -1.00, SPEI ≤ -0.8, RPI $\leq$ 75%, Selianinov index < 1.4. The frequency of dry months was presented for specific decades of the study period, for the whole year, the warm half-year (May–October) and the cold half-year (November–April). The forest drought indices can be calculated with only an annual resolution so the number of years with conditions unfavorable for beech in specific decades has been shown.

4. Results

All the indices used in the study are based on data concerning air temperature, precipitation or both. Therefore, the spatial and temporal variability of air temperature and precipitation is presented first in order to provide background for the analysis of drought indices.

4.1. Air Temperature and Precipitation Variability

For all stations included in the study, an increase in mean annual air temperature has been observed; the regression analysis showed that it was statistically significant at $p < 0.05$. The rate of increase varied from 0.5 to 0.7 °C per 10 years (R^2 from 0.34 to 0.47). The Mann-Kendall test confirmed statistically significant positive trends for all stations, and Sen's slope values confirmed the rate of increase described above (Appendix B). The variability coefficient for all series is <25%, which means a relatively low variability of mean annual air temperature in the study area. Figure 2 shows that the mean decadal air temperature has been gradually increasing at all stations, too. The most striking increase is observed for Kasprowy Wierch, one of the highest peaks of the Tatra Mts, where mean annual air temperature has exceeded 0 °C which means a shift from the "moderately cold" vertical climatic zone (i.e., from mean annual air temperature from −2 to 0 °C) to "very cool" (0 to 2 °C). Such a shift can also be observed for Zakopane (from "moderately cool" to "moderately warm"). Lesko and Limanowa shifted from "moderately warm" to a mean annual air temperatures above 8 °C, not included in the original scheme described in [29].

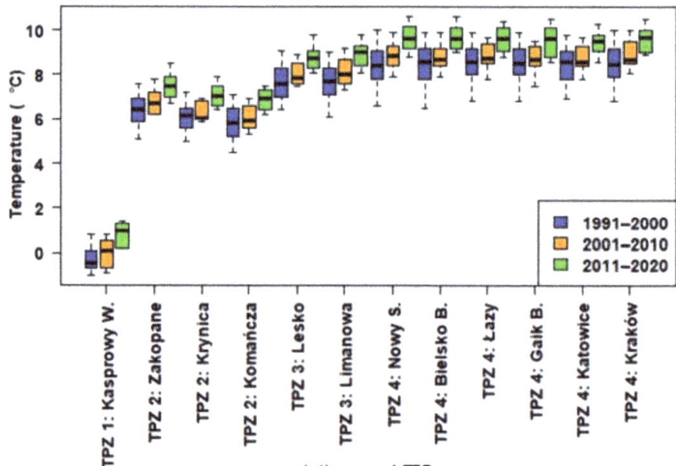

Figure 2. Mean annual air temperature (°C, black horizontal marks inside the boxes) in specific decades at the stations studied. Boxes mark the first and the third quartile, and the whiskers show the highest and the lowest value in a certain decade. Standard errors for the mean values are provided in Appendix C. Stations are ordered following the concept of TPZ explained in Section 4.1.

In the case of precipitation, there are no statistically significant changes for annual totals (according to regression analysis and the Mann–Kendall test; see Appendix B). The comparison of mean annual totals in specific decades confirms this fact (Figure 3); the highest values were noted in the second decade. The values of the variability coefficient for annual totals are below 25% which means low variability. However, the values for particular months reveal that for Kasprowy Wierch, Bielsko-Biała, Limanowa and Kraków, for all months the coefficient values are >45%, which means high variability. For Zakopane, Katowice, Łazy, Nowy Sącz, Gaik-Brezowa and Krynica, from 1 to 3 months show mean variability (25–45%) but for all other months, the coefficient exceeds 45%. For Lesko and Komańcza, 4–5 months show mean variability while all other months have high variability. Additionally, there is no clear dependency between precipitation and altitude (except Kasprowy Wierch) or precipitation and longitude, and this is linked to the strong local impacts of landforms in the mountains on spatial patterns of precipitation.

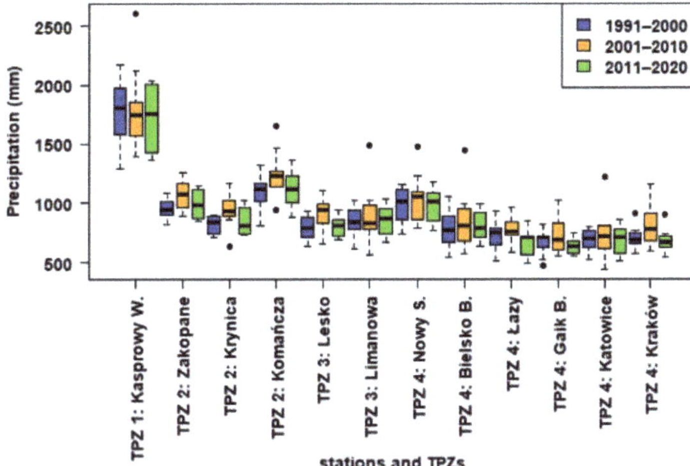

Figure 3. Mean annual total of precipitation (mm, black horizontal marks inside the boxes) in specific decades at the stations studied. Boxes mark the first and the third quartile, and the whiskers show the highest and the lowest value in a certain decade, except the outliers which are shown with black dots. Standard errors for the mean values are provided in Appendix C. Stations are ordered following the concept of TPZ explained in Section 4.1.

Both air temperature and precipitation are key factors controlling drought occurrence and Figure 4 shows their combination for the stations included in the study. The stations can be then assigned to the following temperature-precipitation zones (TPZ):

1. TPZ 1: The highest parts of the Carpathians, located above 1500 m a.s.l., represented by Kasprowy Wierch, where mean annual air temperature is lowest (around 0 °C), and mean annual precipitation is highest (about 1800 mm); therefore, potential drought risk is the lowest.
2. TPZ 2: The Carpathian basins and large valleys located at about 500–900 m a.s.l., represented by Zakopane, Krynica and Komańcza, where annual precipitation exceeds 800 mm, and mean annual air temperature is about 6–7 °C, which allows potential drought risk to be considered as relatively low.
3. TPZ 3: The Carpathian basins and large valleys located at about 400–500 m a.s.l., represented by Lesko and Limanowa, with a mean annual air temperature of 8.2 °C and precipitation above 800 mm. Here the potential drought risk is medium.
4. TPZ 4: The Carpathian foreland and foothills, together with basins located about 200–400 m a.s.l., represented by Bielsko-Biała, Gaik-Brzezowa, Nowy Sącz, Łazy,

Kraków and Katowice, where the mean annual air temperature is the highest in the vertical profile (8.9–9.0 °C), and annual precipitation totals are very diversified, from 670 to 1000 mm, so the potential drought risk is high.

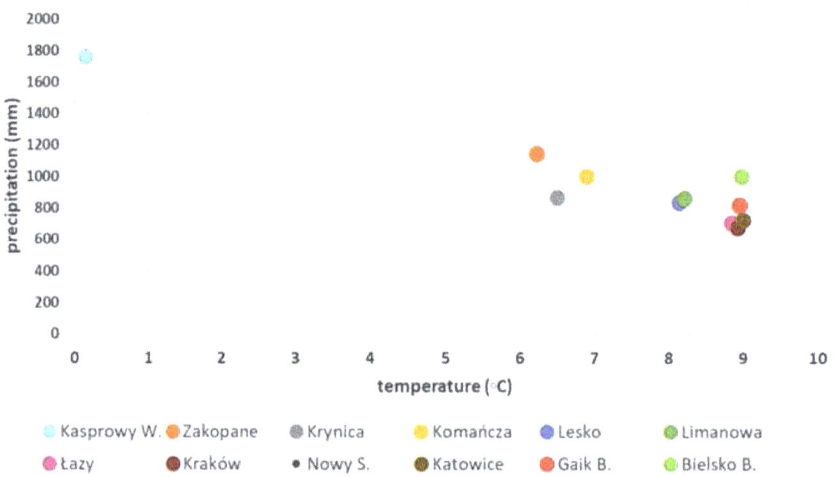

Figure 4. Comparison of mean annual air temperature and precipitation totals for the period 1991–2020 for the stations studied. The value for Nowy Sącz is not visible in the figure as it is almost identical to the value for Katowice and the symbols overlap each other.

4.2. Drought Frequency and Trends in the Polish Carpathians

Drought occurrence was determined by SPI, SPEI, RPI and the Selianinov index. For SPI and SPEI, the percentages of dry months (i.e., with SPI $\leq$ -1.00, SPEI $\leq$ −0.8) for the whole year and for the subperiods May–October and November–April were calculated. All data series show very high variability, that is, the values of Vc exceed 100%.

SPI values for the 1-month time scale vary from 3.74 to −4.28, for the 3-month scale from 3.81 to −3.13, and for the 6-month scale from 3.62 to −2.78. None of the SPI 1- and 3-monthly series shows any statistically significant trend; the results of the Mann–Kendall test are presented in Appendix B. In the case of SPI 6-monthly series for Zakopane, Krynica, Komańcza (TPZ 2) and Gaik-Brzezowa (TPZ 4), the p-values indicate statistically significant increasing trends, but Sen's slope values are as low as 0.001, and low tau values indicate that the trends are weak (Appendix B). A 1-month SPI reflects short-term conditions, related closely to meteorological drought along with short-term soil moisture and crop stress, while a 3-month SPI reflects short- and medium-term moisture conditions and provides a seasonal estimate of precipitation, and a 6-month SPI indicates seasonal to medium-term trends in precipitation [39]. Figure 5 presents the data for the decades and it shows that in a short-term perspective, a clear increase in the frequency of dry months in the cold subperiod can be seen in the last decade in comparison to the previous ones (except Gaik-Brzezowa; Figure 5c). This is also the reason for the increase of dry-month frequency at an annual scale (Figure 5a). The increase is observed throughout the whole study area. In the decade 2011–2020, the frequency of dry months according to SPI reached about 15% on an annual scale at all stations. For medium-term and seasonal perspectives (Figure 5, data for 3- and 6-monthly timescales), there are no clear spatial or temporal patterns.

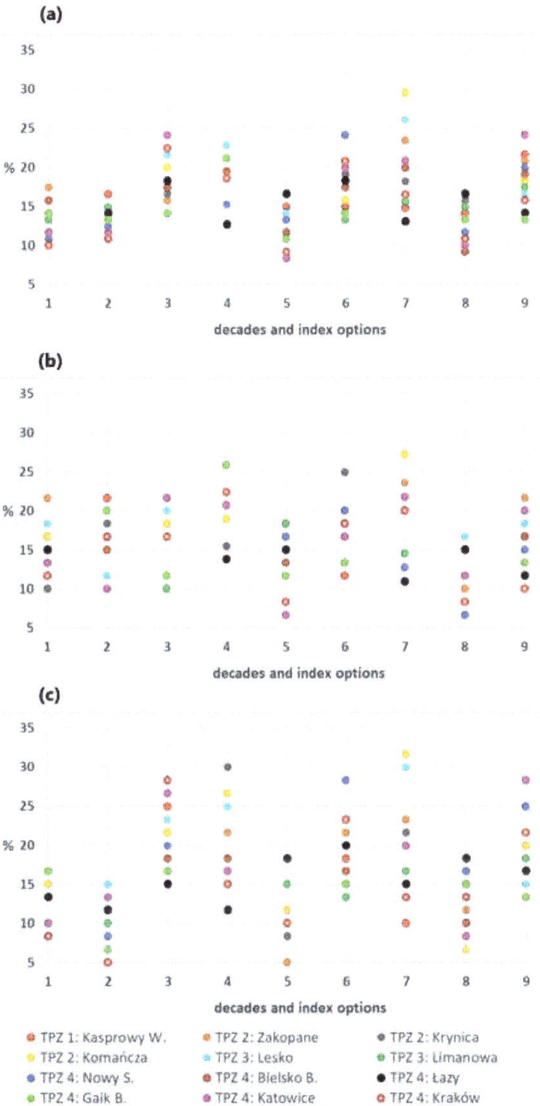

Figure 5. Percentages of dry months according to SPI (SPI ≤ -1.0) for 1-, 3- and 6-monthly timescales, for the whole year (**a**) and for the subperiods May–October (**b**) and November–April (**c**) in the decades of the study period. Stations are ordered following the concept of TPZ explained in Section 4.1. Explanation of numbers on axis x: 1—1991–2000, 1 month; 2—2001–2010, 1 month; 3—2011–2020, 1 month; 4—1991–2000, 3 months; 5—2001–2010, 3 months; 6—2011–2020, 3 months; 7—1991–2000, 6 months; 8—2001–2010, 6 months; 9—2011–2020, 6 months.

For annual values of RPI, the Mann–Kendall test by definition gives the same results as for annual precipitation totals (Appendix B). For the test's results for particular months and particular stations, all p-values exceeded 0.05, so none of the series shows any statistically significant trend. Tau values varied from -0.264 to 0.209. Figure 6 shows that in the decade 2011–2020, for most of the stations, a large increase in the number of dry months per year can be seen which is mainly the effect of the increase in the frequency of such months in the cold half-year. In the last decade of the study period, according to RPI, from 35 to 45% of months per year were dry.

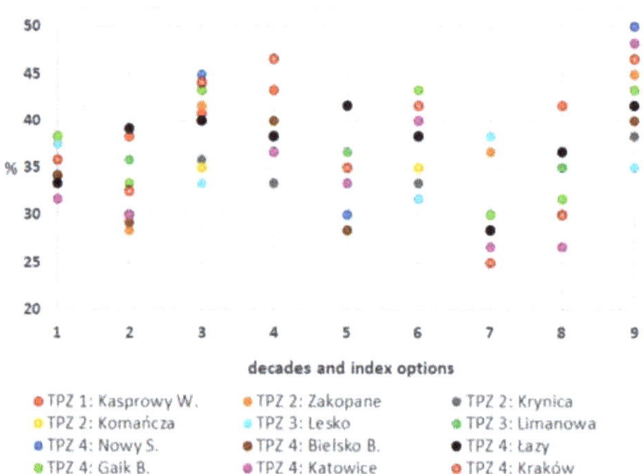

Figure 6. Percentage of dry months according to RPI (RPI $\leq$ 75%-for the whole year) and for the subperiods May-October and November-April in the decades of the study period. Stations are ordered following the concept of TPZ explained in Section 4.1. Explanation of numbers on axis x: 1—1991–2000, year; 2—2001–2010, year; 3—2011–2020, year; 4—1991–2000, May–Oct; 5—2001–2010, May–Oct; 6—2011–2020, May–Oct; 7—1991–2000, Nov–Apr; 8—2001–2010, Nov–Apr; 9—2011–2020, Nov–Apr.

The Selianinov index could be calculated for all stations (except Kasprowy Wierch) and for each year only for June, July and August; for other months the values could be calculated only sporadically. The Mann–Kendall test for those three months showed no statistically significant trend at any station. All *p*-values exceeded 0.05, tau values varied from −0.159 to 0.062. However, the Selianinov index, unlike SPI and RPI, showed a large spatial variability of drought occurrence in the warm part of the year (Figure 7), with a clear increase in drought risk with decrease in altitude (i.e., from less than 10% of dry months at Kasprowy Wierch (TPZ 1) to over 50% in Kraków, TPZ 4). Additionally, the data show that for most stations, the share of dry months is greater in the last decade than in the two previous ones. An increase is especially visible in the highest parts of the Carpathians. Until 2015, the Selianinov index could be calculated for Kasprowy Wierch only once over several years, and only for July, while later it could be calculated also for June and August, as the index is calculated only for the months when the mean daily air temperature exceeds 10 °C. There is no significant difference in the W-E profile concerning drought risk, but the data for Bielsko-Biała are worth attention as the risk is much lower than in other stations of similar locations. Bielsko-Biała and Katowice (TPZ 4) are the westernmost points of the study area, and both of them are exposed to moist oceanic air masses coming from the west; however, Bielsko-Biała is located in the Carpathian foothills, that is, close to an orographic barrier which enhances precipitation. Figure 3 shows that Bielsko-Biała has higher precipitation sums than neighboring stations.

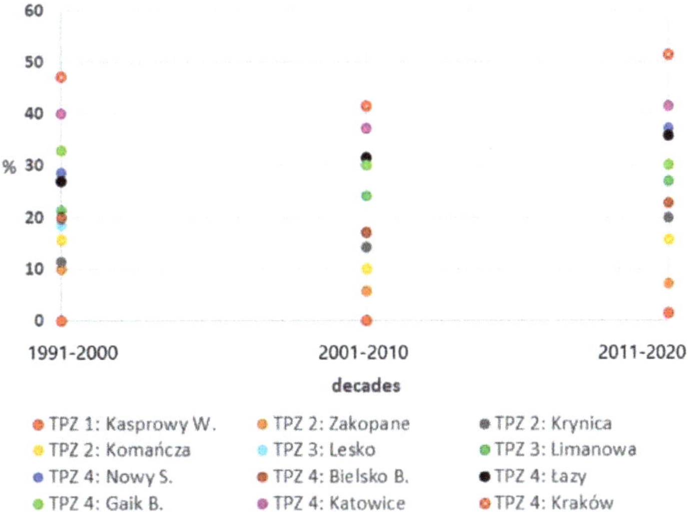

Figure 7. Percentage of dry months according to the Selianinov index for the subperiod April–October in the decades of the study period. Stations are ordered following the concept of TPZ explained in Section 4.1.

SPEI values for the 1-monthly time scale vary from 2.54 to −3.15, for the 3-monthly scale from 2.60 to −2.83, and for the 6-monthly scale from 2.72 to −3.44. For the 1-monthly series, statistically significant trends were found with the Mann–Kendall test for Katowice, Nowy Sącz and Kraków (TPZ 4), and for 3-monthly and 6-monthly series for Kasprowy Wierch (TPZ 1), Katowice, Nowy Sącz and Kraków (TPZ 4) (Appendix D). The trends indicate an increase in drought risk although Sen's slope values are as low as 0.001–0.002 which indicates that the trends are weak. However, most tau values are much higher than for SPI (Appendix B), which shows that those trends, although weak, should be considered important signals of the increasing drought risk at least in some areas of the Polish Carpathians, mainly in foreland areas. SPEI presents combined effects of precipitation and air temperature and concerning the results presented above, it is the increasing temperature that is mainly contributing to those trends. Figure 8 presents the percentage of the dry months (SPEI ≤ -0.8) for the decades and it shows that in the last decade of the study period, there were much more dry months observed at most of the stations than previously. Such tendency is more clear for the cold half-year than for the warm one, especially for the 1-monthly time scale.

4.3. *Drought Risk for Beech in the Polish Carpathians*

The forest drought indices used in the study are based on air temperature and precipitation (EQ, AI and FAI) or on air temperature only (MT). They were calculated for all stations except Kasprowy Wierch, as that station is located far above the tree line. For EQ, FAI and AI, no statistically significant trend in the study period was found for any station, neither by regression analysis nor with the Mann–Kendall test (Appendix B). The indices show high spatial and temporal variability (Figure 9). The Carpathian foreland and foothills, together with basins located about 200–400 m a.s.l., are the areas where, in each decade, there are years with forest drought conditions, and at some stations (i.e., in Katowice, Kraków, Nowy Sącz—TPZ 4) an increasing tendency can be observed. In the remaining part of the study area, drought conditions do not occur at all or they occur only sporadically.

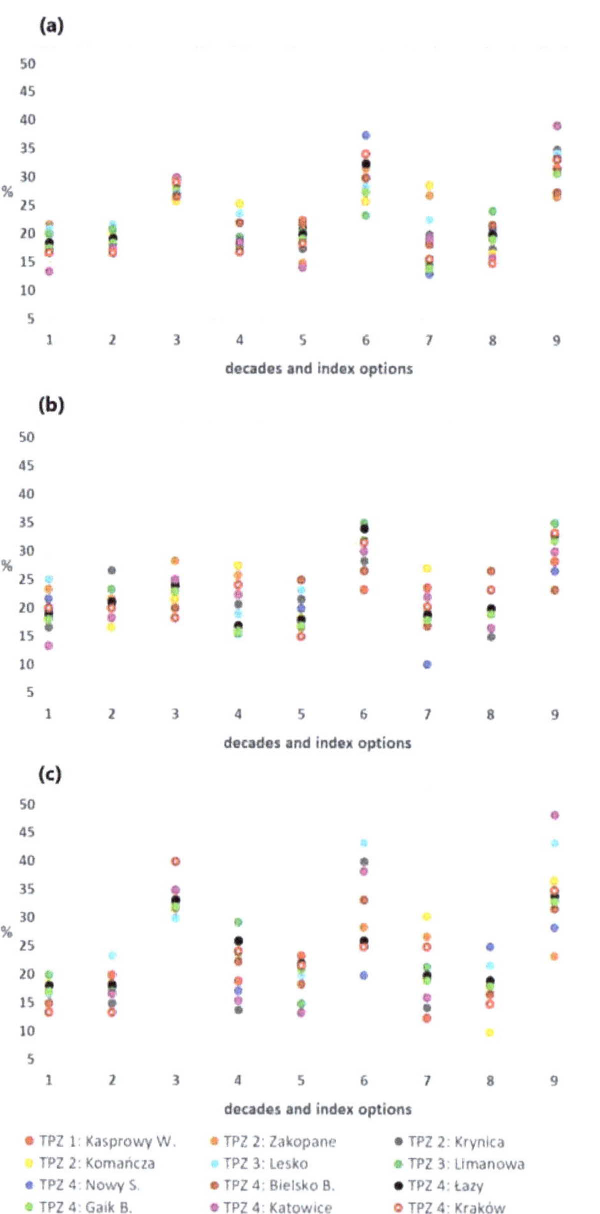

Figure 8. Percentage of dry months according to SPEI (SPEI ≤ -0.8) for 1-, 3- and 6-monthly timescales, for the whole year (**a**) and for the subperiods May–October (**b**) and November–April (**c**) in the decades of the study period. Stations are ordered following the concept of TPZ explained in Section 4.1. Explanation of numbers on axis x: 1—1991–2000, 1 month; 2—2001–2010, 1 month; 3—2011–2020, 1 month; 4—1991–2000, 3 months; 5—2001–2010, 3 months; 6—2011–2020, 3 months; 7—1991–2000, 6 months; 8—2001–2010, 6 months; 9—2011–2020 6 months.

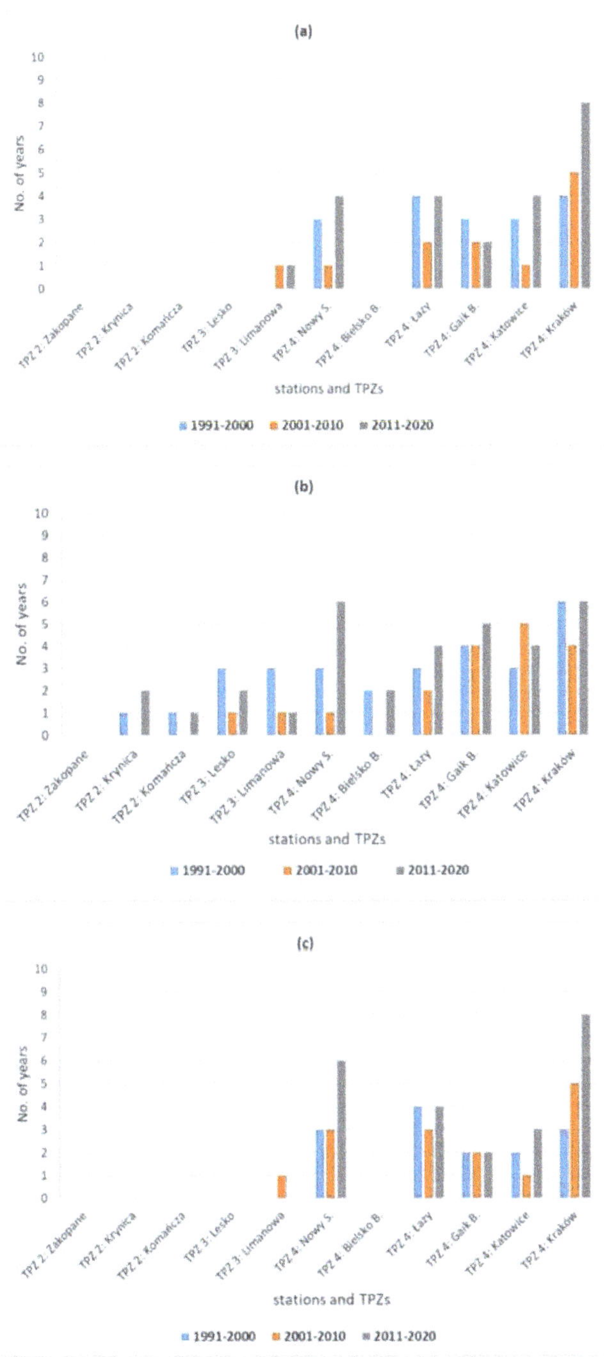

Figure 9. The number of years in specific decades when forest drought conditions unfavorable for beech occurred at the stations studied: (**a**) EQ > 30; (**b**) FAI > 4.75; (**c**) AI < 35. Stations are ordered following the concept of TPZ explained in Section 4.1.

Unlike the indices shown in Figure 9, MT has a statistically significant increasing trend at all stations ($p < 0.05$) at the rate from 0.4 to 0.5 per decade; only for Limanowa did the value reach 0.6 per 10 years. The Mann–Kendall test confirms those results (Appendix B). The threshold value of 18 °C, above which the conditions for beech are estimated to be unfavorable, was crossed for the first time in 2002 at all stations in the zone of the Carpathian foreland and foothills, together with basins located about 200–400 m a.s.l. (TPZ 4), except for Bielsko-Biała. In the second decade, the value was crossed in three years in the zone mentioned, and in the third decade for five years. In 2018, in Katowice and Kraków, the index value reached 19 °C (Figure 10).

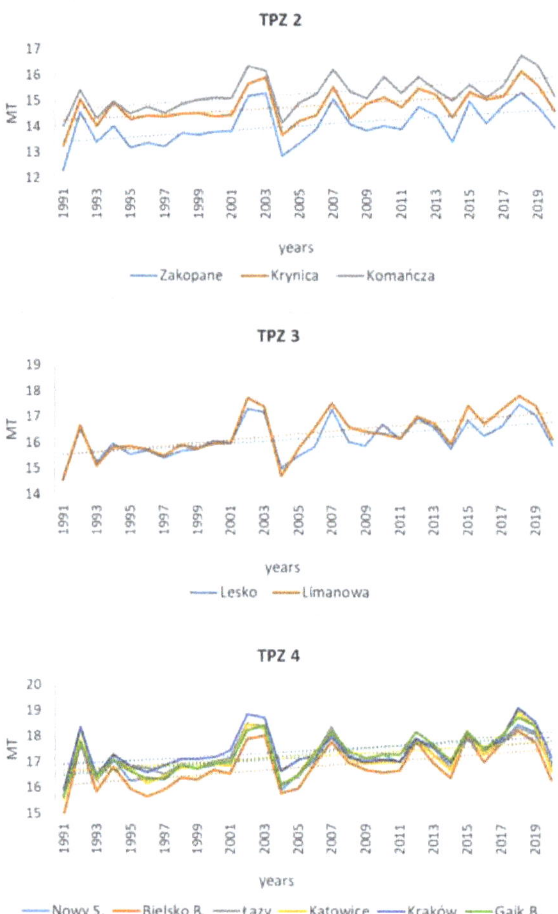

Figure 10. Values of the Mayr Tetratherm Index for particular TPZs and stations in the period 1991–2020, together with linear trends. See Appendix D for regression equations and the values of R^2. Stations are ordered following the concept of TPZ explained in Section 4.1.

5. Discussion

The analyses presented above, based on the application of various indices, show that atmospheric drought risk has increased in the study area, and the main reason is increasing air temperature. It is recommended to study the drought issue with a variety of methods as in many studies, solely precipitation-based indices show only minor changes in drought occurrence, whereas other indices that consider evapotranspiration indicate a significant increase in the area under drought [52]. The results presented for air temperature and precipitation series confirm earlier findings for the Polish Carpathians that there has been significant warming in the area, particularly over recent decades. Climate change is most evident in the foothills; however, it is the highest summits that have experienced the most intensive increases in temperature during the recent period. Precipitation does not demonstrate any substantial trend and has high year-to-year variability. The distribution of annual temperature provides evidence of the upward shift of vertical climate zones in the Polish Carpathians, reaching approximately 350 m, on average, which indicates further ecological consequences [53]. The absence of statistically significant trends in precipitation is in accordance with the results for Czechia, Slovakia and Austria [54] where the main driver of drought is an increase in the evaporative demand of the atmosphere, driven by higher temperatures and global radiation with limited changes to precipitation totals. However, the observed drying trends were most pronounced there during the April–September period and at lower elevations. Conversely, the majority of stations above 1000 m exhibited a significant wetting trend for both the summer and winter (October–March) half-years. The part of the Carpathians included in that study is located on the southern side of the Western Carpathian chain, while the Polish Carpathians, considered in the present paper, are located on the northern side. Therefore, the climatic factors, for example, atmospheric circulation impacts are different for these regions and this is then visible in the values for climatic elements. Conditions for areas above 1000 m a.s.l. can be estimated on the northern side with the data from Kasprowy Wierch only, and they do not show a wetting trend but an increase in drought risk. For stations located at lower altitudes, according to indices based on precipitation only, drought frequency was highest in the warm half-year in the first decade only, while in the last decade it was most frequent in the cold half-year. However, the Selianinov index, based on both air temperature and precipitation, and calculated for the warm half-year, showed that the last decade experienced drought much more often than previous decades, which is in accordance with the results presented in [54]. SPEI is also based on both air temperature and precipitation, but it has been calculated for the whole year, and as shown in Figure 8, the frequency of dry months increased a lot in the last decade all over the study area. The increase was larger in the cold half-year than in the warm one; at some stations, it was more than double in comparison to previous decades. In the study for the whole Carpathian region [5], there are no differences among the mountains and lowlands shown but the general trend confirms an increase in drought frequency. For the Alpine region [55], drought impacts were studied and turned out to be most pronounced in the warm half-year, while the high-altitude region showed this effect the most. The Polish Carpathians can also be compared to the remainder of Poland. Areas classified as dry increased their surface area from 13% in the period 1931–1960 to 20% in the 30-year period of 1971–2000, and 46% in the 30-year period of 1981–2010. However, the northern and western regions of Poland are more endangered by drought than the mountains where the precipitation is always higher [56].

It can be concluded that the present paper shows the frequency and trends of drought in the Polish Carpathians in the most recent 30 years, that is, in the period marked by global warming. The results obtained are in accordance with the outcomes of other studies concerning drought in Central Europe and in the Central European mountains, but they also show some new aspects which have not been analyzed so far. There is no drought frequency variability in the W-E profile. The stations representing the easternmost part of the Polish Carpathians, Komańcza and Lesko, belong to two different TPZs (2 and 3, respectively), and the indices values obtained show that drought risk in that region is similar to that in the remaining part of the study area. However, there is a high impact of local environmental conditions on spatial patterns of precipitation. Unlike the indices based on precipitation only, the index based on both precipitation and air temperature has shown a clear increase in drought risk with decrease in altitude. Additionally, for the highest parts of the mountains represented by Kasprowy Wierch, there are clear indications of increasing drought risk, so both TPZ 1 and TPZ 4 should be considered most endangered with the increase of drought risk due to ongoing air temperature increase. In the Polish Carpathians, agriculture is not the dominating sector, due to more unfavorable environmental conditions than in the lowland part of the country. Much more important is tourism for which the state of the natural environment, including forests, is one of the key factors. Analyses of the indices describing the conditions for beech forest have shown that the zone of the Carpathian foreland and foothills, together with basins located about 200–400 m a.s.l., is the part of the Polish Carpathians where conditions are already unfavorable and are worsening most rapidly. According to [57], Carpathian ecosystems located in water-limited environments of lowland to foothill areas can be particularly exposed to climate change, and the Tatras are climate change hotspots. As the vertical climate-vegetation zones are shifting due to constant warming, it can be expected that the deterioration of climatic conditions for beech will appear at higher altitudes. Other studies show that in the Carpathian Basin, beech has already reached its xeric limit on many sites [58]. Beech total yield production in the Western Carpathians was recently found to be lower by −11% on average compared to beech forests in Central Europe (Germany) [59]. The extraordinary drought and heat in the summers of 2018 and 2019 have demonstrated the climatic vulnerability of European beech in many parts of its Central European distribution range. At its southern and south-eastern range edges, beech is most likely limited by summer drought and probably also by heat [60]. In the high-mountain zone of the Carpathians, populations of cold-adapted species are very vulnerable to climate change, while their habitats tend to shrink. The climate-driven decrease of snow cover often leads to frost damage to vegetation that provides gaps appropriate for the establishment of many rare species [61]. Most probably, many species of conservation concern will irreversibly disappear from the regional flora under the ongoing climate change [62]. The increase of drought risk in the highest parts of the Polish Western Carpathians, shown in the present paper, is another factor that can contribute to those negative processes. In Europe, a strong spatial pattern in the beech growth responses to summer temperature and to drought was found; radial growth of the species generally did not respond to summer drought in Central Europe (Germany, Slovakia and Romania), but it became highly responsive in the Balkan Peninsula (Bosnia and Herzegovina). However, beech shows a wide variety of growth patterns driven by several factors, and beech growth has been declining over the last two decades [63]. The results shown in the present paper suggest that the drought risk has been increasing during the last 30 years in the Polish Western Carpathians, especially in the foothills zone, so we can expect the beech forests growing there to be effects and even decline in the next decades.

It should be mentioned that the trends from observed data might, for two main reasons, not be suitable for extrapolation into the future [64,65]. First, they could be related to climate variability and not too persistent changes over time. Second, an investigated trend depends on the observation period, so it could differ if the observation period was extended.

Author Contributions: Conceptualization, A.B., M.K., P.K. and W.K.; methodology, A.B., M.K. and P.K.; software, not applicable; validation, M.K., P.K. and W.K.; formal analysis, A.B., M.K. and P.K.; investigation, A.B., M.K. and P.K.; resources, M.K., P.K. and W.K.; data curation, M.K., P.K.; writing—original draft preparation, A.B., M.K. and P.K.; writing—review and editing, A.B., M.K. and P.K.; visualization, A.B., M.K., P.K.; supervision, A.B.; project administration, not applicable; funding acquisition, not applicable. All authors have read and agreed to the published version of the manuscript.

Funding: This research received no external funding.

Institutional Review Board Statement: Not applicable.

Informed Consent Statement: Not applicable.

Data Availability Statement: Not applicable.

Acknowledgments: The authors wish to thank the reviewers for their valuable comments and suggestions, and Piotr Sekuła, for the assistance in SPEI calculations and preparation of Figures 2 and 3.

Conflicts of Interest: The authors declare no conflict of interest.

Appendix A

Formulas for calculation of the indices used

RPI:

$$RPI = (P/P_{mean})*100\%$$

where:
P—precipitation sum in a certain period (e.g., a month; in mm);
P_{mean}—mean multi-annual precipitation sum for the same period (e.g., a certain month; in mm).

Selianinov index:

$$K = 10P/\Sigma t,$$

where:
P—monthly precipitation sum
Σt—sum of mean daily air temperatures; the index is calculated only for the months when mean daily air temperature exceeds 10 °C

EQ:

$$EQ = TW/P \times 1000$$

where:
TW—temperature of the warmest month in a year
P—annual precipitation total

FAI:

$$FAI = 100 \times {}_{(Jul\text{-}Aug)}/(P_{(May\text{-}Jul)} + P_{(Jul\text{-}Aug)}$$

where:
$T_{(Jul–Aug)}$—mean air temperature in July and August;
$P_{(May–Jul)}$—precipitation total from May to July;
$P_{(Jul–Aug)}$—precipitation total from July to August.

MT:

$$MT = (T_{May} + T_{Jun} + T_{Jul} + T_{Aug})/4$$

where:
T_{May}, T_{Jun} etc.—mean monthly air temperature for May, June etc.

AI:

$$AI = P/(TA + 10)$$

where:
P—annual precipitation total
TA—mean annual air temperature

Appendix B

Table A1. Values of test statistics S (upper value) and tau (middle/bottom value) of the Mann–Kendall test for the series of mean annual air temperature, annual sums of precipitation, 1-, 3- and 6-monthly SPI, and annual values of AI, FAI, EQ and MT. For mean annual air temperature, the bottom value presents Sen's slope value (provided only if the p-value calculated was lower than 0.05).

Station	Mean Annual Air Temper.	Annual Total of Precip.	SPI 1 Month	SPI 3 Months	SPI 6 Months	AI	FAI	EQ	MT
Kasprowy W.	199 0.457 0.056	−19 −0.044	−1865 −0.029	−2855 −0.045	−2731 −0.044	na	na	na	na
Zakopane	201 0.462 0.059	69 0.159	1818 0.028	3862 0.061	4739 0.076	−15 −0.034	3 0.007	−7 −0.016	173 0.398 0.048
Krynica	205 0.471 0.055	81 0.186	1824 0.028	4197 0.066	5081 0.081	23 0.053	−9 −0.021	−33 −0.076	155 0.356 0.043
Komańcza	213 0.489 0.061	51 0.117	2183 0.034	4266 0.067	5088 0.081	−21 −0.048	−33 −0.076	−3 −0.007	195 0.448 0.045
Lesko	221 0.508 0.062	35 0.080	2632 0.041	3518 0.055	3382 0.054	−31 −0.071	−25 −0.057	9 0.021	175 0.402 0.048
Limanowa	221 0.508 0.067	37 0.085	1076 0.017	3335 0.052	4306 0.069	−13 −0.030	−21 −0.048	1 0.002	201 0.462 0.063
Nowy S.	187 0.429 0.059	−11 −0.025	−2015 −0.031	−1992 −0.031	−1925 −0.031	−45 −0.103	17 0.039	37 0.085	177 0.407 0.053
Bielsko B.	203 0.466 0.063	35 0.080	414 0.006	2524 0.040	3079 0.049	−1 −0.002	5 0.011	11 0.025	159 0.366 0.057
Łazy	185 0.425 0.057	33 0.076	878 0.014	3253 0.051	3255 0.052	−9 −0.021	−3 −0.007	−5 −0.011	162 0.375 0.041
Gaik B.	190 0.467 0.065	48 0.118	2227 0.037	4150 0.070	5110 0.087	6 0.015	−4 −0.010	−10 −0.025	196 0.483 0.063
Katowice	190 0.436 0.055	−13 −0.030	−1584 −0.025	−1408 −0.022	−3117 −0.050	−59 −0.136	−1 −0.002	29 0.067	143 0.329 0.042
Kraków	170 0.390 0.052	−5 −0.011	−1857 −0.029	−1972 −0.031	−2947 −0.047	−57 −0.131	1 0.002	35 0.080	143 0.329 0.039

Explanation: na: non-applicable.

Appendix C

Table A2. Standard error for the mean annual air temperatures and precipitation totals shown in Figures 2 and 3.

Decade	Kasprowy W.	Zakopane	Krynica	Komańcza	Lesko	Limanowa	Nowy S.	Bielsko B.	Łazy	Gaik B.	Katowice	Kraków
						Air temperature						
1991–2000	0.186	0.253	0.213	0.234	0.256	0.259	0.295	0.310	0.267	0.308	0.276	0.283
2001–2010	0.188	0.180	0.141	0.181	0.165	0.214	0.194	0.199	0.205	0.204	0.189	0.189
2011–2020	0.165	0.160	0.175	0.197	0.186	0.189	0.180	0.188	0.175	0.196	0.193	0.210
						Precipitation						
1991–2000	89,065	44,822	25,336	26,587	31,309	40,597	30,344	50,045	28,757	47,517	37,713	33,755
2001–2010	112,533	67,436	43,544	36,754	42,832	77,618	53,026	66,311	65,414	78,120	35,040	48,688
2011–2020	83,909	47,339	36,619	36,037	27,844	43,163	32,619	46,110	35,948	42,072	34,658	20,368

Appendix D

Table A3. Values of test statistics S (upper value), tau (middle value) and Sen's slope (bottom value) of the Mann-Kendall test for those series of SPEI for which *p*-value calculated was <0.05.

Station	1-Month SPEI	3-Month SPEI	6-Month SPEI
Kasprowy W.	na	−4498 −0.070 −0.001	−4639 −0.074 −0.001
Katowice	−4754 −0.074 −0.001	−6267 −0.098 −0.002	−8855 −0.141 −0.002
Nowy S.	−4861 −0.075 −0.001	−6567 −0.103 −0.002	−7649 −0.122 −0.002
Kraków	−4796 −0.074 −0.001	−6767 −0.106 −0.002	−9261 −0.147 −0.002

Explanation: na: non-applicable.

Appendix E

Table A4. Linear regressin equations and the values of R^2 for the series of MT.

Station	Equation	R^2
Zakopane	y = 0.0443x + 13.376	0.2593
Krynica	y = 0.0394x + 14.199	0.2718
Komańcza	y = 0.045x + 14.604	0.3366
Lesko	y = 0.0435x + 15.504	0.2792
Limanowa	y = 0.0575x + 15.473	0.3495
Nowy Sącz	y = 0.0447x + 16.504	0.2747
Bielsko-Biała	Y = 0.0492x + 16.067	0.2572
Łazy	y = 0.0377x + 16.655	0.2285
Gaik-Brzezowa	y = 0.0603x + 16.408	0.4182
Katowice	y = 0.0406 + 16.604	0.1912
Kraków	y = 0.037x + 16.882	0.1868

References

1. Mishra, A.K.; Singh, V.P. A review of drought concepts. *J. Hydrol.* **2010**, *391*, 202–216. [CrossRef]
2. IPCC. *Climate Change 2014: Synthesis Report. Contribution of Working Groups I, II and III to the Fifth Assessment Report of the Intergovernmental Panel on Climate Change*; IPCC: Geneva, Switzerland, 2014; p. 151. Available online: https://www.ipcc.ch/report/ar5/syr/ (accessed on 5 August 2021).
3. Haslinger, K.; Schöner, W.; Anders, I. Future drought probabilities in the Greater Alpine Region based on COSMO-CLM experiments-spatial patterns and driving forces. *Meteor. Zeit.* **2015**, *25*, 137–148. [CrossRef]
4. Spinoni, J.; Naumann, G.; Vogt, J.; Barbosa, P. *Meteorological Droughts in Europe: Events and Impacts—Past Trends and Future Projections*; Publications Office of the European Union: Luxembourg, 2016. [CrossRef]
5. Spinoni, J.; Antofie, T.; Barbosa, P.; Bihari, Z.; Lakatos, M.; Szalai, S.; Szentimey, T.; Vogt, J. An overview of drought events in the Carpathian Region in 1961–2010. *Adv. Sci. Res.* **2013**, *10*, 21–32. [CrossRef]
6. IPCC. Sixth Assessment Report, Working Group I—The Physical Science Basis, Regional Fact Sheet—Europe. 2021. Available online: https://www.ipcc.ch/report/ar6/wg1/#Regional (accessed on 5 August 2021).
7. János Vágó-András Hegedűs-Endre Dobos. Analysis of drought conditions in the Carpathians using standardized precipitation index (SPI) 1951–2050. *Geosci. Eng.* **2013**, *2*, 66–80.
8. Antofie, T.; Naumann, G.; Spinoni, J.; Vogt, J. Estimating the water needed to end the drought or reduce the drought severity in the Carpathian region. *Hydrol. Earth Syst. Sci.* **2015**, *19*, 177–193. [CrossRef]
9. Škvarenina, J.; Tomlain, J.; Hrvoľ, J.; Škvareninová, J.; Nejedlík, P. Progress in dryness and wetness parameters in altitudinal vegetation stages of West Carpathians: Time-series analysis 1951–2007. *IDŐJÁRÁS Q. J. Hung. Meteorol. Serv.* **2009**, *113*, 47–54.
10. Mezősi, G.; Blanka, V.; Ladányi, Z.; Bata, T.; Urdea, P.; Frank, A.; Meyer, B.C. Expected mid- and long-term changes in drought hazard for the south-eastern Carpathian basin. *Carpathian J. Earth Environ. Sci.* **2016**, *11*, 355–366.
11. Crocetti, L.; Forkel, M.; Fischer, M.; Jurečka, F.; Grl, A.; Salentinig, A.; Trnka, M.; Anderson, M.; Ng, W.-T.; Kokalj, Z.; et al. Earth Observation for agricultural drought monitoring in the Pannonian Basin (southeastern Europe): Current state and future directions. *Reg. Environ. Chang.* **2020**, *20*, 123. [CrossRef]
12. Skvarenina, J.; Tomlain, J.; Hrvol', J.; Skvareninova, J. Occurrence of Dry and Wet Periods in Altitudinal Vegetation Stages of West Carpathians in Slovakia: Time-Series Analysis 1951–2005. In *Bioclimatology and Natural Hazards*; Strelcová, K., Matyas, C., Kleidon, A., Lapin, M., Matejka, F., Blazenec, M., Skvarenina, J., Holecy, J., Eds.; Springer: Amsterdam, The Netherlands, 2009; pp. 97–106. [CrossRef]
13. Vido, J.; Nalevanková, P. Drought in the Upper Hron Region (Slovakia) between the Years 1984–2014. *Water* **2020**, *12*, 2887. [CrossRef]
14. Cheval, S.; Busuioc, A.; Dumitrescu, A.; Birsan, M.-V. Spatiotemporal variability of meteorological drought in Romania using the standardized precipitation index (SPI). *Clim. Res.* **2014**, *60*, 235–248. [CrossRef]
15. Vido, J.; Tadesse, T.; Šustek, Z.; Kandrík, R.; Hanzelová, M.; Škvarenina, J.; Škvareninová, J.; Hayes, M. Drought Occurrence in Central European Mountainous Region (Tatra National Park, Slovakia) within the Period 1961–2010. *Adv. Meteorol.* **2015**, 248728. [CrossRef]
16. Kholiavchuk, D.; Cebulska, M. Precipitation Shortage in the High Ukrainian and Polish Carpathians. In Proceedings of the Air and Water—Components of the Environment Conference, Cluj-Napoca, Romania, 25–26 March 2021; pp. 33–42. [CrossRef]
17. Kalbarczyk, R.; Kalbarczyk, E.; Leshchenko, O. Występowanie suszy atmosferycznej w półroczu ciepłym 2018 roku w Polsce—ocena intensywności i zasięgu [Occurrence of Atmospheric Drought in the Warm Half-Year in 2018 in Poland—Assessment of Intensity and Range]. In *Zmienność Klimatu Polski i Europy oraz jej Cyrkulacyjne Uwarunkowania [Variability of the Climate of Poland and Europe and Its Circulation Background]*; Kolendowicz, L., Bednorz, E., Tomczyk, A.M., Eds.; Bogucki Wyd. Naukowe: Poznan, Poland, 2019; pp. 105–116.
18. Łabędzki, L.; Bąk, B. Prognozowanie suszy meteorologicznej i rolniczej w systemie monitorowania suszy na Kujawach i w Dolinie Górnej Noteci [Predicting meteorological and agricultural drought in the system of drought monitoring in Kujawy and the upper Notec valley]. *Infrastruct. Ecol. Rural. Areas* **2011**, *5*, 19–28. Available online: http://www.infraeco.pl/pl/art/a_16106.htm?plik=904 (accessed on 5 August 2021).
19. Bartoszek, K. Występowanie susz atmosferycznych w okolicy Lublina i ich uwarunkowania cyrkulacyjne [Occurrence of atmospheric droughts in the Lublin area and atmospheric circulation conditions]. *Ann. Univ. Mariae Curie-Skłodowska Sec. E* **2014**, *69*, 49–61. Available online: https://czasopisma.up.lublin.pl/index.php/as/article/view/679/518 (accessed on 5 August 2021).
20. Twardosz, R. Charakterystyka okresów bezopadowych w Krakowie 1863–1995 [The characteristic of dry spells in Cracow in period 1863–1995]. *Geogr. Rev.* **1999**, *71*, 435–446.
21. Cebulska, M. Okresy bez opadów i ze słabymi opadami w polskich Karpatach (1984–2013) [Periods without precipitation and with low precipitation in the Polish Carpathians in the years of 1984–2013]. *Pol. J. Agron.* **2013**, *34*, 52–61. [CrossRef]
22. Cebulska, M. Niedobór opadów atmosferycznych w okresie wegetacyjnym w zlewni Małej Wisły (1984–2013) [Deficiencies of precipitation in the growing season in the Mała Wisła catchment (1984–2013)]. *Acta Sci. Pol. Form. Circumiectus* **2016**, *15*, 13–26. Available online: http://www.formatiocircumiectus.actapol.net/pub/15_2_13.pdf (accessed on 5 August 2021). [CrossRef]
23. Somorowska, U. Wzrost zagrożenia suszą hydrologiczną w różnych regionach geograficznych Polski w XX wieku [Increase in the hydrological drought risk in different geographical regions of Poland in the 20th century]. *Geogr. Works Stud.* **2009**, *43*, 97–114. Available online: https://wgsr.uw.edu.pl/wgsr/wp-content/uploads/2018/11/Somorowska.pdf (accessed on 5 August 2021).

24. Doroszewski, A.; Jóźwicki, T.; Wróblewska, E.; Kozyra, J. *Susza Rolnicza w Polsce w Latach 1961–2010 [Agricultural Drought in Poland in the Years 1961–2010]*; Institute of Soil Science and Plant Cultivation-State Research Institute: Puławy, Poland, 2014; ISBN 978-83-7562-171-6.
25. Spinoni, J.; Szalai, S.; Szentimrey, T.; Lakatos, M.; Bihari, Z.; Nagy, A.; Németh, A.; Kovács, T.; Mihic, D.; Dacic, M.; et al. Climate of the Carpathian Region in the period 1961–2010: Climatologies and trends of 10 variables. *Int. J. Climatol.* **2015**, *35*, 1322–1341. [CrossRef]
26. Gurung, A.B.; Bokwa, A.; Chełmicki, W.; Elbakidze, M.; Hirschmugl, M.; Hostert, P.; Ibisch, P.; Kozak, J.; Kuemmerle, T.; Matei, E.; et al. Global change research in the Carpathian Mountain Region. *Mt. Res. Dev.* **2009**, *28*, 282–288. [CrossRef]
27. Izmaiłow, B.; Kaszowski, L.; Krzemień, K.; Święchowicz, J. Rzeźba [Relief]. In *Karpaty Polskie [The Polish Carpathians]*; Warszyńska, J., Ed.; Institute of Geography, Jagiellonian University: Krakow, Poland, 1995; pp. 23–30.
28. Kondracki, J. *Geografia Polski [Geography of Poland]*; PWN: Warsaw, Poland, 1978; p. 463.
29. Hess, M. Piętra klimatyczne w Polskich Karpatach Zachodnich [Vertical climatic zones in the Polish Western Carpathians]. *Prace Geogr. Uniw. Jagiell.* **1965**, *11*, 266.
30. Obrębska-Starklowa, B. *Typologia i Regionalizacja Fenologiczno-Klimatyczna na Przykładzie Dorzecza Górnej Wisły [Phenological-Climatological Typology and Regionalization with the Upper Vistula River Basin as an Example]*; Series: Habilitation Dissertations of the Jagiellonian University, 11; Jagiellonian University: Kraków, Poland, 1977.
31. Niedźwiedź, T. *Sytuacje Synoptyczne i ich Wpływ na Zróżnicowanie Przestrzenne Wybranych Elementów Klimatu w Dorzeczu Górnej Wisły [Synoptic Situations and Their Impact on Spatial Variability of Selected Climate Elements in the Upper Vistula River Basin]*; Series: Habilitation Dissertations of the Jagiellonian University, 58; Jagiellonian University: Kraków, Poland, 1981.
32. Niedźwiedź, T.; Obrębska-Starklowa, B. Klimat [Climate]. In *Dorzecze Górnej Wisły [The Upper Vistula River Basin]*; Dynowska, I., Maciejewski, M., Eds.; PWN: Warsaw-Kraków, Poland, 1991.
33. Bokwa, A.; Wypych, A.; Ustrnul, Z. Climate Changes in the Vertical Zones of the Polish Carpathian in the Last 50 Years. In *The Carpathians: Integrating Nature and Society towards Sustainability*; Kozak, J., Ostapowicz, K., Bytnerowicz, A., Wyżga, B., Eds.; Springer: Berlin/Heidelberg, Germany, 2013; pp. 89–110, ISBN 978-3-642-12724-3.
34. Micu, D.M.; Dumitrescu, A.; Cheval, S.; Birsan, M.-V. *Climate of the Romanian Carpathians, Variability and Trends*; Springer: Cham, Switzerland, 2015; p. 221, ISBN 978-3-319-02886-6. [CrossRef]
35. Kobiv, Y. Response of rare alpine plant species to climate change in the Ukrainian Carpathians. *Folia Geobot.* **2017**, *52*, 217–236. [CrossRef]
36. Obrębska-Starklowa, B.; Hess, M.; Olecki, Z.; Trepińska, J.; Kowanetz, L. Klimat [Climate]. In *Karpaty Polskie [The Polish Carpathians]*; Warszyńska, J., Ed.; Institute of Geography, Jagiellonian University: Krakow, Poland, 1995; pp. 31–47.
37. Towpasz, K.; Zemanek, B. Szata roślinna [Plants]. In *Karpaty Polskie [The Polish Carpathians]*; Warszyńska, J., Ed.; Institute of Geography, Jagiellonian University: Krakow, Poland, 1995; pp. 77–94.
38. European Environment Agency. *Global and European Temperatures*; EEA: København, Denmark, 2021. Available online: https://www.eea.europa.eu/data-and-maps/indicators/global-and-european-temperature-10/assessment (accessed on 5 August 2021).
39. World Meteorological Organization. *Standardized Precipitation Index User Guide*; WMO-No. 1090; World Meteorological Organization: Geneva, Switzerland, 2012; p. 24, ISBN 978-92-63-11090-9.
40. Gasiorek, E.; Musiał, E. Porównanie i klasyfikacja warunków opadowych na podstawie wskaźnika standaryzowanego opadu i wskaźnika względnego opadu [Comparison and classification of precipitation conditions based on standardized SPI and relative RPI precipitation indices]. *Water-Environ.-Rural Areas* **2011**, *11*, 107–119. Available online: https://www.itp.edu.pl/old/wydawnictwo/woda/zeszyt_36_2011/artykuly/Gasiorek%20Musial.pdf (accessed on 5 August 2021).
41. Gałęzewska, M.; Kapuściński, J. Próba określenia prawdopodobieństwa posuchy i suszy w Wielkopolsce na przykładzie Poznania [An attempt of defining the probability of drought in Wielkopolska region with Poznań as an example]. *Ann. Agric. Acad. Poznań* **1978**, *105*, 3–11.
42. Vicente-Serrano, S.M.; Begueria, S.; Lopez-Moreno, J.I. A multi-scalar drought index sensitive to global warming: The Standardized Precipitation Evapotranspiration Index. *J. Clim.* **2010**, *23*, 1696–1718. [CrossRef]
43. Ellenberg, H.; Weber, H.E.; Düll, R.; Wirth, V.; Werner, W.; Paulissen, D. Zeigerwerte von Pflanzen in Mitteleuropa. *Scr. Geobot.* **1991**, *18*, 1–248.
44. Führer, E.; Horvath, L.; Jagodics, A.; Machon, A.; Szabados, I. Application of a new aridity index in Hungarian forestry practice. *Időjárás* **2011**, *115*, 205–216. Available online: https://www.met.hu/en/ismeret-tar/kiadvanyok/idojaras/index.php?id=24 (accessed on 5 August 2021).
45. Mayr, H. *Waldbau auf Naturgesetzlicher Grundlage*; Verlag Paul Perey: Berlin, Germany, 1909.
46. Satmari, A. Lucrari Practice de Biogeografie (Practical Applications of Biogeography). Available online: http://www.academia.edu/9909429/05_indici_ecometrici (accessed on 28 July 2021).
47. De Martonne, E. Une nouvelle fonction climatologique: L'indice d'aridité. *La Meteorol.* **1926**, *2*, 449–458.
48. National Drought Mitigation Center—UNL, SPI Generator. 2018. Available online: https://drought.unl.edu/droughtmonitoring/SPI/SPIProgram.aspx (accessed on 5 August 2021).
49. World Meteorological Organization. *Guide to Hydrological Practices, Vol. II: Management of Water Resources and Application of Hydrological Practices*; WMO-No. 168; World Meteorological Organization: Geneva, Switzerland, 2009; p. 302, ISBN 978-92-63-10168-6.

50. Kendall, M.G. *Rank Correlation Measures*; Charles Griffin: London, UK, 1975.
51. Mann, H.B. Non-parametric tests against trend. *Econometrica* **1945**, *13*, 245–259. [CrossRef]
52. IPCC. Managing the Risks of Extreme Events and Disasters to Advance Climate Change Adaptation. In *A Special Report of Working Groups I and II of the Intergovernmental Panel on Climate Change*; Field, C.B., Barros, V., Stocker, T.F., Qin, D., Dokken, D.J., Ebi, K.L., Mastrandrea, M.D., Mach, K.J., Plattner, G.-K., Allen, S.K., et al., Eds.; Cambridge University Press: Cambridge, UK; New York, NY, USA, 2012; 582p.
53. Wypych, A.; Ustrnul, Z.; Schmatz, D. Long-term variability of air temperature and precipitation conditions in the Polish Carpathians. *J. Mt. Sci.* **2018**, *15*, 237–253. [CrossRef]
54. Trnka, M.; Balek, J.; Štěpánek, P.; Zahradniček, P.; Možny, M.; Eitzinger, J.; Žalud, Z.; Formayer, H.; Turňa, M.; Nejedlik, P.; et al. Drought trends over part of Central Europe between 1961 and 2014. *Clim. Res.* **2016**, *70*, 143–160. [CrossRef]
55. Stephan, R.; Erfurt, M.; Terzi, S.; Žun, M.; Kristan, B.; Haslinger, K.; Stahl, K. An Inventory of Alpine Drought Impact Reports to explore past droughts in a mountain region. *Nat. Hazards Earth Syst. Sci.* **2021**, *21*, 2485–2501. [CrossRef]
56. Ziernicka-Wojtaszek, A.; Kopcińska, J. Variation in Atmospheric Precipitation in Poland in the Years 2001–2018. *Atmosphere* **2020**, *11*, 794. [CrossRef]
57. Hlasny, T.; Trombik, J.; Dobor, L.; Barcza, Z.; Barka, I. Future climate of the Carpathians: Climate change hot-spots and implications for ecosystems. *Reg. Environ. Chang.* **2016**, *16*, 1495–1506. [CrossRef]
58. Berki, I.; Rasztovits, E.; Móricz, N.; Mátyás, C. Determination of the drought tolerance limit of beech forests and forecasting their future distribution in Hungary. In Proceedings of the VIII. Alps-Adria Scientific Workshop, Neum, Bosnia-Herzegovina, 27 April–2 May 2009; pp. 613–616. [CrossRef]
59. Bosela, M.; Štefančík, I.; Petráš, R.; Vacek, S. The effects of climate warming on the growth of European beech forests depend critically on thinning strategy and site productivity. *Agric. For. Meteorol.* **2016**, *222*, 21–31. [CrossRef]
60. Hohnwald, S.; Indreica, A.; Walentowski, H.; Leuschner, C. Microclimatic Tipping Points at the Beech-Oak Ecotone in the Western Romanian Carpathians. *Forests* **2020**, *11*, 919. [CrossRef]
61. Kobiv, Y. Trends in Population Size of Rare Plant Species in the Alpine Habitats of the Ukrainian Carpathians under Climate Change. *Diversity* **2018**, *10*, 62. [CrossRef]
62. Svitková, I.; Svitok, M.; Petrík, A.; Bernátová, D.; Senko, D.; Šibík, J. The Fate of Endangered Rock Sedge (Carex rupestris) in the Western Carpathians—The Future Perspective of an Arctic-Alpine Species under Climate Change. *Diversity* **2019**, *11*, 172. [CrossRef]
63. Bosela, M.; Lukac, M.; Castagneri, D.; Sedmák, R.; Biber, P.; Carrer, M.; Konôpka, B.; Nola, P.; Nagel, T.A.; Popa, I.; et al. Contrasting effects of environmental change on the radial growth of co-occurring beech and fir trees across Europe. *Sci. Total. Environ.* **2018**, *615*, 1460–1469. [CrossRef]
64. Blöschl, G.; Hall, J.; Viglione, A.; Perdigão, R.A.P.; Parajka, J.; Merz, B.; Lun, D.; Arheimer, B.; Aronica, G.T.; Bilibashi, A.; et al. Changing climate both increases and decreases European river floods. *Nature* **2019**, *573*, 108–111. [CrossRef]
65. De Luca, D.L.; Petroselli, A.; Galasso, L.A. Transient Stochastic Rainfall Generator for Climate Changes Analysis at Hydrological Scales in Central Italy. *Atmosphere* **2020**, *11*, 1292. [CrossRef]

Article

Monitoring of Expansive Clays over Drought-Rewetting Cycles Using Satellite Remote Sensing

André Burnol [1,*], Michael Foumelis [1,2,*], Sébastien Gourdier [1], Jacques Deparis [1] and Daniel Raucoules [1]

1 BRGM, Risks and Risk Prevention Division (DRP), 45060 Orléans, France; s.gourdier@brgm.fr (S.G.); j.deparis@brgm.fr (J.D.); d.raucoules@brgm.fr (D.R.)
2 Department of Physical and Environmental Geography, Aristotle University of Thessaloniki (AUTh), 54124 Thessaloniki, Greece
* Correspondence: a.burnol@brgm.fr (A.B.); mfoumelis@geo.auth.gr (M.F.)

Abstract: New capabilities for measuring and monitoring are needed to prevent the shrink-swell risk caused by drought-rewetting cycles. A clayey soil in the Loire Valley at Chaingy (France) has been instrumented with two extensometers and several soil moisture sensors. Here we show by direct comparison between remote and in situ data that the vertical ground displacements due to clay expansion are well-captured by the Multi-Temporal Synthetic Aperture Radar Interferometry (MT-InSAR) technique. In addition to the one-year period, two sub-annual periods that reflect both average ground shrinking and swelling timeframes are unraveled by a wavelet-based analysis. Moreover, the relative phase difference between the vertical displacement and surface soil moisture show local variations that are interpreted in terms of depth and thickness of the clay layer, as visualized by an electrical resistivity tomography. With regard to future works, a similar treatment relying fully on remote sensing observations may be scaled up to map larger areas in order to better assess the shrink-swell risk.

Keywords: Copernicus Sentinel-1; electrical resistivity tomography; expansive clay; InSAR; shrink-swell risk; SMOS surface soil moisture; wavelet analysis

Citation: Burnol, A.; Foumelis, M.; Gourdier, S.; Deparis, J.; Raucoules, D. Monitoring of Expansive Clays over Drought-Rewetting Cycles Using Satellite Remote Sensing. *Atmosphere* **2021**, *12*, 1262. https://doi.org/10.3390/atmos12101262

Academic Editors: Andrzej Walega and Agnieszka Ziernicka-Wojtaszek

Received: 21 July 2021
Accepted: 24 September 2021
Published: 28 September 2021

1. Introduction

Among the various natural risks in France, the risk due to shrinking and swelling of subsurface clays is the second most important cause of financial compensation from insurance companies behind the flooding risk. In 2010, a first shrink/swell hazard map of metropolitan France, based on 1:50,000 geological maps, geotechnical data and spatial distribution of building damages has been published by the French Geological Survey (BRGM). So far, in situ monitoring of soil moisture and ground movements and ex situ clay characterization are traditionally used to assess this risk (e.g., the Mormoiron site with a Mediterranean climate [1] or at the Pessac site with an oceanic climate [2]). Since 2016, a new site characterized by a high shrink/swell hazard, level and located at Chaingy (France) has been instrumented by BRGM.

Standard ground displacement monitoring techniques (e.g., extensometers and GNSS) provide information on a very limited number of points within an area. Allowing a higher density of measurement points, the monitoring using Multi-Temporal Synthetic Aperture Radar Interferometry (MT-InSAR) techniques has been intensively developed in the last decades to track land subsidence or uplift related to groundwater extraction or recharge of aquifers around large cities [3–5]. Conversely, the monitoring of expansive clays based on InSAR has been the subject of very few studies until now [6–8]. The major limiting factor is the non-availability of relatively high-temporal resolution remote sensing datasets. There is indeed the requirement for fine temporal sampling due to the non-linear behavior of the shrink/swell cycles [9]. The launch by the European Space Agency (ESA) of the

Copernicus Sentinel-1A/1B satellites enables the systematic data provision, with a 6-day repeat cycle at the equator.

The aim of this article is to present an investigation of the shrink/swell behavior of a clay soil in relation to drought-rewetting cycles using both in situ and satellite remote sensing monitoring techniques over an instrumented site. The ground displacement measured by InSAR is compared to the in situ measurements in the studied zone during a three-year period. We will analyze how accurate Sentinel-1 data can measure the ground displacement due to the shrink/swell process, when the Parallel Small Baseline Subset (P-SBAS) technique is applied, and how well P-SBAS results and precise extensometers agree in validation for our study area. Our hypothesis is that the vertical displacement captured by the P-SBAS technique using a 90 m by 90 m cell is an average vertical displacement in that cell. Moreover, we show the time lag between the in-ground soil moistures at 1.2 m depth and the surface soil moisture acquired by the SMOS satellite. Finally, we propose for a first time a methodology for evaluating the depth and the thickness of subsurface clay layers relying fully on remote sensing observations, namely the use of the relative phase difference between Sentinel-1 InSAR displacement and SMOS surface soil moisture time series. In order to further validate this new approach, we deployed an electric tomography survey providing insights on the subsurface structure.

2. Materials and Methods

2.1. Studied Area

Our investigation site, an urban area located at Chaingy (France) with a semi-oceanic (i.e., slightly continental) climate, is characterized by a high level of shrink/swell hazard (Figure 1). Since 2016, two in situ extensometers (EXT1 and EXT2), spaced about 12 m apart, and a battery of soil moisture sensors at 1.2 m depth are deployed (Figure 1).

2.2. Synthetic Aperture Radar (SAR) Data and Interferometic Processing

Copernicus Sentinel-1 SAR data was utilized for the investigation of induced ground displacements in the vicinity of the extensometers EXT1 and EXT2 (see Table 1). Interferometric SAR (InSAR) processing was performed on the Geohazards Exploitation Platform (GEP) (https://geohazards-tep.eu, accessed on 23 September 2021). GEP is a platform originated by ESA as part of the Thematic Exploitation Platforms (TEP) initiative, aiming to support the exploitation of Earth Observation (EO) satellites to assess geohazards and their impact [10].

For the InSAR processing the Parallel Small Baseline Subset (P-SBAS) algorithm [11–13], as implemented on the GEP, was exploited. P-SBAS method is based on the processing of temporal series of co-registered SAR images acquired over the same target area for the generation of Line-of-Sight (LoS) ground displacement time series and average velocity maps. The technique allows for the extraction of both linear and non-linear motion components without a priori assumption on the displacement model. The service is provided at 90 m spacing distributed on a regular grid covering the user defined area of interest.

Between September 2016 and December 2019, the entire Sentinel-1A/1B archive data from both ascending and descending orbit geometries, 183 and 178 scenes, respectively, were processed. Supposing a 1D vertical displacement, the LOS motions are projected to vertical, whereas combination of different viewing geometries provide us the actual vertical motion component (the equation below was adapted from Hanssen (2001) [14]):

$$\mathrm{VERT\_ASC\_DES} = (\mathrm{LOS_{ASC}} + \mathrm{LOS_{DES}})/(\cos\theta_{\mathrm{ASC}} + \cos\theta_{\mathrm{DES}}) \quad (1)$$

VERT is the vertical displacement, LOS the motion in the Line-Of-Sight direction, θ the incidence angle, and the subscript ASC and DES for the ascending (θ_{ASC} = 34.4°) and descending (θ_{DES} = 42.8°) tracks, respectively.

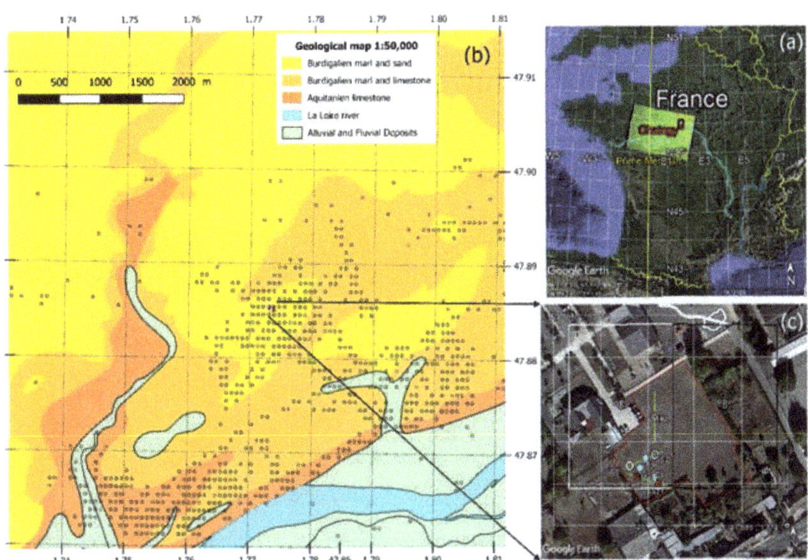

Figure 1. Map of the Chaingy experimental site. (**a**) Regional setting in France (Map data: Google, Landsat/Copernicus, SIO, NOAA, US Navy, NGA, and GEBCO) that contains modified Copernicus Sentinel data (2016) and a ~25 km cell of the EASE equal-area grid used by the SMOS satellite (black rectangle). (**b**) Simplified superficial geology of the studied zone modified from the BRGM geological map of France at the 1:50,000 scale showing the P-SBAS grid (black circles) and the location of the studied zone (red filled polygon). (**c**) Local setting (Map data: Google, Landsat/Copernicus, SIO, NOAA, US Navy, NGA, and GEBCO) of the studied zone (red polygon) showing two extensometers (EXT1 and EXT2) (red placemarks), three soil moistures sensors at E1, E5, and W5 (blue placemarks), three electric profiles (green lines), two core sampling SC1 and SC2 (white polygons), and two P-SBAS grid cells (white and black rectangle) around two grid points West Point WP and East Point EP (white circle). Maps (**a**–**c**) were created using Google Earth Pro and map b by QGIS (Version 3.4.14, Bern and Chur, Switzerland, http://qgis.org, accessed on 23 September 2021).

Table 1. Sentinel-1A/1B GEP InSAR processing parameters.

Parameters	Ascending Orbit	Descending Orbit
Number of scenes	183	178
Date of measurement start	4 September 2016	8 September 2016
Date of measurement end	30 December 2019	28 December 2019
Track number	59	110
Repeat cycle	6 days	6 days
Look angle	37.4 degrees	42.8 degrees
Applied algorithm	Parallel SBAS Interferometry Chain	
Software version	CNR-IREA P-SBAS 28	
Date of production	23 January 2020	21 January 2020
Geographic Coordinate System	EPSG 4326	
Number of looks azimuth	5	5
Number of looks range	20	20

Table 1. *Cont.*

Parameters	Ascending Orbit	Descending Orbit
Polarization	VV	VV
Temporal Coherence Threshold	0.85	0.85
Reference date	4 September 2016	8 September 2016
Reference point	Global average of zero-mean points	

2.3. *SMOS Level 3 Surface Soil Moisture (SSM) Products*

SMOS satellite was successfully launched on 2 November 2009 by ESA. We use here the term Surface Soil moisture (SSM) to refer to the volumetric soil moisture in the first few centimeters (0–5 cm) of the soil. L-band radiometry is achieved resulting in a ground resolution of 50 km. SMOS Level 0 (L0) to Level 2 (L2) data products are designed by ESA for scientific and operational use. The products are divided in half orbits, from pole to pole, ascending or descending, spanning about 50 min of acquisition 3. Level 3 products are geophysical variables with improved characteristics through temporal resampling or processing. In order to prevent any inconsistency resulting from interpolation over highly heterogeneous surfaces, no spatial averaging is operated in the algorithms [15–17]. It must also be noted that ascending and descending overpasses are bound to show different values of the retrieved parameters that may not be always comparable, and they are, thus, retrieved separately. The performance of each satellite SSM product depends on many factors such as, but not limited to, soil type, climate, presence of noise (Radio Frequency Interference), and land cover. It is therefore difficult to predict the performance of SSM products over a region, without performing a quality assessment using in situ measurements. The SMOS Level 3 SSM products were accessed through the CATDS Data Processing Center [16] (https://www.catds.fr, accessed on 23 September 2021). The data are presented over the Equal-Area Scalable Earth (EASE grid 2) [18] with a sampling of about 25 km $\times$ 25 km and the studied area is included in one grid cell (Figure 1). We used these 3-day aggregated SMOS-CATDS SSM products for ascending and descending overpasses between September 2016 and December 2019 for each Sentinel-1 acquisition (6 day-repeat cycle).

2.4. *Signal Processing Using Fourier Analysis*

Fourier analysis is well suited for the quantification of constant periodic components in time series. To perform filtering of satellite and in situ signals, time series of measurements are smoothed using a one-dimensional convolution approach with a HANNING window [19]. To carry out the spectral analysis, the filtered time series is first padded with trailing zeros to a length of 100 yr before computing the Discrete Fourier Transform (DFT) using the Fast Fourier Transform (FFT) Matlab (Matrix Laboratory, the MathWorks, Natick MA, USA) function [20]. The frequency maxima of Fourier power spectrum are therefore computed with a precision of $1/100\ yr^{-1}$. The magnitude and the phase angles of complex FFT values are calculated by Matlab ABS and ANGLE operators, respectively. Whenever the jump between consecutive angles is greater than or equal to π radians, UNWRAP function shifts the angles by adding multiples of $\pm 2\pi$ until the jump is less than π.

2.5. *Signal Processing Using Wavelet Analysis*

Fourier analysis does not provide any information about when the frequencies are present during the time-span covered by the time series. Conversely, the wavelet transform is especially suited to identify localized intermittent periodicities from low signal-to-noise ratio time-series [21,22]. The Morlet wavelet [23] is adapted to geophysical time series as described by [24] who developed the software provided at http://paos.colorado.edu/research/wavelets (accessed on 23 September 2021). We used the Matlab wavelet coherence toolbox as adapted by [25] and provided at http://www.glaciology.net/wavelet-coherence

(accessed on 23 September 2021). The Continuous Wavelet Power Spectrum (CWT) expands time-series records into time/frequency space. The time-series input data must be equally spaced in time. Although Copernicus satellites have a regular revisit interval (6 days for Sentinel-1A/B and 3 days for SMOS), some acquisitions may be missing or excluded from processing. These missing values are linearly interpolated using a constant time interval. The other time-series data of extensometers and soil moistures sensors are down-sampled using the same time interval. Two individual CWTs can be combined by using the Cross Wavelet Transform (XWT) tool which is computed by multiplying the CWT of one time-series by the complex conjugate of the CWT of the second time-series. XWT image is the 2-D representation of the absolute value and the phase of the complex number in the time-frequency space. For many geophysical phenomena, an appropriate background spectrum is the red/Brownian noise (increasing power with decreasing frequency) [26]. This background spectrum was recently used in a wavelet analysis for land subsidence [27] and clay expansion [7]. ANGLEMEAN function calculates the mean angle found by XWT during a time period, and the associated sigma which corresponds to a standard deviation.

2.6. Clay Layer Characterization Using Electric Method

The data were collected using the IRIS Syscal Pro Plus multi-electrode imaging system with internal multiplexer. Three profiles were acquired using 96 stainless electrodes spaced every 30 cm. Length of the profile was 28.5 m. The measurement was carried out using Dipole–Dipole and Wenner configuration. Due to good coupling, no data was removed during pre-processing. Inversion will be carried out using Res2DINV software [28,29] with a L1 regulation norm. Root means square values was lower than 1% after fives iterations for the three profiles.

3. Results

3.1. Intercomparison of InSAR and In Situ Displacement Time Series

The InSAR LoS measurements are independently projected to the vertical as well as combined to calculate the actual vertical motion component, using the ascending track and descending track (VERT-ASC-DES) at the West Point (WP) (see Materials and Methods section). Qualitatively, there is a positive correlation between the vertical displacement VERT-ASC-DES at WP, as calculated by InSAR using the satellite observations and as measured by ground-based extensometers EXT1 and EXT2 (Figure 2). Quantitatively, the least squares correlation coefficient (cc) is much higher for EXT2 (cc = 0.69) than for EXT1 (cc = 0.47). The relatively low cc values are expected given the non-linearity phenomenon and the heterogeneities of the clay layer as described in Section 3.2.

During the three-year period, the ground displacement time-series may be decomposed with the addition of a linear trend component T, a seasonal component S and an irregular residual component. There is a linear trend with a shrinking of about −2.2 mm/yr at EXT1, while the measured motion at EXT2 is merely a cyclic component with a negligible shrinking of about −0.3 mm/yr (Figure 2). During the same period, the linear trend measured by Sentinel-1A/B lies within in situ rates at EXT1 and EXT2, with a global shrinking of about −0.55 mm/yr. Concerning the cyclic part, the expansion is up to three times higher at EXT2 than at EXT1, with average swelling of 9.4 ± 2.0 mm and 3.1 ± 1.0 mm, respectively (Figure 2). Over the same period, the swelling magnitude measured by VERT-ASC-DES is 4.5 ± 0.5 mm.

3.2. Electrical Tomography Survey

Electrical resistive tomography has been carried out in order to image the lithology stratification and the heterogeneity of the clay layer (i.e., depth, thickness, and fraction of clay) (see Materials and Methods section). Along three 28.5 m long profiles in the WP cell (Figure 1), the resistivity values range from 8 to 100 Ω.m and a three-layer stratified subsurface is highlighted (Figures 3 and 4).

Figure 2. Comparison of the in situ vertical displacement of both extensometers (EXT1 and EXT2) to VERT-ASC-DES, the vertical displacement using the ascending and the descending track of Sentinel-1 at West Point (WP). The linear trendlines of the displacements are also shown (dotted lines).

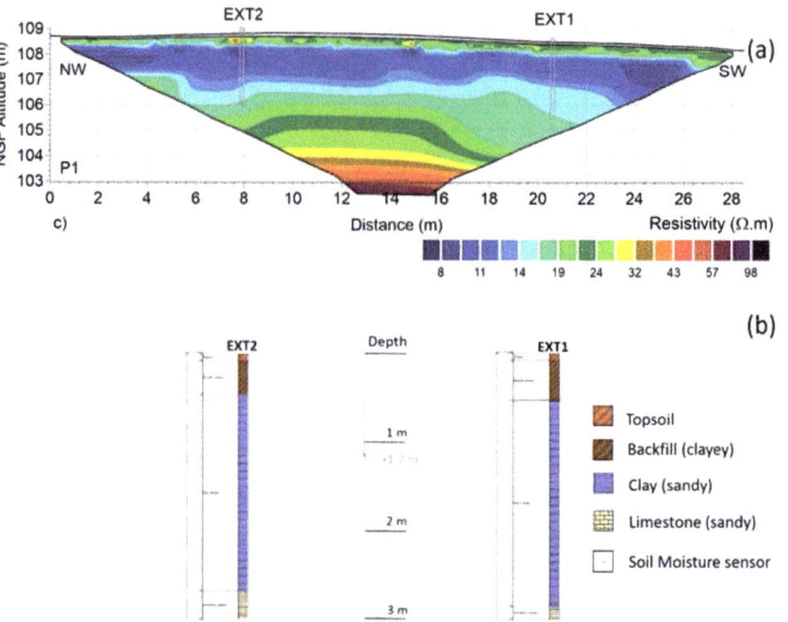

Figure 3. (**a**) Vertical cross section of resistivity along P1 profile (EXT1-EXT2). (**b**) Clay depth and thickness at EXT1 and EXT2 deduced from the resistivity profile (see text). E1 sensor (near EXT1, Figure 1) and W5 (near EXT2) are both located at the same 1.2 m depth inside the clay layer.

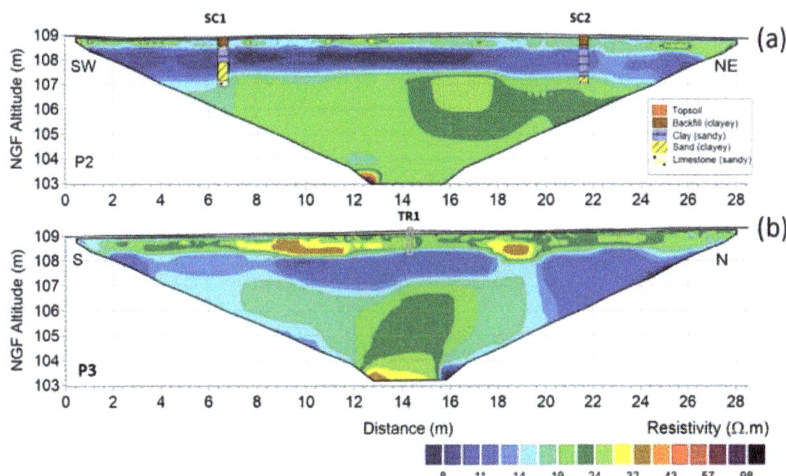

Figure 4. (**a**) Vertical cross section of resistivity along P2 profile (SC1-SC2) showing the SC1 and SC2 geological logs. (**b**) Vertical cross section of resistivity along P3 profile (South-North) showing TR1 drilling.

We used two drill cores along the SC1-SC2 profile to define the clay subsurface material by the resistivity value [30] (Figure 4). A first layer above the depth of 0.59 m ± 0.22 m corresponds to topsoil and backfill with some clay lenses (Figures 3 and 4). A second layer with a mean thickness of 1.5 m ± 0.42 m corresponds to the clay material with a resistivity lower than 17 Ω.m ± 1 Ω.m. Below 2.09 m ± 0.42 m, a third layer corresponds to sandy limestones with a resistivity gradient due to the weathering process.

Along the EXT1-EXT2 resistivity profile (Figure 3), the mean depth and thickness of the clay material layer is 0.46 m ± 0.05 m (min-max, 0.34–0.55) and 1.97 m ± 0.18 m (min-max, 1.70–2.39), respectively. The mean depth of the carbonate layer is 2.43 m ± 0.16 m (min-max, 2.15–2.73). The depth (approximately at 0.5 m) and thickness (about 2.3 m) of the clay layer at EXT1 and EXT2 are very similar (Figure 3). However, the resistivity of the clay layer at EXT2 is much lower compared to EXT1, indicating that the clay fraction is much higher beneath EXT2 [20]. That is also consistent with the expansion difference between the extensometers showing a much higher swelling magnitude at EXT2.

Along the three profiles, the mean depth and thickness of the clay material layer in the West Point (WP) cell is 0.59 m ± 0.22 m and 1.5 m ± 0.42 m.

3.3. Intercomparison of In Situ Soil Moistures and SMOS Satellite Surface Soil Moistures

We investigated here the soil moisture variations acquired by E1 sensor (near EXT1) and W5 (near EXT2) that are both located at the same 1.2 m depth inside the clay layer (Figure 3). The temporal variations of E1 and W5 are strongly correlated (cc = 0.73) during the three-year period (Figure 5). For the seasonal one-year period, there is a slight leading of E1 moisture relative to W5 (about 0.75 months) as calculated by the Cross Wavelet Transform (XWT) (see Materials and Methods Section 2.5). Indeed, the infiltration time of the meteoric water inside the clay layer at 1.2 m depth is lower in the E1 case, since the clay fraction of low permeability is shallower at EXT1 compared to EXT2 (Figure 3).

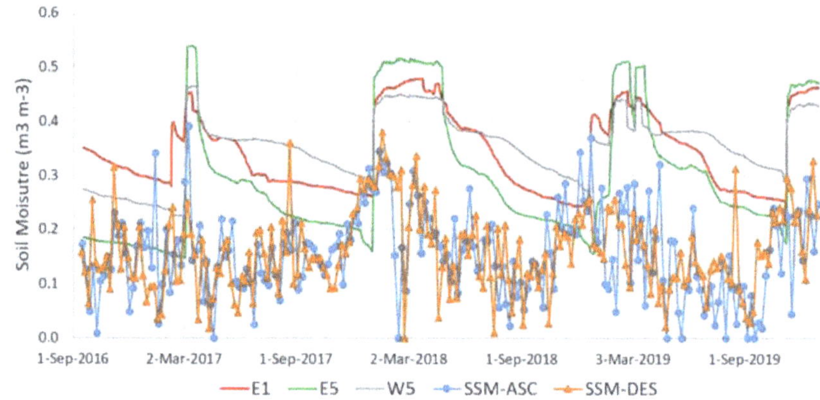

Figure 5. Comparison of three soil moistures at about 1.2 m depth (E1, E5, W5) to the Surface Soil Moisture SSM-ASC and SSM-DES using the SMOS ascending and descending track, respectively.

We use here the term Surface Soil Moisture (SSM) to refer to the volumetric soil moisture in the first few centimeters (0–5 cm) of the soil. We investigated also the SSM acquired by the ascending (SSM-ASC) and descending (SSM-DES) orbits of the SMOS satellite, showing rather positive correlation (cc = 0.48). For the one-year period, SSM-ASC and SSM-DES are both in-phase and there is a slight leading of SSM-ASC relative to SSM-DES for the 5-month period (about 0.9 months). Hence, the SSM-ASC dataset was considered more appropriate for calculating the phase differences to other time series in the discussion. For the one-year period, the phase lead of the surface soil moisture SSM-ASC relative to the in-ground soil moistures E1 and W5 calculated by XWT is about 0.16 yr and 0.22 yr, respectively.

4. Discussion

4.1. Intercomparison of InSAR and In Situ Displacement Time Series

Displacements measured by InSAR are in the direction of the Line-of-Sight (LoS) of the satellite, while displacements measured by extensometers depict motion in the vertical direction. However, the combination of ascending and descending InSAR measurements for the calculation of the actual vertical motion component (VERT-ASC-DESC) allows the direct inter-comparison of satellite and in situ observations.

When exposed to moisture, the magnitude of the vertical expansion of soils will depend on the amount of expansive clay minerals in the subsurface. In addition to the vertical displacement, the subsurface heterogeneities (i.e., clay depth and thickness and percentage of sand) may introduce spatial variability to the observed ground displacements. The above-mentioned factors need to be considered when examining the consistency of our measurements. In our case, the vertical displacements VERT-ASC-DESC from remote sensing are within the motion values of both extensometers, not only for linear displacement rates but also in terms of observed seasonality (see Section 3.1). This result is in agreement with the assumption that the effect of the spatial resolution of the remote sensing measurements (90 m by 90 m cell) represents an average of the vertical displacements within that resolution cell.

4.2. Shrinking and Swelling Periods Using Fourier Power Spectra and Continuous Wavelet Transform

We use first the Fourier spectra of vertical displacements to calculate the main frequency components of time series of both extensometers as well as at WP and East Point (EP) cells of the P-SBAS grid. The one-year seasonal period is found in the power spectrum of the four displacement time series at EXT1, EXT2, WP, and EP (Figure 6). In the frequency band between 2 and 4 yr^{-1} (corresponding to a period between 0.5 and 0.25 yr), there are three other frequencies which are noted F1, F21, and F22 (order of increasing frequency) and P1, P21, and P22 periods (order of decreasing period). It should be noted that the frequency values for all displacement time series are close to the first three sub-annual components of the Fourier transform of the soil moistures E1 and W5 at 1.2 m depth (Figure 6a) and the two ascending and descending time series of the surface soil moisture (SSM-ASC and SSM-DES) as measured by SMOS (Figure 6b).

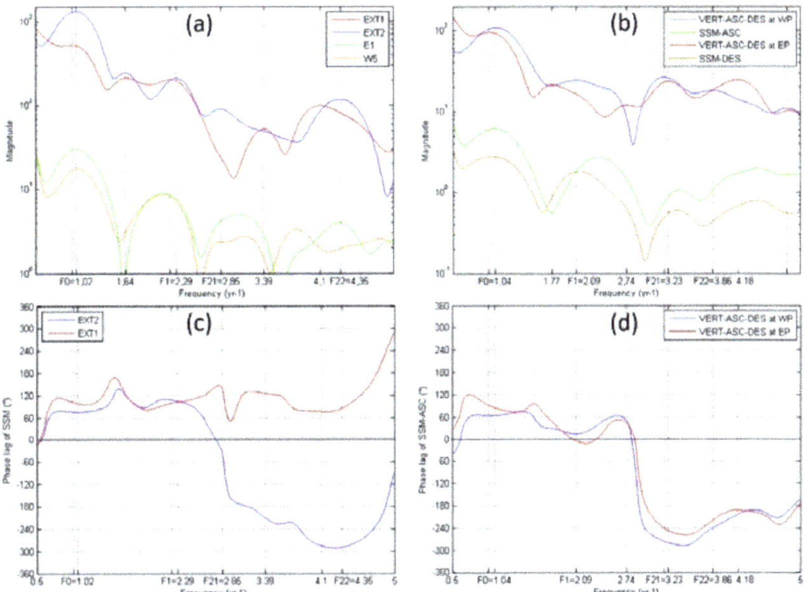

Figure 6. In the top, the magnitude of the Fast Fourier Transform (FFT) of different time series between frequencies 0.5 and 5 year^{-1} are shown: (**a**) vertical displacement of both extensometers (EXT1 and EXT2) and the Soil Moisture at E1 and W5 (about 1.2 m depth), (**b**) VERT-ASC-DES for vertical displacement using ascending and descending S-1A/1B tracks at West Point (WP) and East Point (EP), and SSM-ASC and SSM-DES for the Surface Soil Moisture using the ascending and descending SMOS track, respectively. In the bottom, the phase angle difference in degrees between the FFT of the SSM-ASC using SMOS ascending track and the FFT of the displacement of four time series are shown: (**c**) EXT1 and EXT2 and (**d**) VERT-ASC-DES at WP and EP.

The Continuous Wavelet Transform (CWT) tool permits the recognition of power in time-frequency space, along with assessing confidence levels against red noise backgrounds (see Materials and Methods section). We use now the continuous wavelet power spectrum in order to unravel the intermittent physical mechanisms of the shrink/swell process at EXT2 and WP cell during 3.3 years (September 2016–December 2020):

- September 2018 is the end of the shrinking period (Figure 2) corresponding to the maximum of spectrum power at the P1 period (at 2 yr in Figure 7);
- December 2017 is the start of the swelling period for EXT2 corresponding to the maximum of power at the P21 and P22 periods (at 1.25 yr in Figure 7). Both swelling periods at WP cell are unraveled at different times, before 1.25 yr for P21 period and after 1.5 yr for P22 period.

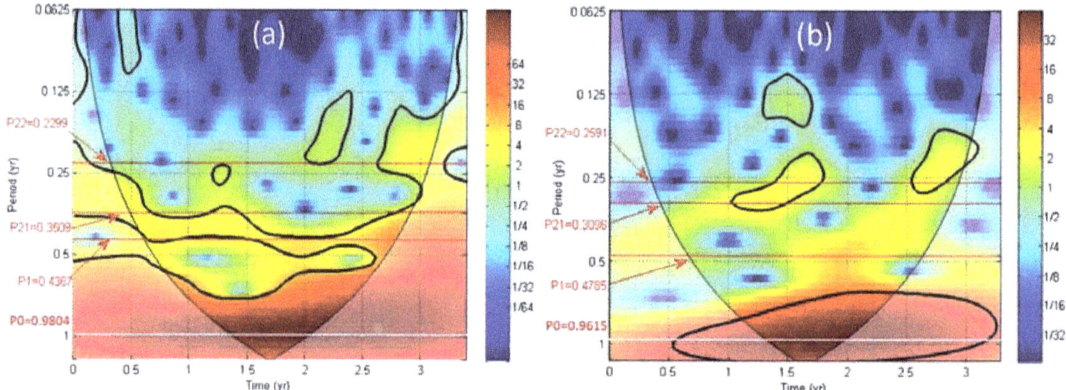

Figure 7. The continuous wavelet transform (CWT) during 3.3 years (September 2016–December 2020) are shown for two vertical displacements: (**a**) EXT2 and (**b**) VERT-ASC-DES at WP. The thick contour designates the 5% significant level against red noise (see Materials and Methods section). The Cone of Influence (COI) where edge effects might distort the picture is shown as a lighter shadow.

In conclusion, the CWT analysis highlights two sub-annual periods that reflect both average ground shrinking (P1) and swelling timeframes (P21 and P22). A similar behavior is observed at EXT1 extensometer and within EP cell (data not shown).

4.3. Estimating Variations of Expansive Clay Depth and Thickness Using the Time Series Phase Difference

The phase difference between the Surface Soil Moisture (SSM) and InSAR displacement time series indicates the time lag between the cause and the effect in the shrink/swell process. We test the assumption that depth and thickness of expansive clays are linked to the phase difference between both time series for the shrinkage and swelling periods, respectively. We use and compare the Fourier and XWT analyses to calculate this phase difference (see Materials and Methods section). The XWT spectra will be high in the time-frequency areas where both CWTs display high values, so this helps identify common time patterns in the two data sets. The XWT permits also the recognition of relative phase lags in time-frequency space (noted $\Delta\Phi$). While P1 and P21 are the shrinking and swelling periods found by the Fourier method (as described above), P1* and P21* are the same similar periods we found using the thick contours of XWT that designate the 5% significant level against red noise (Figure 8). XWT tool is used as well to calculate the circular standard deviation of this phase difference $\Delta\Phi$ (see Materials and Methods section) and all these values are reported in Table 2.

No significant difference is found between $\Delta\Phi$ values of both extensometers EXT1 and EXT2 for the shrinking period P1* and the swelling period P21* (Table 2). This result is consistent with the tomography results which show that both depth and thickness of the clay is indeed comparable at the location of the extensometers (Figure 3b). For the shrinking period P1*, the phase difference in degrees is three times higher at EXT2 ($\Delta\Phi = 107^\circ$) than at WP ($\Delta\Phi = 34^\circ$) (Table 2). The same phase difference for the shrinking period is observed between EXT1 and EP (Table 2 and Figure 9). This result is further consistent with the tomographic findings indicating clays lenses in the first 0.5 m subsurface layer of the three 28m-wide profiles in the WP cell (Figures 3 and 4). Conversely, $\Delta\Phi$ of the swelling periods is higher at EP cell ($\Delta\Phi = 110^\circ$) in comparison to the WP cell ($\Delta\Phi = 72^\circ$) (Table 2 and Figure 9). A higher clay thickness in the EP cell in comparison to the WP cell may be an explanation of this result (there is a need of electrical tomography data to check this assumption).

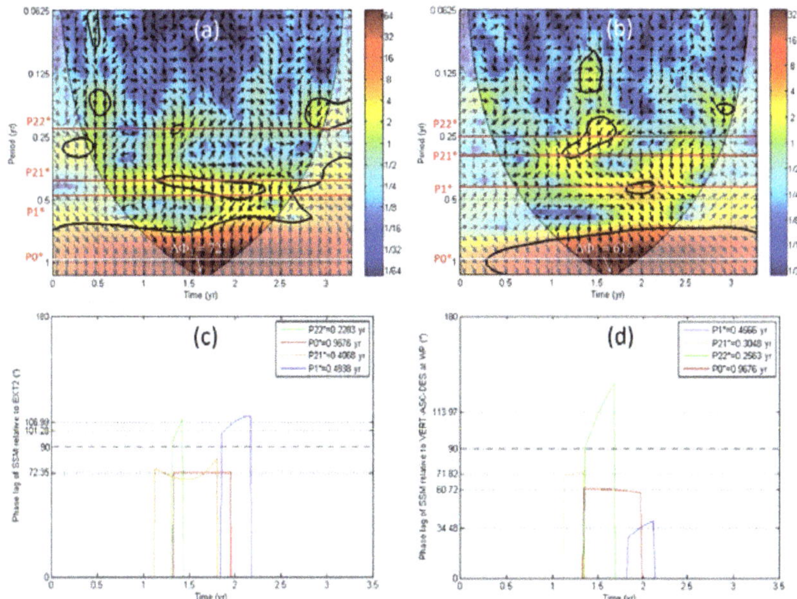

Figure 8. In the top, the cross wavelet transform (XWT) between SSM-ASC and two times series is shown: (**a**) EXT2 and (**b**) VERT-ASC-DES at WP. The relative phase relationship is shown as arrows, with in-phase pointing right and anti-phase pointing left and the Surface Soil Moisture leading by 90° pointing straight down. The Cone of Influence (COI) where edge effects might distort the picture is shown as a lighter shadow. In the bottom, the phase lags ΔΦ of SSM-ASC relative to the same times series are shown: (**c**) EXT2 and (**d**) VERT-ASC-DES at WP. The shrinking (P1*) and swelling (P21* and P22*) periods are found using the thick contours of XWT that designate the 5% significant level against red noise (see Materials and Methods section).

Table 2. Comparison of Fourier FFT and cross-wavelet XWT analysis of phase angles and time lags of the surface soil moisture (SSM-ASC) relative to the vertical displacement at EXT1, EXT2, West Point (WP), and East Point (EP) for P0, P1, P21, and P22 periods.

Method	1-Year Period (P0)		Shrinking Period (P1)		Swelling First Period (P21)		Swelling Second Period (P22)	
	Angle ΔΦ (°)	Time (month)	Angle ΔΦ (°)	Time (month)	Angle ΔΦ (°)	Time (month)	Angle ΔΦ (°)	Time (month)
FFT at EXT1	102.0	3.51	99.63	1.50	125.86	1.24	77.25	0.63
XWT at EXT1	89.09 ± 0.4	3.04 ± 0.01	101.20 ± 7.94	1.63 ± 0.13	70.41 ± 7.95	1.07 ± 0.12	94.82 ± 9.78	0.76 ± 0.08
FFT at EXT2	76.76	2.51	105.09	1.53	−38.69	0.45	70.76	0.54
XWT at EXT2	72.35 ± 0.02	2.33 ± 0.00	106.99 ± 3.68	1.73 ± 0.06	71.13 ± 3.76	0.96 ± 0.05	101.28 ± 5.02	0.77 ± 0.04
FFT at WP cell	64.75	2.08	14.68	0.23	83.81	0.86	119.70	1.03
XWT at WP cell	60.72 ± 0.96	1.96 ± 0.03	34.48 ± 3.40	0.52 ± 0.05	71.82 ± 0.98	0.73 ± 0.01	113.97 ± 13.30	0.97 ± 0.11
FFT at EP cell	92.16	3.20	42.85	0.52	112.48	1.14	168.36	1.34
XWT at EP cell	90.10 ± 0.32	3.08 ± 0.01	34.90 ± 3.75	0.56 ± 0.06	109.66 ± 0.56	1.11 ± 0.01	144.28 ± 5.42	1.15 ± 0.04

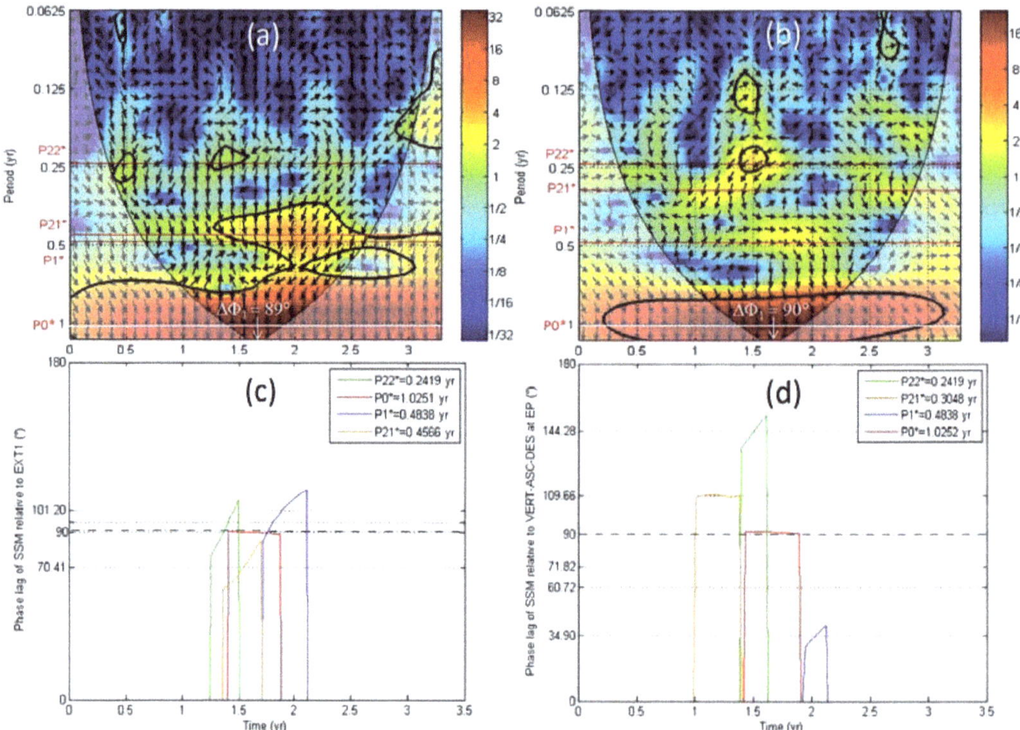

Figure 9. In the top, the cross wavelet transform (XWT) between SSM-ASC using SMOS ascending track and two times series is shown: (**a**) EXT1 and (**b**) VERT-ASC-DES at East Point (EP). The relative phase relationship is shown as arrows, with in-phase pointing right and anti-phase pointing left and the Surface Soil Moisture leading by 90° pointing straight down. The Cone of Influence (COI) where edge effects might distort the picture is shown as a lighter shadow. In the bottom, the phase lags $\Delta\Phi$ of SSM-ASC relative to the same times series are shown: (**c**) EXT1 and (**d**) VERT-ASC-DES at EP. The shrinking (P1*) and swelling (P21* and P22*) periods are found using the thick contours of XWT that designate the 5% significant level against red noise (see Materials and Methods section).

5. Conclusions

By comparing space-borne interferometric ground motion measurements with ground-based data for a well instrumented site at Chaingy (France), we provide evidence that Sentinel-1 InSAR data enables accurate measurement of small cyclic motions (in the order of a few millimeters) related to the shrink/swell process of expansive clays. A wavelet-based method has been applied to these MT-InSAR time series to better characterize the shrinking and swelling timeframes. Moreover, we show how coupling SMOS surface soil moisture and Sentinel-1 InSAR data may reveal apart from the presence of expansive clays additional information on their spatial characteristics. The relative phase differences between SMOS and InSAR time series are used for assessing the variations of the clay layer in terms of depth and thickness, as confirmed by an electrical resistivity tomography campaign. With regard to future works, the proposed methodology could be applied coupling different satellite observations to scale up over wide areas for tracking the presence of expansive clays, with the ultimate goal to better estimate the shrink-swell risk.

Author Contributions: Conceptualization, A.B.; Data curation, S.G. and J.D.; Funding acquisition, S.G. and D.R.; Methodology, M.F. and D.R.; Software, M.F.; Visualization, J.D.; Writing—original draft, A.B. All authors have read and agreed to the published version of the manuscript.

Funding: The instrumentation of the new Chaingy experimental site was funded by the French Ministry of Environment (research program DGPR-BRGM 2016, Action C8).

Institutional Review Board Statement: Not applicable.

Informed Consent Statement: Not applicable.

Data Availability Statement: Acquisitions of Sentinel-1 satellite were accessed and processed through GEP (https://geohazards-tep.eu, accessed on 23 September 2021) and SMOS satellite data via the French ground segment for the Level 3 data (CATDS, https://www.catds.fr/, accessed on 23 September 2021).

Acknowledgments: We would like to thank Céline Roux for her contribution to the graphical abstract.

Conflicts of Interest: The authors declare no conflict of interest.

References

1. Vincent, M.; Le Roy, S.; Dubus, I.; Surdyk, N. Experimental monitoring of water content and vertical displacements in clayey soils exposed to shrinking and swelling. *Rev. Française Géotechnique* **2007**, *120–121*, 45–48. [CrossRef]
2. Fernandes, M.; Denis, A.; Fabre, R.; Lataste, J.-F.; Chrétien, M. In situ study of the shrinkage-swelling of a clay soil over several cycles of drought-rewetting. *Eng. Geol.* **2015**, *192*, 63–75. [CrossRef]
3. Declercq, P.-Y.; Walstra, J.; Gérard, P.; Pirard, E.; Perissin, D.; Meyvis, B.; Devleeschouwer, X. A Study of Ground Movements in Brussels (Belgium) Monitored by Persistent Scatterer Interferometry over a 25-Year Period. *Geosciences* **2017**, *7*, 115. [CrossRef]
4. Foumelis, M.; Papageorgiou, E.; Stamatopoulos, C. Episodic ground deformation signals in Thessaly Plain (Greece) revealed by data mining of SAR interferometry time series. *Int. J. Remote Sens.* **2016**, *37*, 3696–3711. [CrossRef]
5. Foumelis, M. Human induced groundwater level declination and physical rebound in northern Athens Basin (Greece) observed by multi-reference DInSAR techniques. In Proceedings of the 2012 IEEE International Geoscience and Remote Sensing Symposium, Munich, Germany, 22–27 July 2012; pp. 820–823.
6. Bonì, R.; Bosino, A.; Meisina, C.; Novellino, A.; Bateson, L.; McCormack, H. A Methodology to Detect and Characterize Uplift Phenomena in Urban Areas Using Sentinel-1 Data. *Remote Sens.* **2018**, *10*, 607. [CrossRef]
7. Burnol, A.; Aochi, H.; Raucoules, D.; Veloso, F.M.L.; Koudogbo, F.N.; Fumagalli, A.; Chiquet, P.; Maisons, C. Wavelet-based analysis of ground deformation coupling satellite acquisitions (Sentinel-1, SMOS) and data from shallow and deep wells in Southwestern France. *Sci. Rep.* **2019**, *9*, 8812. [CrossRef]
8. Fryksten, J.; Nilfouroushan, F. Analysis of Clay-Induced Land Subsidence in Uppsala City Using Sentinel-1 SAR Data and Precise Leveling. *Remote Sens.* **2019**, *11*, 2764. [CrossRef]
9. Raucoules, D.; Bourgine, B.; De Michele, M.; Le Cozannet, G.; Closset, L.; Bremmer, C.; Veldkamp, H.; Tragheim, D.; Bateson, L.; Crosetto, M. Validation and intercomparison of Persistent Scatterers Interferometry: PSIC4 project results. *J. Appl. Geophys.* **2009**, *68*, 335–347. [CrossRef]
10. Foumelis, M.; Papadopoulou, T.; Bally, P.; Pacini, F.; Provost, F.; Patruno, J. Monitoring Geohazards Using On-Demand and Systematic Services on Esa's Geohazards Exploitation Platform. In Proceedings of the IGARSS 2019—2019 IEEE International Geoscience and Remote Sensing Symposium, Yokohama, Japan, 28 July–2 August 2019; pp. 5457–5460.
11. Casu, F.; Elefante, S.; Imperatore, P.; Zinno, I.; Manunta, M.; Luca, C.D.; Lanari, R. SBAS-DInSAR Parallel Processing for Deformation Time-Series Computation. *IEEE J. Sel. Top. Appl. Earth Obs. Remote Sens.* **2014**, *7*, 3285–3296. [CrossRef]
12. Berardino, P.; Fornaro, G.; Lanari, R.; Sansosti, E. A new algorithm for surface deformation monitoring based on small baseline differential SAR interferograms. *IEEE Trans. Geosci. Remote Sens.* **2002**, *40*, 2375–2383. [CrossRef]
13. Manunta, M.; Luca, C.D.; Zinno, I.; Casu, F.; Manzo, M.; Bonano, M.; Fusco, A.; Pepe, A.; Onorato, G.; Berardino, P.; et al. The Parallel SBAS Approach for Sentinel-1 Interferometric Wide Swath Deformation Time-Series Generation: Algorithm Description and Products Quality Assessment. *IEEE Trans. Geosci. Remote Sens.* **2019**, *57*, 6259–6281. [CrossRef]
14. Hanssen, R.F. *Radar Interferometry: Data Interpretation and Error Analysis*; Springer Science & Business Media: New York, NY, USA; Boston, MA, USA; Dordrecht, The Netherlands; London, UK; Moscow, Russian, 2001; Volume 2, p. 308.
15. Al Bitar, A.; Mialon, A.; Kerr, Y.H.; Cabot, F.; Richaume, P.; Jacquette, E.; Quesney, A.; Mahmoodi, A.; Tarot, S.; Parrens, M.; et al. The global SMOS Level 3 daily soil moisture and brightness temperature maps. *Earth Syst. Sci. Data* **2017**, *9*, 293–315. [CrossRef]
16. Cabot, F. CATDS-PDC L3SM Aggregated—3-Day, 10-Day and Monthly Global Map of Soil Moisture Values from SMOS Satellite 2016. Available online: https://www.catds.fr/Publications (accessed on 23 September 2021).
17. Kerr, Y.H.; Waldteufel, P.; Richaume, P.; Wigneron, J.P.; Ferrazzoli, P.; Mahmoodi, A.; Al Bitar, A.; Cabot, F.; Gruhier, C.; Juglea, S.E.; et al. The SMOS Soil Moisture Retrieval Algorithm. *IEEE Trans. Geosci. Remote Sens.* **2012**, *50*, 1384–1403. [CrossRef]
18. Brodzik, M.J.; Billingsley, B.; Haran, T.; Raup, B.; Savoie, M.H. EASE-Grid 2.0: Incremental but Significant Improvements for Earth-Gridded Data Sets. *ISPRS Int. J. Geo-Inf.* **2012**, *1*, 32–45. [CrossRef]
19. Oppenheim, A.V.; Ronald, W.S. *Discrete-Time Signal Processing*; Prentice-Hall: Hoboken, NJ, USA, 1999.
20. *MATLAB Version 8.1 (R2013a)*; The MathWorks Inc.: Natick, MA, USA, 2013.

21. Cazelles, B.; Chavez, M.; Berteaux, D.; Ménard, F.; Vik, J.O.; Jenouvrier, S.; Stenseth, N.C. Wavelet analysis of ecological time series. *Oecologia* **2008**, *156*, 287–304. [CrossRef] [PubMed]
22. Bloomfield, D.S.; McAteer, R.J.; Lites, B.W.; Judge, P.G.; Mathioudakis, M.; Keenan, F.P. Wavelet phase coherence analysis: Application to a quiet-sun magnetic element. *Astrophys. J.* **2004**, *617*, 623. [CrossRef]
23. Goupillaud, P.; Grossmann, A.; Morlet, J. Cycle-octave and related transforms in seismic signal analysis. *Geoexploration* **1984**, *23*, 85–102. [CrossRef]
24. Torrence, C.; Compo, G.P. A practical guide to wavelet analysis. *Bull. Am. Meteorol. Soc.* **1998**, *79*, 61–78. [CrossRef]
25. Grinsted, A.; Moore, J.C.; Jevrejeva, S. Application of the cross wavelet transform and wavelet coherence to geophysical time series. *Nonlinear Process. Geophys.* **2004**, *11*, 561–566. [CrossRef]
26. Kasdin, N.J. Discrete simulation of colored noise and stochastic processes and 1/f/sup/spl alpha//power law noise generation. *Proc. IEEE* **1995**, *83*, 802–827. [CrossRef]
27. Tomás, R.; Pastor, J.L.; Béjar-Pizarro, M.; Bonì, R.; Ezquerro, P.; Fernández-Merodo, J.A.; Guardiola-Albert, C.; Herrera, G.; Meisina, C.; Teatini, P.; et al. Wavelet analysis of land subsidence time-series: Madrid Tertiary aquifer case study. *Proc. IAHS* **2020**, *382*, 353–359. [CrossRef]
28. Loke, M.; Barker, R. Least-squares deconvolution of apparent resistivity pseudosections. *Geophysics* **1995**, *60*, 1682–1690. [CrossRef]
29. Loke, M.H.; Barker, R.D. Rapid least-squares inversion of apparent resistivity pseudosections by a quasi-Newton method1. *Geophys. Prospect.* **1996**, *44*, 131–152. [CrossRef]
30. Long, M.; Donohue, S.; L'Heureux, J.-S.; Solberg, I.-L.; Ronning, J.S.; Limacher, R.; O'Connor, P.; Sauvin, G.; Romoen, M.; Lecomte, I. Relationship between electrical resistivity and basic geotechnical parameters for marine clays. *Can. Geotech. J.* **2012**, *49*, 1158–1168. [CrossRef]

Article

Summer Drought in 2019 on Polish Territory—A Case Study

Agnieszka Ziernicka-Wojtaszek

Department of Ecology, Climatology and Air Protection, Faculty of Environmental Engineering and Land Surveying, University of Agriculture in Kraków, al. Mickiewicza 24/28, 30-059 Kraków, Poland; agnieszka.ziernicka-wojtaszek@urk.edu.pl

Abstract: The summer 2019 drought in Poland, i.e., the warmest year in observation history, was characterized. Meteorological, agricultural, hydrological, and hydrogeological aspects were taken into account. Meteorological drought in the light of regionally differentiated days of low precipitation frequency lasted the longest, i.e., over 3 months in central-western Poland. In the period between June–August 2019, in the belt of South Baltic Lakes and Central Polish Lowlands, the lowest precipitation sums of 30–60% of the norm were recorded. The values of the climatic water balance (CWB) calculated by the Institute of Soil Science and Plant Cultivation (IUNG) method for individual months of June–August for the Polish area were −129, −64, and −53 mm, respectively. The most threatened were fruit bushes, spring cereals, maize for grain and silage, and leguminous plants. In central-western and south-western Poland, the drought accelerated the date of the lowest flows by two months on average from the turn of September and October to the turn of July and August. In the lowland belt, where the drought was the most intensive, the average monthly groundwater level, both of free and tight groundwater table, was lower than the monthly averages for the whole hydrological year.

Keywords: precipitation; precipitation deficit; climatic water balance; drought

Citation: Ziernicka-Wojtaszek, A. Summer Drought in 2019 on Polish Territory—A Case Study. *Atmosphere* **2021**, *12*, 1475. https://doi.org/10.3390/atmos12111475

Academic Editors: Alfredo Rocha and Ognjen Bonacci

Received: 24 September 2021
Accepted: 3 November 2021
Published: 8 November 2021

1. Introduction

Spatial and temporal variability of precipitation amounts in Poland is very high. The diversity of relief means that the areas with the lowest precipitation covering the central part of the country receive less than 500 mm of precipitation annually. On the other hand, on the upper border of the moderately warm story in the Western Carpathians, precipitation of 1000 mm should be expected, while on the upper border of the moderately cool story, which is the limit of agricultural use, it was 1400 mm [1–3].

The location of Poland in moderate latitudes of Central Europe determines high variability of weather in particular years. The values of the lowest and highest precipitation in Warsaw may vary from 60% to 150% of the standard multi-year average for the year, 27% to 250% for seasons, and even 5% and 505% for October [4,5]. In the summer, the maximum daily sums may exceed the multi-year average monthly sums. Such high variability of precipitation with simultaneous variability of values of other meteorological elements results in different meteorological conditions of crops in particular years manifested by the occurrence of dry periods and periods with an excess of precipitation.

In recent years, a growing interest among climatologists, hydrologists, environmentalists, farmers, and political activists has been the observed and projected climate change [6]. The trends of precipitation changes in a warming climate have not yet found a clear assessment. In the area of Poland lying between northern Europe, with marked and predicted increasing trends of annual precipitation sums and southern Europe where decreasing annual precipitation sums are observed and predicted, no major changes in precipitation are observed and predicted, with only a very slight increasing trend in some areas [7,8]. The largest increases in precipitation with different significance of changes were observed in the northern part of Poland with different levels of significance and in a small area of south-eastern Poland [9]. In the light of the analysis of observational series averaged for the area of Poland, only the increasing variability of precipitation sums is undeniable [10,11]. Rising air temperatures, especially since the second decade of the 20th century, have increased evapotranspiration and may be a significant cause of increased drought caused by insufficient precipitation among other things [12–14].

In contrast to floods, the effects of droughts are not immediate. The phenomenon increases slowly and its consequences become visible over a longer period of time. Furthermore, they are initially less visible and extend over a larger area than in the case of other extreme weather events [15–17]. The impact on the economy in a drought-affected region depends not only on the duration, intensity, and spatial extent of the phenomenon, but also on the vulnerability of the environment to the negative effects of droughts. On soils with deep groundwater levels and low useful retention, which prevail in Poland, with relatively low and variable precipitation and the observed increase in air temperature as well as the signaled increase in the frequency of meteorological extreme events, an increase in both the frequency and intensity of drought phenomena should be expected [18,19].

In the extensive literature on droughts, and particularly on droughts in Poland, one can distinguish several research trends or thematic sections covering the phenomenon in question. Drought as a meteorological phenomenon unfavorable for agriculture does not occur suddenly but it shows a specific cycle of development. Dębski [20] breaks down this cycle into four phases: distinguishing atmospheric drought, soil drying out transforming into soil drought, lowering of groundwater level, occurrence of deep lows in rivers and drying out of springs and small watercourses–hydrological drought, and long-term lowering of groundwater resources defined as hydrogeological drought. Extended atmospheric drought may develop into soil drought, often referred to as agricultural drought. It occurs when a lack of precipitation, usually combined with high air temperatures, causes the soil to dry out, severely restricting the growth and development of crops and resulting in a significant decrease in crop yields. The time scale over which soil drought can occur is 1–3 months [15]. The prolonged period of low precipitation, often in combination with increased air temperature, leads to hydrological drought manifested by decreased water flow in rivers and water level in lakes, and at a further stage to hydrogeological drought manifested by decreased groundwater resources. The phases distinguished by the author correspond to the divisions into atmospheric, soil, and hydrological droughts often used in the literature [21,22].

Each of the mentioned phases is characterized by a different course and requires different research methods. One of them is the method of rain-free sequences. In Poland, Schmuck [23] was the first to analyze droughts on the basis of rain-free sequences. The author together with Koźmiński [24] presented a spatial distribution of frequencies of droughts lasting over 8 and 17 days in Poland. Drought monitoring uses many indices based on precipitation alone or precipitation and other meteorological elements and indices, often taking into account evapotranspiration of plants and soil water reserves, as well as groundwater. A review of this is provided by Przedpełska [21] and later by Łabędzki [15,25]. The simplest and the most widely used are indices using precipitation in such modifications as the relative precipitation index RPI [4] and the standardized precipitation index SPI [26–29].

Drought monitoring in agricultural areas should take into account meteorological evaporation conditions in addition to precipitation. For example, we can mention the Sielianinov hydrothermal index, the De Martonne dryness index, index evapotranspiration, or standardized climatic water balance [30–33]. The Institute of Soil Science and Plant Cultivation–State Research Institute (IUNG-PIB) in Puławy, at the request of the Ministry of Agriculture and Rural Development, developed and launched an agricultural drought monitoring system (PL abbr. SMSR). The importance of the problem of drought monitoring is also emphasized by the fact that it has found appropriate legal empowerment [34]. Based on the Act, the Ministry of Agriculture and Rural Development carries out the task of drought monitoring by specifying that the IUNG-PIB determines the current values of climatic water balance (CWB) "in the period from 1 June to 20 October, within 10 days after the end of the six-day period, indicators of climatic water balance for individual crop species and soils, broken down by voivodeship, on the basis of data provided by the Institute of Soil Science and Plant Cultivation–State Research Institute" [35].

Drought is a relative phenomenon and its assessment should be related to the current agricultural area and a specific crop. Therefore, by analyzing the current state of research, a number of studies on drought can be listed for particular regions, such as Schmuck [36] for Lower Silesia, Prawdzic and Koźmiński [37] for the Szczecin province, Konopka [38] for the Bydgoszcz region, and Łabędzki [39] for the Bydgoszcz-Kujawy region, as well as publications on selected crops, e.g., winter wheat [40], medium–late and late potatoes [41], or spring cereals [42].

Characteristics of particular droughts or drought periods, such as the 1959 drought [43], the drought of 1969 [44], the dry period of 1982–1992 by Bobiński and Meyer [45], the drought of 1992 [46], or the drought of 2005 [47], have been developed. A few sentences on the course and effects of droughts in 2003 and 2005 may be found in the monograph by Łabędzki [15] devoted to agricultural droughts in Poland.

As for drought problems discussed directly in this paper and concerning precipitation deficits and drought cases in Poland in the first years of the 20th century, it is worth mentioning the first study by Hohendorf [48] on precipitation deficits and excesses for the period 1891–1930 and a more recent one by Dzieżyc et al. [49] for the realities of the period from 1952–1980, Farat et al. [50] for the 40-year period from 1951–1990, Ziernicka-Wojtaszek and Zawora [51] for the 30-year period from 1971–2000, Doroszewski et al. [14] for the period from 1961–2010, and Przybylak et al. [52] for the period of over 1000 years from the establishment of the Polish state in the year 1996 to 2015.

In recent years, studies from the stream of contemporary climate change taking into account the impact of rising temperature on the occurrence of droughts have become increasingly frequent [53–56]. Drought is increasingly monitored using satellite data [57,58].

In the warmest year in the history of observation, 2019, especially in the summer, another intense drought occurred in Poland. This study uses preliminary documentation and characterization, and is a continuation of such studies made in the 20th and 21st centuries for dry years, such as 1959, 1969, 1982, 1983, 1989, 1992, 2000, 2003, 2005, 2008, and 2013. The study includes analysis of the causes, the course, and the consequences of the summer drought of 2019, and characterizes consecutive drought-meteorological, agricultural, hydrological, and hydrogeological phases. It shows, by comparison with extreme thermal droughts of 2003, 2018, and 2019 in Central Europe, its uniqueness. It proves that, in terms of climatic water balance values in summer, it is a case of maximum maximorum, which means that there is no other value in the series that is above or at least at the level of this maximum. It is a typical case of thermal drought caused not so much by precipitation deficit as by intensive evapotranspiration, following extremely high air temperatures. Hence, the methods which were atypical or modified for drought analysis, such as the analysis of months with precipitation frequency lower than the average rather than the analysis of classical sequences of days without precipitation, or the additional analysis of the simultaneous influence of precipitation and air temperature on the value of climatic water balance using the method of stepwise multiple regression, were used to show the leading role of temperature in generating the drought phenomenon.

2. Materials and Methods

The research material consists of verified, homogeneous monthly mean values of insolation, air temperature, and precipitation totals from 47 meteorological stations evenly distributed across Poland for the periods June–August 2019 and June–August 1981–2010 (Figure 1). Due to the insufficient number of stations, mountainous, and especially high-mountainous, areas are poorly represented. Data on sunshine (hours), temperature (°C), and precipitation (mm) were taken from the database of the Institute of Meteorology and Water Management. The study material was evaluated in terms of the degree of data homogeneity [59].

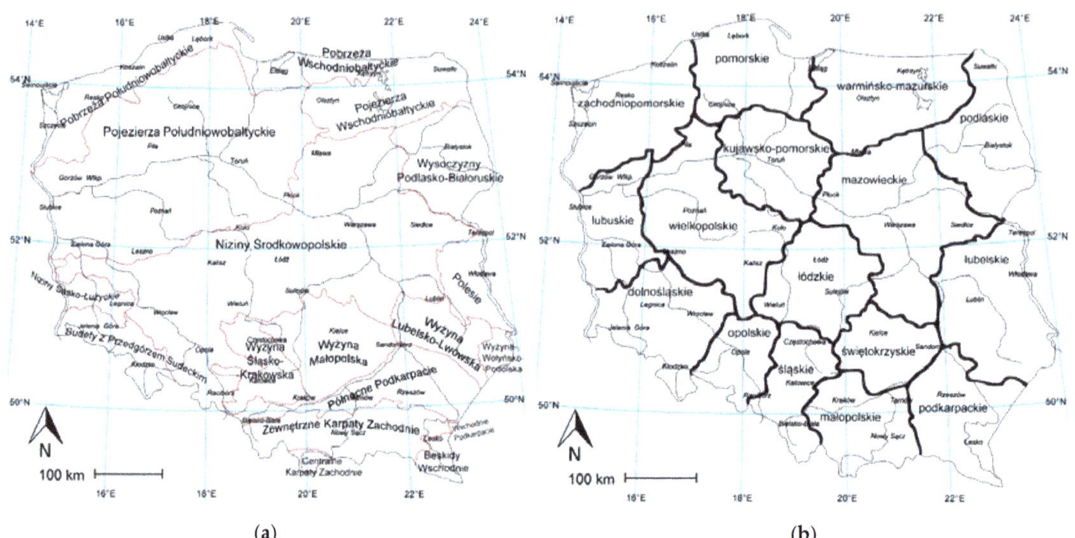

Figure 1. Distribution of weather stations in Poland: (**a**) geographical regions; (**b**) voivodeships. PL-EN: Beskidy Wschodnie-Eastern Beskids, Centralne Karpaty Zachodnie-Central Western Carpathians, Niziny Sasko-Łużyckie-Lowlands Sasko-Łużyckie, Niziny Środkowopolskie-Central Polish Lowlands, Pobrzeża Południowobałtyckie-South Baltic Coast, Pobrzeża Wschodniobałtyckie-Eastern Baltic Coast, Podkarpacie Wschodnie-Eastern Subcarpathia, Pojezierza Południowobałtyckie-South Baltic Lakes, Pojezierza Wschodniobałtyckie-Eastern Baltic Lakes, Polesie–Polesie, Północne Podkarpacie-Northern Subcarpathia, Sudety z Przedgórzem Sudeckim-Sudety Mountains with the Sudeten Foreland, Wysoczyzny Podlasko-Białoruskie-Podlasie-Byelarus Uplands, Wyżyna Lubelsko-Lwowska-Lublin-Lviv Upland, Wyżyna Małopolska-Małopolska Upland, Wyżyna Śląsko-Krakowska-Silesia-Cracow Upland, Wyżyna Wołyńsko-Podolska-Volyn-Podolia Upland, Zewnętrzne Karpaty Zachodnie-Outer West Carpathians.

In addition to meteorological data, the following were used: maps published within the agricultural drought monitoring system of IUNG–CWB, maps with comments from 14 reporting periods, summaries of drought-prone areas for the mentioned 14 reporting periods, all 14 studied plants and 16 voivodeships, and summaries of drought occurrence for selected communes of the Lubuskie voivodeship on particular soil categories for the year 2019.

The characteristics of weather patterns, i.e., temperature, precipitation, and synoptic situations against multi-year averages contained in the Bulletin of Climate Monitoring of Poland of the Institute of Meteorology and Water Management and the Bulletin of the National Hydrological and Meteorological Service for 2019; data on the outflow from the Bulletin of the National Hydrological and Meteorological Service for 2019; and data on the groundwater table level from the Hydrogeological Annual Report Polish Hydrogeological Survey for 2019 were also used.

The scope of the work included: (1) Characteristics of the weather pattern during the 2019 drought: (a) pluviothermic December 2018 to October 2019 including days with maximum temperature equal to and above 25 °C on the background of the multi-year (1981–2010); (b) characteristics of the occurrence of probability of extreme values of temperature and precipitation on the background of the multi-year (1951–2018); (c) frequency of synoptic situations on the background of the multi-year (1951–2018); (d) characteristics of the driest months June to August 2019 on the background of the multi-year (1951–2019). (2) The characteristics of meteorological drought: (a) the low rainfall frequency method for the entire summer drought period; and (b) the relative precipitation method RPI for the period of June–August. (3) The characteristics of agricultural drought: (a) on the basis of the climatic water balance (CWB) values for the following months

June, July, and August 2019 and in the period 1981–2010; (b) the temporal dynamics of drought (during the growing season); (c) a ranking of the provinces affected by the drought; and (d) the vulnerability of particular crops to drought. (4) Elements of hydrological drought. (5) Elements of hydrogeological drought.

Due to the extraordinary nature of the phenomenon, very different methods were used.

The method for periods of low precipitation frequency.

It was not possible to distinguish classical sequences of rain-free days, even in the areas regarded as the driest, comprising south-western and central-western parts of Poland. Precipitation-free sequences of several days at the most, often lasting several days, were separated by one, two, or even several days of local precipitation of thunderstorm character. Attempts were made to separate such drought sequences interrupted by rainfall in June–August by the predominance of rain-free days in the selected periods. At the same time, meteorological stations were found at which precipitation dominated separated by rainless days and which could not be regarded as drought periods. These were high-mountain stations and the Lesko station and stations in north-central and north-eastern Poland, such as Lębork, Elbląg, Olsztyn, Kętrzyn, and Suwałki. The limiting value was the average multi-year number of days with precipitation amounting to 41% of the duration of the summer period from June–August [60]. On the basis of this criterion, drought periods with rare precipitations were distinguished, in which the average number of days with precipitation amounted to 24%. In the remaining periods, not included in the drought periods, the number of days with precipitation amounted to 43%, which slightly exceeded the average number of days with precipitation—41%.

The relative precipitation method *RPI*.

The relative precipitation index RPI [4] is defined as the ratio of the total precipitation in a given period to the average multi-year sum, taken as the norm:

$$RPI = \frac{P}{\overline{P}} \cdot 100\% \tag{1}$$

where P—the total precipitation in the study period (mm); and $\overline{P}$—the average precipitation in the studied multi-year period (mm). The 1981–2010 norm was assumed.

The method of climatic water balance (CWB).

The climatic water balance (*CWB*) was calculated as the difference between precipitation (*P*) and potential evapotranspiration (*PET*):

$$CWB = P - PET \tag{2}$$

where P—precipitation (mm); and PET—potential evapotranspiration (mm).

To calculate the potential evapotranspiration, the simplified formula developed by Doroszewski and Górski [61], based on Penmann's algorithm [62], was used:

$$PET = -89.6 + 0.0621 \cdot t^2 + 0.00448 \cdot h^{1.66} + 9.1 \cdot f \tag{3}$$

where: PET—the monthly potential evapotranspiration (mm · [month]^(−1));

f—the length of the middle day of the month (hour);

h—the monthly insolation (hour);

t—the average monthly air temperature 2 m above the ground surface (°C).

The Institute of Soil Science and Plant Cultivation (IUNG) method of determining climatic water balance (CWB) in 6-decade periods with a step every decade was modified to characterize the CWB in monthly periods. The calculated values of climatic water balance for the months of June, July, and August were compared with the same months from the multi-year period from 1981–2010 and with the corresponding IUNG reporting periods for the estimated monthly periods.

The temporal dynamics of drought.

The dynamics of drought over time in the growing season were characterized by presenting—for the entire area of Poland without regional differentiation (discussed in other chapters)—the values of climatic water balance (mm) and the area at risk of drought (%) in the following 14 reporting periods from 21 March to 30 September 2019: reporting period 1 (from 21 March to 20 May), reporting period 2 (from 1 April to 31 May), reporting period 3 (from 11 April to 10 June), reporting period 4 (from 21 April to 20 June), reporting period 5 (from 1 May to 30 June), reporting period 6 (from 11 May to 10 July), reporting period 7 (from 21 May to 20 July), reporting period 8 (from 1 June to 31 July), reporting period 9 (from 11 June to 10 August), reporting period 10 (from 21 June to 20 August), reporting period 11 (from 1 July to 31 August), reporting period 12 (from 11 July to 10 September), reporting period 13 (from 21 July to 20 September), and reporting period 14 (from 1 August to 30 September).

Ranking of voivodeships by area at risk of drought.

A ranking of voivodeships by drought risk, illustrating spatial diversification of drought phenomenon, in addition to values of climatic water balance, was presented for two time periods: the whole reporting period covering the vegetation period from 21 March to 30 September, and the reporting period 8 of the highest drought intensity covering the months of June and July. The corresponding summaries for voivodeships published in IUNG reports for the individual 14 reporting periods were used. Voivodeships were ranked in terms of the area at risk of drought based on the average value of the 14 reporting periods (as above) and the eighth period with the highest drought intensity.

Sensitivity of crops to drought.

Sensitivity of crops to drought was determined by the percentage of cases of drought occurrence in all 14 reporting periods for the entire Poland, in all reporting periods for the three voivodeships where drought was the most intense (i.e., lubuskie, wielkopolskie, and łódzkie voivodeships), and for three voivodeships for the driest reporting period covering June and July. The 14 crops studied by IUNG were ranked from the most drought-sensitive to the least drought-sensitive on the basis of the average risk for the whole of Poland, within the three voivodeships most threatened by drought and within those voivodeships in the eighth period with the highest drought intensity.

Hydrological aspects of drought.

Hydrological aspects of drought were characterized by the acceleration of the timing of minimum flows and comparison of minimum flows with multi-year averages.

Hydrogeological aspects of drought.

Hydrogeological aspects of drought were characterized on the basis of the difference in the level of the groundwater table with respect to multi-year average values in individual months.

The results on the background of drought patterns in different European countries are presented against a historical reconstruction of a 254-year climate database for Europe and drought projections in Europe for the period 2041–2070 compared to 1981–2010 for two emission scenarios: RCP4.5 and RCP8.5.

The uniqueness of the 2019 drought phenomenon against the background of (compared to) the hottest and driest years of 2003 and 2018 was demonstrated by comparing the values of the climatic water balance in individual months of the summer (June–August) and for the whole summer period in the stated three years. The results of the comparison are included in the final "Discussion" chapter.

The maps of spatial distribution of drought were made in the Surfer 10 program. The kriging method using spherical function fitting was used for their interpolation. Taking into consideration small scale of the maps, it was evident that they had illustrative character. To describe spatial variability of precipitation, the names of mesoregions according to physico-geographical regionalization by Kondracki [63] were used.

3. Results

3.1. Synthesis of Weather Patterns December 2018–October 2019 for Air Temperature, Precipitation, Atmospheric Circulation

The months leading up to the 2019 growing season were treated more generally, more specifically the driest months of June, July, and August.

The winter period and the April–October 2019 growing season in question can be characterized as follows:

December 2018 was anomalously warm with a deviation of 2.0 °C from the 1981–2010 norm and very wet with precipitation of 126–150% of normal with snow cover lingering for several to a dozen days.

January 2019 was slightly cool with a temperature deviation from normal of −0.5 °C on the borderline between humid and very humid with precipitation of the order of 125% of normal with snow cover persisting throughout the month, in warmer regions a few to a dozen days.

February was very warm with temperature deviation from the norm of 3.0–4.0 °C and was dry with precipitation of 80% of the norm with snow cover lasting several days, especially in colder regions.

March was anomalously warm with temperature deviation from the norm of 2.0–3.0 °C and was slightly dry with precipitation of 80% of the norm.

April, in terms of temperature, was anomalously warm with temperature deviations from the norm exceeding 2.0 °C in the central-western part of Poland, in the Mazowiecka Basin and near Suwałki. Only the coast, i.e., the Sudety Mountains and the Carpathians, saw very warm conditions. In terms of precipitation, the month was normal in the south of the country and, locally, it was humid and very humid there. In the remaining part of Poland, April was mostly extremely dry. Anticyclonic situations prevailed: 57% over cyclonic 20% and zero 23%. Advections from the north-east direction prevailed for 30% of days, followed by advections from the east, the south-east, and the south for 20% each.

May was the only month of the year with temperatures lower than normal within the limits of −0.7 °C and was very cold in the prevailing area of the country. In the prevailing area of Poland, it exceeded the precipitation norm and was the month with the highest relative precipitation in the year within 145% of the norm. Anticyclonic situations prevailed at 54%, and the prevailing north-west direction of advection was 28%. This frequency was more than twice the frequency of the 1951–2018 multi-year average of 12%.

June was extremely warm over most of Poland, with temperatures within the range of 5.0–6.0 °C. Across about 80 percent of the country, there were days with a maximum temperature of >25.0 °C every day. In 18 days on almost half of the Polish territory, the maximum temperature exceeded 30.0 °C. The probability of such a warm June can be estimated at less than 1%. June was extremely dry on most of the country, and only Eastern Pomerania and Warmia and Mazury received normal or even wet precipitation. The probability of occurrence of such low precipitation at representative meteorological stations can be estimated as 1% at the Kraków station, 27% at the Słubice station, 31% at the Toruń station, 6% at the Warszawa station, and 11% at the Wrocław station. Anticyclonic situations prevailed, accounting for 59%. The dominant direction of advection was southerly. Advection from the south direction was 2.5 times higher than the average of 8%. The frequency of circulation from the south-east direction was twice as high as the average for the 1951–2018 period.

July was normal in terms of temperature, only slightly warm in the Carpathians, and very warm in the Sudetes. The probability of such a warm July can be estimated at 40%. July in the prevailing area of Poland was dry or extremely dry. After June and April, it was the third relatively driest month of the year. The probability of occurrence of such low precipitation at representative meteorological stations can be estimated as 59% at the Kraków station, 45% at the Słubice station, 16% at the Toruń station, 20% at the Warszawa station, and 22% at the Wrocław station. Cyclonic situations accounted for almost half of the cases. Advection from the north-east direction prevailed strongly, accounting for 33% of all cases in this month.

August was extremely warm over the prevailing area, with anomalies exceeding 2.0 °C in the warmest places. The probability of such a warm August can be estimated at 5%. August was humid in the east and south-east of Poland, while the remaining areas were dry and very dry, and, in the Lubuskie Land, it was extremely dry. The probability of occurrence of such a dry August can be estimated as 59% at the Kraków station, 6% at the Słubice station, 2% in Toruń, 20% in Warszawa, and 30% in Wrocław. Anticyclonic situations prevailed with a frequency of 44%. South-west-oriented advection prevailed with a frequency of 29%, followed by south-oriented advection with a frequency of 16%.

September was warm and sometimes slightly warm. September on the prevailing area of Poland was in the range between humid and extremely humid, only in the south-east, and, locally, in Silesia and the Sudeten basins, September was dry or very dry. Advection from the north-west direction prevailed with a frequency of 17%, which was 8% higher than the norm. Anticyclonic, cyclonic, and null situations occurred with an equal frequency of 33%.

October was generally very warm. The month was dry and very dry over the prevailing area of Poland, and, locally, in the center of the country, in the east, and in the south, it was even extremely dry. Cyclonic systems prevailed, accounting for 41%, with zero circulation 38%. Advection from a southernly direction with a frequency of 20% was dominant, which was 7% higher than normal. Advection from west direction was frequent—19%, also 7% more frequent than the norm, north-east direction—16% which was 9% more frequent than the norm.

Driest months (June, July, and August).

June with a temperature deviation of 5.0 °C for the whole Poland was the warmest month in the period since 1951. Although in the period 1951–2019 there were extreme years with very high temperatures as in 1964 with a deviation of 2.2 °C and in 1979 with a deviation of 1.9 °C but they were within 2.0 °C or below.

July was no longer so warm but normal. Years with temperature deviations above 2.0 °C occurred in the period since 1951 in 2006 with a deviation of 3.5 °C, with a deviation of 2.6 °C in 1994 and with a deviation of 2.4 °C in 2010.

In August, which was an extremely warm month in the mentioned period 1951–2019, there were years even warmer, such as 2015 with a deviation of 3.5 °C, 2.7 °C in 1992, 2.6 °C in 2018, and 2.1 °C in 2002.

As far as the moisture characteristics of the year 2019 are concerned, in the light of the flows for the whole area of Poland for the period 1951–2019, it ranks fifth after the driest years of 1954, 2015, 2016, and 1952 [Bulletin of the National Hydrological and Meteorological Service No. 13/215 2019].

3.2. Periods of Low Precipitation Frequency

The summer drought in terms of periods of low precipitation frequency began on 29 May with the change from cyclonal to anti-cyclonal circulation, which lasted for very different periods of time reaching 100 days in Warsaw and 101 days in Zielona Góra, i.e., until the first days of September. The longest period, i.e., exceeding 90 days, occurred at stations located in the Central Poland Lowlands (up to Warsaw) and the southern part of the South Baltic Lake District. On the outskirts of the mentioned area and in the Lublin Upland, the length of this period was shorter, of the order of 60–90 days. In the Sandomierz Basin and lower parts of the Sudety Mountains, it was a period of 30–60 days, and, in the Carpathian, the period lasted from several to twenty days. On the South Baltic Coast in its western and central part, this period lasted from 30 to 30 several days, while, in the Podlasie–Byelarus Uplands and Podlasie, the period lasted in between 35–37 days. In the surroundings of the Gulf of Gdańsk, the Eastern Baltic Coast, and the Lake Districts, the period of low precipitation frequency did not occur (Figure 2).

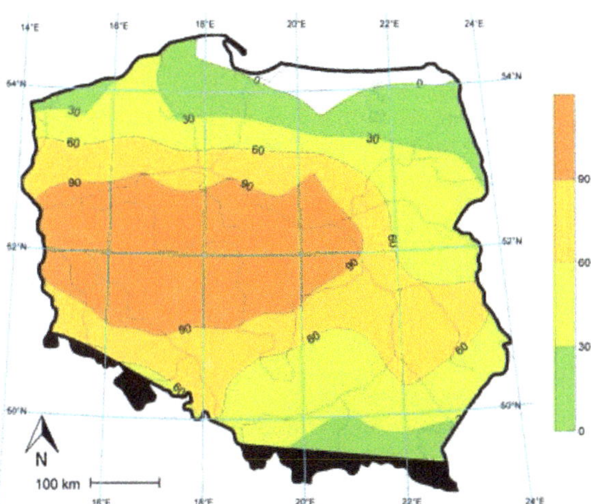

Figure 2. Duration of sequences of days with low precipitation frequency during summer (June–August) 2019. White color—phenomenon does not occur. Black color areas above 500 m above sea level.

3.3. Percentage of Normal Rainfall June–August 2019

From June to August 2019 in the South Baltic Lakes belt and the Central Polish Lowlands, precipitation totals were 30–60% of normal with a minimum of 30.1% in Poznań. Slightly higher values of precipitation occurred in the western part of the South Baltic Coast and Lake Districts belt within 70–90% of the norm. Values exceeding 90% of the precipitation norm occurred in the southern part of the Sudety Mountains with the Sudeten Foreland, in the northern part of the Małopolska Upland, in the western part of the Outer West Carpathians, and in the eastern part of Poland. The highest values of precipitation were recorded in the eastern part of the South Baltic Coast, the Eastern Baltic Coast, and the Podlasie–Byelarus Uplands within 100–125% of the norm. The average precipitation in Poland was only 66.5% of the norm (Figure 3).

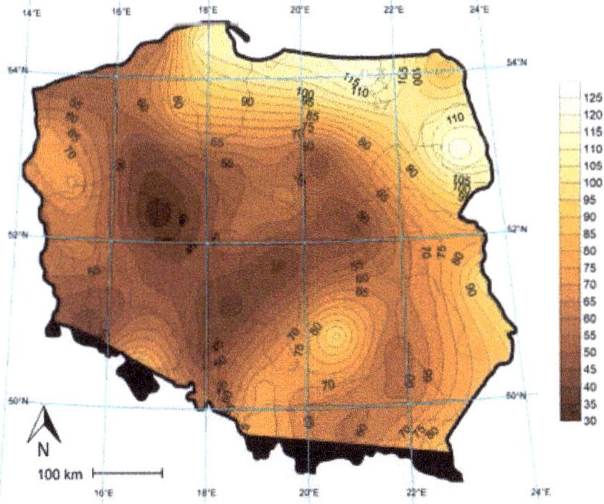

Figure 3. Percentage of normal rainfall for June–August 2019. Explanation same as in Figure 2.

3.4. Climate Water Balance (CWB) for June, July, August with Aspects of Regional Variation

In June 2019, the greatest water deficit occurred in the central Poland belt, especially its central and western parts with a maximum in Poznań at 176.6 mm. The water deficit decreased towards the north, south, and east. The lowest values of water deficit occurred in Eastern Baltic Coast and Lake Districts with the minimum of −71.3 mm in Kętrzyn. Lower deficit values were recorded in southern Poland with a minimum of −34.1 mm near Lesko. The mean value of CWB in Poland was −129 mm (Figure 4a). In June 1981–2010, the highest water deficit occurred in central Poland, especially in its central and western part with its maximum in Poznań at 106.7 mm. Water deficit decreased towards the north and south. The smallest deficit values occurred in the Eastern Baltic Coast and the Lake Districts with the minimum of −69.5 mm in Koszalin. Much lower values of deficit were recorded in southern Poland with a minimum of −12.9 mm near Lesko. The mean value of CWB in Poland amounted to −79 mm (Figure 4b).

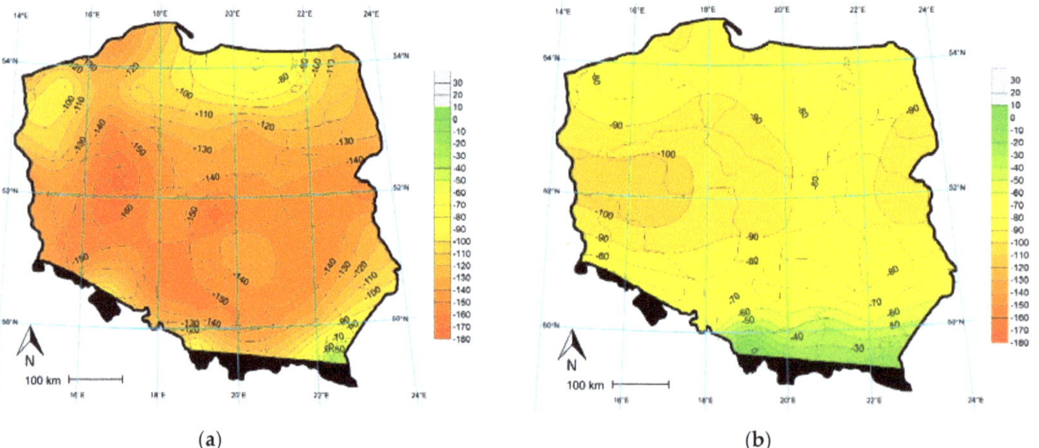

Figure 4. Climatic water balance (CWB) in June: (**a**) 2019; (**b**) 1981–2010. Explanation same as in Figure 2.

In July 2019, the largest water deficits occurred in the western part of the South Baltic Coast with a maximum of −109.8 mm in Świnoujście. Slightly lower values were observed in central Poland ranging from −60 mm to −90 mm. The lowest values of water deficit were observed in the eastern part of South Baltic Coast, Eastern Baltic Coast and Lake Districts and locally in the south of Poland. In the vicinity of Katowice, an excess of 6.1 mm was recorded. The average CWB over Poland was −64 mm (Figure 5a). In July 1981–2010, the highest water deficits occurred in western part of the South Baltic Coast with a maximum of −69.9 mm recorded in Ustka. Slightly lower values were observed in central Poland ranging from −30 to −60 mm. The lowest values of water deficit occurred in the southern part of Poland. In the vicinity of Bielsko–Biała and Lesko, an excess of water was recorded, at 19.1 and 13.5 mm, respectively. The mean CWB value over Poland was −36 mm (Figure 5b).

In August 2019, the largest water deficit values were characterized by the South Baltic Lakes region, especially around Słubice, Gorzów Wlk., and Poznań from −92.4 mm to −100.6 mm. As one moved away from these regions, a decrease in water deficit values in all directions was observed. Smaller values of water deficit were recorded in the north-eastern part of Poland and in the south of the country from −1.5 mm to −19 mm. Locally, in Kielce, Bielsko–Biała, and Lesko, there was an excess of water, with recordings of 23.8 mm, 18 mm, and 1.4 mm, respectively. The mean CWB value over Poland was −53 mm (Figure 6a). In August 1981–2010, the highest values of water deficit were characterized by the eastern part of Poland, especially near Terespol and Włodawa, which saw values of −55.8 mm and

−55.7 mm, respectively. Smaller values of water deficit occurred in northern part of Poland and in the south of the country, ranging from −1.5 mm to −30 mm. The mean value of CWB in Poland was −38 mm (Figure 6b).

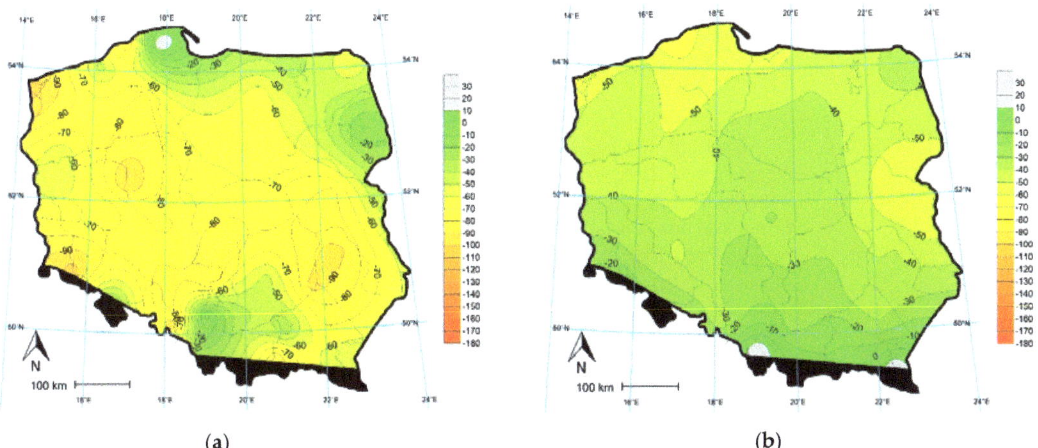

Figure 5. Climatic water balance (CWB) in July: (**a**) 2019; (**b**) 1981–2010. Explanation same as in Figure 2.

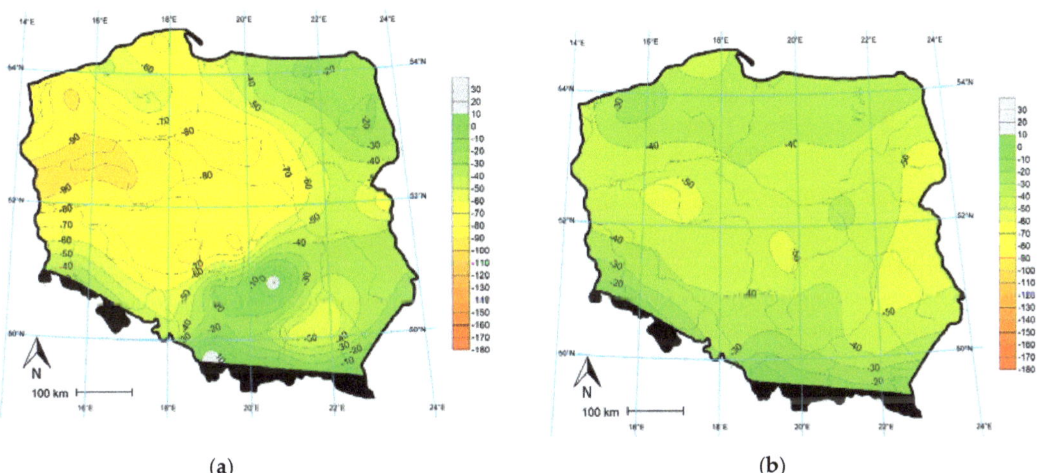

Figure 6. Climatic water balance (CWB) in August: (**a**) 2019; (**b**) 1981–2010. Explanation same as in Figure 2.

In light of the comparison presented, June saw with the highest temperature deviation from normal and low rainfall, and was the month with the highest CWB deviation of 50 mm. The values of these deviations from normal decreased to −28 mm in July and −15 mm in August (Table 1).

Table 1. Comparison of mean climatic water balance (CWB) values (mm) over Poland in the months June–August in 2019 with values from the 1981–2010 multi-year period.

Monthly	Year 2019	Multi-Year Period 1981–2010	Difference
June	−129	−79	−50
July	−64	−36	−28
August	−53	−38	−15

3.5. The Temporal Dynamics of Drought

Drought was practically absent in the first four reporting periods, i.e., 21 March to 20 May up to 21 April to 20 June. CWB values in these periods were several tens of mm with the highest value of −48 mm in the first period, and the area at risk of drought ranged from a fraction of a percent to 2.6% in the fourth reporting period. From the fifth reporting period, i.e., 1 May to 30 June, the mean CWB value for Poland decreased from −87 mm, through subsequent periods −94 mm and −133 mm to reach the highest value of −173 mm in the eighth reporting period from June to July. In the following reporting periods, CWB values systematically decreased from −134 mm, −106 mm, −91 mm, −54 mm, and −44 mm to reach −20 mm in the last reporting period. The percentage of area at risk of drought increased at a similar rate from a fraction of percent in the first three reporting periods through several to a dozen or so percent in periods 4–6, i.e., 21 April–20 June to 11 May–10 July, to reach 28% in period 7 and the highest value of 42% in period 8. In periods 9 and 10, i.e., 11 June to 10 August and 21 June to 20 August, it was already 18% and 6%, respectively, in the following periods a few percent and fractions of a percent. The highest CWB values occurred in the Lubuskie Lake District and systematically increased from −120 mm in the first reporting period to −240 mm to −279 mm in the eighth period of the highest drought intensity. In subsequent periods, they decreased systematically to reach the value from 100 mm to −159 mm in the last reporting period. The percentage of the area endangered by drought in the three lubuskie, łódzkie, and wielkopolskie voivodeships (most endangered by drought) increased from a fraction of a percent to several percent in the first three reporting periods to reach the values of 52%, 73%, 97%, and 56% in periods 6–9, respectively (Figure 7).

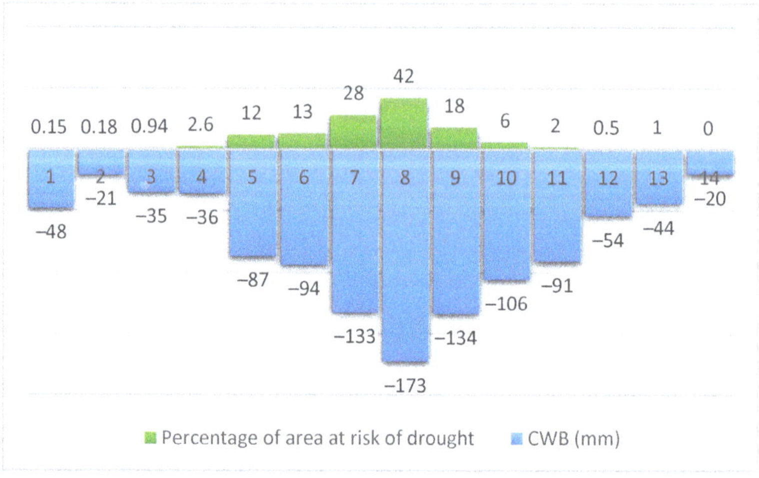

Figure 7. Temporal dynamics of drought in 14 IUNG reporting periods based on climatic water balance (CWB) values and percentage of area at risk of drought average values for Poland without regional differentiation.

Explanations: reporting period 1 (from 21 March to 20 May), reporting period 2 (from 1 April to 31 May), reporting period 3 (from 11 April to 10 June), reporting period 4 (from 21 April to 20 June), reporting period 5 (from 1 May to 30 June), reporting period 6 (from 11 May to 10 July), reporting period 7 (from 21 May to 20 July), reporting period 8 (from 1 June to 31 July), reporting period 9 (from 11 June to 10 August), reporting period 10 (from 21 June to 20 August), reporting period 11 (from 1 July to 31 August), reporting period 12 (from 11 July to 10 September), reporting period 13 (from 21 July to 20 September), and reporting period 14 (from 1 August to 30 September).

3.6. Ranking of Voivodeships by Area at Risk of Drought

Despite slight differences in the order of voivodeships for average data from all reporting periods and from the eighth period with the highest drought intensity, a clear regularity was observed. It was possible to distinguish four groups of voivodeships with the highest, medium, and lowest drought risk, and one voivodeship in which drought practically did not occur. Voivodeships with the highest drought risk included the lubuskie, łódzkie, and wielkopolskie voivodeships. The voivodeships with a medium drought threat were the dolnośląskie, kujawsko-pomorskie, lubelskie, mazowieckie, opolskie, śląskie świętokrzyskie, and zachodnio-pomorskie voivodeships. The voivodeships with a low drought risk included małopolskie, podkarpackie, podlaskie, and pomorskie. In the warmińsko-mazurskie voivodeship, drought practically did not occur (Table 2).

Table 2. Ranking of provinces in terms of area at risk of drought (%) at the area of the voivodeship at draught risk.

Item	Voivodeship	Average over 14 Study Periods (%)	Voivodeship	Peak Period 8 Value (%)
1	lubuskie	39	wielkopolskie	92
2	wielkopolskie	25	lubuskie	88
3	łódzkie	20	łódzkie	81
4	mazowieckie	8	mazowieckie	49
5	zachodniopomorskie	8	śląskie	45
6	dolnośląskie	7	opolskie	43
7	opolskie	7	kujawsko-pomorskie	40
8	kujawsko-pomorskie	6	lubelskie	40
9	lubelskie	6	zachodniopomorskie	40
10	śląskie	5	dolnośląskie	33
11	podlaskie	3	świętokrzyskie	32
12	świętokrzyskie	3	podkarpackie	17
13	pomorskie	1.6	podlaskie	16
14	podkarpackie	1.4	pomorskie	16
15	małopolskie	0.6	małopolskie	7
16	warmińsko-mazurskie	0.07	warmińsko-mazurskie	1

Source: Agricultural drought monitoring system in Poland IUNG (2019).

3.7. Sensitivity of Crops to Drought

Analysis of areas covered by drought, both in the whole of Poland and in the three voivodeships most threatened by drought, i.e., the lubuskie, wielkopolskie, and łódzkie voivodeships, revealed clear differences in the drought sensitivity of individual crops. The crops most threatened by drought are fruit bushes, spring cereals, maize, and legumes. For Poland, the percentage of these crops threatened by drought ranges from 16–18%, while, on the area of the three voivodeships most exposed to drought, it is as high as 42–49%. Tobacco, winter cereals, field vegetables, and strawberries are among the crops which are moderately sensitive to drought. On the other hand, the crops least exposed to drought are rape and colza seed, hops, potatoes, fruit trees, and sugar beet, and the percentage of the area of these crops affected by drought is 2–8%, respectively, on the area of Poland as a whole, and 11–26% on the area of the three voivodeships mentioned. In the eighth driest reporting period, from 1 June to 31 July, the majority of crops in the lubuskie, wielkopolskie, and łódzkie voivodeships were almost completely threatened by drought (Table 3).

Table 3. Sensitivity of crops to drought. Percentage of drought incidents in 14 consecutive reporting periods and during the period of maximum drought intensity.

Item	Crop and Percentage of Area at Risk of Drought Across Poland	Crops and Percentage of Area at Risk of Drought in the Lubuskie, Wielkopolskie, and Łódzkie Voivodeships	Crops and Percentage of Area at Risk of Drought in the Lubuskie, Wielkopolskie, and Łódzkie Voivodeships during the Period of 8 Highest Intensity
1	Fruit bushes: 18	Spring cereals: 49	Spring cereals: 92
2	Spring cereals: 17	Fruit bushes: 47	Fruit bushes: 99
3	Grain maize: 17	Silage maize: 46	Silage maize: 96
4	Silage maize: 17	Grain maize: 45	Grain maize: 96
5	Leguminous plants: 16	Leguminous plants: 42	Leguminous plants: 98
6	Tobacco: 13	Winter cereals: 39	Winter cereals: 97
7	Winter cereals: 12	Strawberries: 39	Strawberries: 82
8	Ground vegetables: 12	Ground vegetables: 38	Field vegetables: 91
9	Strawberries: 11	Tobacco: 37	Tobacco: 97
10	Rapeseed and colza seed: 8	Potatoes: 26	Potatoes: 86
11	Hops: 7	Rapeseed and colza seed: 25	Rapeseed and colza seed: -
12	Potatoes: 5	Hops: 25	Hops: 84
13	Fruit trees: 5	Fruit trees: 19	Fruit trees: 76
14	Sugar beet: 2	Sugar beet: 11	Sugar beet: 45

Source: agricultural drought monitoring system in Poland IUNG (2019).

3.8. Hydrological Aspects of Drought

According to the classification of Kaczorowska [4], which assesses the deficiency or excess of precipitation in relation to the perennial norm (1971–2000), the year 2019 was classified as normal [Bulletin of the National Hydrological and Meteorological Service No. 13/215 2019]. Annual precipitation on a national scale, determined by measurements from 52 synoptic stations, amounted to 556 mm, corresponding to 91.7% of the multi-year value (1971–2000). In the central part of Poland, 2019 was classified as dry, locally even as very dry, and in the rest of the country as normal, only locally in the north and south as wet. The hydrological year 2019 with a total outflow of Polish rivers equal to 41.9 km^3 (with an average in the multi-year period 1951–2018 equal to 60.4 km^3) was classified as a dry year [Bulletin of the National Hydrological and Meteorological Service No. 13/215 2019]. The outflow in 2019 was, therefore, about 2/3 of the multi-year average outflow. In the distribution sequence from the period 1951–2019, which contains 69 years, where the years are arranged in the order of increasing total annual outflow of Polish rivers, the year 2019 occupies the 5th place after the driest the years, i.e., 1954, 2015, 2016, and 1952. In the first and second hydrological half-years, lower values of outflow occurred in the Odra basin than in the Vistula basin. The outflow of the Pomeranian rivers was, like other rivers, lower than the norm, and was generally relatively higher than the outflow of the rivers of the Vistula and Oder basins. Outflow of the Pomeranian rivers, like other rivers, was lower than the regular values; however, in general, it was relatively higher than the outflow of the rivers of the Vistula and Odra basins [Bulletin of the National Hydrological and Meteorological Service No. 13/215 2019] (Table 4).

Table 4. Months with the lowest flow in 2019 compared to the 1951–2018 multi-year period.

River and Water Level Gauge	Flow 2019 in m^3/s and Months	Average Flow 1951–2018 in m^3/s and Months
Vistula–Warsaw	228.0 (July)	435.0 (September)
Wieprz–Kośmin	12.0 July)	26.5 (August)
Pilica–Sulejów	7.53 (August)	17.5 (September)
Odra–Racibórz-Miedonia	16.3 (July)	41.0 (October)
Odra–Ścinawa	48.4 (July)	129.0 (October)
Odra–Nowa Sol	56.2 (August)	157.0 (October)
Nysa Kłodzka–Skorogoszcz *	8.6 (November)	24.6 (October)
Bóbr–Żagań	9.1 (August)	24.9 (October)
Warta–Sieradz	14.4 (August)	32.9 (September)
Warta–Poznań	26.5 (July)	69.8 (September)

* Flow values are affected by water management in the reservoir. Source: Bulletin of the National Hydrological and Meteorological Service No. 13/176 2016, 13/189 2017, 13/202 2018, 13/215 2019

After analyzing 20 river flows of very different river basins [Bulletin of the National Hydrological and Meteorological Service No. 13/215 2019], in almost half of them, i.e., in nine, the drought accelerated the date (month) of the lowest flow. Thus, this acceleration was observed in three cases by one month (two cases from September to August and one case from August to July). In four cases, it was an acceleration by two months (in two cases from October to August and in two cases from September to July). In two cases, the acceleration of the lowest flow was by three months, i.e., in all cases from October to July. These flows constituted on average of 41% in average perennial flows, a maximum of 52% (Vistula–Warszawa), and a minimum of 36% (Odra–Nowa Sól). It is noteworthy that all cases of minimum flow acceleration only occurred in central-western and south-western Poland with the highest air temperature. In seven cases, apart from those analyzed above, minimum flows occurred in the same months as the multi-year averages, but their values were, on average, smaller by half. In the case of the Skorogoszcz gauge on the Nysa Kłodzka River, flow values were affected by water management in the existing reservoir. Indeed, at the gauge mentioned, minimal flows did not occur in July and August, like most of the analyzed gauges, but in November. Low flows also occurred in October, December, and August. Water management in the reservoir, in this case, disturbed the natural rhythm of flows.

3.9. Hydrogeological Aspects of Drought

In the hydrological year of 2019, the level of the groundwater table was lower than the monthly average for many years (1991–2015). In waters with a free well, the largest deviation was recorded in August (about 31 cm below the average for many years). In waters with a tight well, the highest values below the average were recorded in the period of August–October where it was about 32–34 cm below the multi-year average. In July, the state hydrogeological service declared a state of hydrogeological emergency due to the phenomenon of a very intense hydrogeological low. In August, an extension of the area of the phenomenon was found, and, by the end of the hydrological year, there were no grounds for revoking the state of hydrogeological emergency [Hydrogeological Annual Report Polish Hydrogeological Survey].

In the lowland belt, where the drought was the most intensive, the mean monthly groundwater levels of both free and tight groundwater tables were lower than the monthly averages for the whole hydrologic year. The highest deviation of free-bore waters was recorded in August (36 cm below the average), and, by the end of the year, they were over 30 cm below the average for individual months of the studied period. Similarly, it was the case for waters with a tight well, the average monthly groundwater level was at a level lower than the monthly averages (this difference was 9 cm below the average in January), and then increased from month to month to 45 cm below the average in October, representing the largest deviations of waters with a tight well [Hydrogeological Annual Report Polish Hydrogeological Survey].

4. Discussion

A direct comparison of the IUNG method (CWB in 6-decade periods, step every decade) with the method used in the study (CWB in monthly periods), i.e., concerning the months of the most intensive drought in June, July, and August, is not possible due to different comparison periods. The first two decades of June in the IUNG method fall within the 6th reporting period, ie 11 May to 10 July. The second and third decades of June fall within the seventh reporting period, covering the period from 21 May to 21 July. Therefore, the average of the reporting periods for June (from reporting periods 6, 7, and 8), for July (from reporting periods 9, 10, and 11), and for August (from reporting periods 12, 13, and 14) were conventionally taken for the estimation of CWB values.

The results vary slightly, though there was the least variation in June. June was the warmest month since at least 1951 with a temperature deviation from the national norm of 5.0–6.0 °C and the driest relatively month of the year in 2019, and was very dry on the verge of extremely dry which was the reason for the very low CWB values [Bulletin of the National Hydrological and Meteorological Service No. 6/208 2019]. Smaller differences occurred in August and the largest in July. The differences may be caused, on the one hand, by the different way of estimating values for individual months as an average of the three reporting periods and, on the other hand, by the extension of large negative CWB values from June to the July (post-June) using the IUNG method to determine CWB for 6-week periods (Table 5).

Table 5. Comparison of climatic water balance (CWB) values (mm) estimated from IUNG studies and those obtained in this study.

Monthly	Results Presented in the Study	Estimated IUNG Results
June	−129	−133
July	−64	−110
August	−53	−39

The drought of 2019 was preceded by the drought of 2018. The analysis of consecutive communiqués of the IUNG agricultural drought monitoring system from 2020 shows that, in that year, drought was indicated, especially in north-western Poland voivodeships; in the wielkopolskie voivodeship during all 14 reporting periods; and in the lubuskie, pomorskie, and zachodniopomorskie voivodeships during 13 periods (except the last reporting period). In 2021, in the fifth reporting period from 1 May to 30 June, drought of various intensity was observed in all Polish voivodeships. The analysis of the causes, the course, and the consequences of the dry years of 2018–2021 requires a separate study.

The drought of 2019 has marked itself with varying intensity across Europe. The manifestations and impacts of drought are reported from countries, such as Belgium, Czechia, Finland, France, Germany, Greenland, Italy, Luxembourg, Netherlands, Norway, Poland, Spain, Sweden, Switzerland, and the United Kingdom [64]. A historical reconstruction of the 254-year climate database [65,66] indicates that many years with similar precipitation anomalies occurred in the summer months, but 2018–2019 saw two of the three warmest summer periods in the cited period. The third year when average summer temperature anomalies over Central Europe reached record extreme conditions of over 2.0 °C was 2003. Of the three cited years with the highest air temperature anomaly in Europe, in 2019, the strongest impact of the thermal anomaly was marked in the Central European region including Poland where the increase in temperature was accompanied by a simultaneous significant reduction in summer precipitation, leading to extreme drought conditions.

The comparison of the values of climatic water balance in the months of June, July, and August in the analyzed three extreme years in terms of temperature in the area of Poland shows that, in 2003, these values were −100, −21, and −79 mm, respectively. In 2018, the values averaged for Poland were −77, −36, and −80 mm, respectively. In 2019, the year under analysis, these values were, −129, −64, and −53 mm, respectively, so the highest were in June and July, and were slightly lower only in August. If the values of climatic water balance for the analyzed three months were averaged, then in the next extreme years, in terms of temperature, the values would be −67 mm in 2003, −64 mm in 2018, and −82 mm in 2019. Thus, it can be concluded that, in the light of the values of climatic water balance in Poland representing a significant area of central Europe, this was the most intense drought in the reconstruction of the last more than 250 years of climatic data.

Analysis of drought frequency over a multi-year period shows that the frequency has increased since 1950 throughout southern Europe and most parts of central Europe, while it has decreased in many parts of northern Europe [67,68]. Other drought indices, including drought severity indices, also show significant increases in the Mediterranean region and parts of central and south-eastern Europe, and decreases in northern Europe and parts of eastern Europe [69–72]. Projections for the period 2041–2070 compared to 1981–2010 for two emission scenarios, i.e., RCP4.5 and RCP8.5, indicate an increase in meteorological droughts in most of Europe, especially in southern Europe, while decreases in droughts are projected only for a limited part of northern Europe. The changes are most pronounced for the high emissions scenario (RCP8.5) and somewhat smaller for the moderate scenario (RCP4.5) [73].

5. Conclusions

The summer drought of 2019 was not only caused by a shortage of precipitation, but above all by extremely high temperatures, especially in June with a record deviation of 5.0–6.0 °C and in August when deviations exceeded 2.0 °C. Anticyclonic situations prevailed especially in June (59%) and August. In June, the advection from south direction was 2.5 times higher than the average of 8%.

In the light of days with low precipitation frequency (frequency lower than average), 41% of the duration of summer drought lasted from 29 May to the first days of September. The longest, i.e., 90 days and more, occurred at stations located in the Central Poland Lowlands (up to Warsaw) and the southern part of the South Baltic Lake District. On the outskirts of this driest area, the number of such days with a below-average precipitation frequency was systematically lower. Such a period with lower precipitation frequency basically did not occur only in the area of the Gulf of Gdańsk, the Eastern Baltic Coast and the Lake Districts.

In the summer season of June–August, the lowest precipitation amounts, constituting 30–60% of the norm were recorded in the South Baltic lakes belt and Central Polish Lowlands with the minimum of 30.1% in Poznań. The highest values of precipitation were noted in the eastern part of the South Baltic Coast, the Eastern Baltic Coast, and the Podlasie–Byelarus Uplands with 100–125% of the norm. The average precipitation in Poland in summer was only 66.5% of the norm.

The values of climatic water balance (CWB) calculated by the Institute of Soil Science and Plant Cultivation (IUNG) method for particular months of June–August for the area of Poland amounted successively to −129 mm, −64 mm, and −53 mm with minima, respectively, of −176.6 mm in Poznań, −109.8 mm in Świnoujście, and from −92.4 mm to −100.6 mm near Słubice, Gorzów Wielkopolski, and Poznań.

Summer drought in the light of the area of Poland threatened by drought increased gradually exceeding the values of several percent of the area of Poland in the periods from the beginning of June to 10 July, with the maximum at the end of June—42% of the area of Poland covered by drought. Most threatened by drought were the wielkopolskie, lubuskie, and łódzkie voivodeships, in which the percentage of the area threatened by drought during its highest intensity in the eighth reporting period (1 June to 31 July) amounted to 92%, 88%, and 81%, respectively.

The sensitivity of individual crops to drought is evident. The least resistant to drought are: fruit bushes, spring cereals, maize for silage and grain, and legumes. Medium drought tolerant are: winter cereals, tobacco, field vegetables, and strawberries. The most resistant are hops, potatoes, rapeseed and canola, fruit trees, and sugar beet.

The summer drought accelerated the timing of flow minima, which typically fall in the autumn months, by 1 to 3 months or halved the minimum flows without changing the timing of their occurrence. In the hydrological year 2019, the level of the groundwater table was lower than the average monthly level in many years. In waters with a free well, the largest deviations were recorded in August, and, in waters with a tight well, the largest values below the average were recorded in the August–October period.

Although contemporary trends and scenarios of precipitation changes at the end of the 21st century for Poland do not predict a decrease in precipitation and even a small increase of about 5–10%, it cannot be said that we will not be threatened by droughts. These will be droughts caused not so much by a lack of precipitation but by high air temperature combined with low precipitation (the increase of which in Poland has been documented). The phenomenon of drought will also be aggravated by the observed increase in the coefficient of variation of precipitation in Poland.

Funding: The research was partially financed by Ministry of Science and Higher Education in Poland.

Institutional Review Board Statement: Not applicable.

Informed Consent Statement: Not applicable.

Data Availability Statement: The data presented in this study are available on request from the corresponding author.

Conflicts of Interest: The authors declare no conflict of interest.

References

1. Kożuchowski, K. Zmienność opadów atmosferycznych w Polsce w XX i XXI wieku. In *Skala, Uwarunkowania i Perspektywy Współczesnych Zmian Klimatycznych w Polsce*, 1st ed.; Kożuchowski, K., Ed.; Łódź. Wydaw. Biblioteka: Łódź, Poland, 2004; pp. 47–58. (In Polish)
2. Mager, P.; Kasprowicz, T.; Farat, R. Change o air temperature and precipitation in Poland in 1966–2006. In *Climate Change and Agriculture in Poland—Impacts, Mitigation and Adaptation Measures*, 1st ed.; Leśny, J., Ed.; Acta Agrophys. Rozprawy i Monografie: Lublin, Poland, 2009; Volume 169, pp. 19–38.
3. Czarnecka, M.; Nidzgorska-Lencewicz, J. Wieloletnia zmienność sezonowych opadów w Polsce. *Woda-Sr. -Obsz. Wiej.* **2012**, *12*, 45–60. (In Polish)
4. Kaczorowska, Z. Opady w Polsce w przekroju wieloletnim. *Pr. Geogr. IG PAN* **1962**, *33*, 1–109. (In Polish)
5. IMGW-PIB: Climate Monitoring Bulletin of Poland, ISSN 2391-6362. Available online: http://klimat.pogodynka.pl/biuletyn-monitoring/ (accessed on 14 July 2021).
6. IPCC. *2018: Global Warming of 1.5 °C. An IPCC Special Report on the Impacts of Global Warming of 1.5 °C above Pre-Industrial Levels and Related Global Greenhouse Gas Emission Pathways, in the Context of Strengthening the Global Response to the Threat of Climate Change, Sustainable Development, and Efforts to Eradicate Poverty*; Masson-Delmotte, V., Zhai, P., Pörtner, H.-O., Roberts, D., Skea, J., Shukla, P.R., Pirani, A., Moufouma-Okia, W., Péan, C., Pidcock, R., et al., Eds.; IPCC: Geneva, Switzerland, 2019.
7. Peseta. Available online: http://peseta.jrc.es (accessed on 5 July 2021).
8. Jacob, D.; Petersen, J.; Eggert, B.; Alias, A.; Christensen, O.B.; Bouwer, L.M.; Braun, A.; Colette, A.; Déqué, M.; Georgievski, G.; et al. EUROCORDEX: New high-resolution climate change projections for European impact research. *Reg. Environ. Chang.* **2014**, *14*, 563–578. [CrossRef]
9. Szwed, M. Variability of precipitation in Poland under climate change. *Theor. Appl. Climatol.* **2018**, *135*, 1003–1015. [CrossRef]
10. Kożuchowski, K. Współczesne zmiany klimatyczne w Polsce na tle zmian globalnych. *Przegl. Geogr.* **1996**, *68*, 79–98. (In Polish)
11. Ziernicka-Wojtaszek, A.; Kopcińska, J. Variation in atmospheric precipitation in Poland in the years 2001–2018. *Atmosphere* **2020**, *11*, 794. [CrossRef]
12. Kożuchowski, K.; Żmudzka, E. Ocieplenie w polsce: Skala i rozkład sezonowy zmian temperatury powietrza w drugiej połowie XX wieku. *Prz. Geof.* **2001**, *46*, 81–90. (In Polish)
13. Żmudzka, E. Współczesne zmiany klimatu Polski. *Acta Agrophys.* **2009**, *13*, 555–568. (In Polish)
14. Doroszewski, A.; Jóźwicki, T.; Wróblewska, E.; Kozyra, J. *Susza Rolnicza w Polsce w Latach 1961–2010*; IUNG-PIB: Puławy, Poland, 2014; pp. 1–144. (In Polish)
15. Łabędzki, L. Susze rolnicze-zarys problematyki oraz metody monitorowania i klasyfikacji. *Woda-Sr. -Obsz. Wiejskie. Rozp. Nauk. I Monogr.* **2006**, *17*, 1–107. (In Polish)
16. Seneviratne, S.I.; Nicholls, N.; Easterling, D.; Goodess, C.M.; Kanae, S.; Kossin, J.; Luo, Y.; Marengo, J.; McInnes, K.; Rahimi, M.; et al. Changes in climate extremes and their impacts on the natural physical environment. In *Managing the Risks of Extreme Events and Disasters to Advance Climate Change Adaptation*; Field, C.B., Barros, V., Stocker, T.F., Qin, D., Dokken, D.J., Ebi, K.L., Mastrandrea, M.D., Mach, K.J., Plattner, G.-K., Allen, S.K., et al., Eds.; A Special Report of Working Groups I and II of the Intergovernmental Panel on Climate Change (IPCC); Cambridge University Press: Cambridge, UK; New York, NY, USA, 2012; pp. 109–230. Available online: https://www.ipcc.ch/site/assets/uploads/2018/03/SREX-Chap3_FINAL-1.pdf (accessed on 14 July 2021).
17. Krausmann, E.; Raimond, E.; Wood, M. Understanding disaster risk: Hazard related risk issues-Section IV. Technological risk. In *Science for Disaster Risk Management 2017: Knowing Better and Losing Less*; Poljanšek, K., Marín Ferrer, M., De Groeve, T., Clark, I., Eds.; EUR 28034 EN; Publications Office of the European Union: Luxembourg, 2017. [CrossRef]
18. Bartczak, A.; Glazik, R.; Tyszkowski, S. Identyfikacja i ocena intensywności okresów suchych we wschodniej części Kujaw. *Nauka Przyr. Technol.* **2014**, *8*, 1–22. Available online: http://www.npt.up-poznan.net/pub/art_8_46.pdf (accessed on 15 July 2021). (In Polish)
19. Szyga-Pluta, K. Zmienność występowania susz w okresie wegetacyjnym w Polsce w latach 1966–2015. *Przegl. Geof.* **2018**, *1–2*, 51–67. (In Polish)
20. Dębski, K. Rozważania na temat metod przewidywania suszy. *Prz. Meteor. Hydrol.* **1953**, *3–4*, 96–113. (In Polish)
21. Przedpełska, W. Zagadnienie susz atmosferycznych w Polsce i metody ich określania. *Pr. PIHM* **1971**, *103*, 3–24. (In Polish)
22. Wilhite, D.A.; Glantz, M.H. Understanding the drought phenomenon: The role of definitions. *Water Int.* **1985**, *10*, 111–120. [CrossRef]

23. Schmuck, A. Częstość występowania dni bezopadowych we Wrocławiu w latach 1883-1960. *Studia Geogr.* **1963**, *1*, 157–167. (In Polish)
24. Schmuck, A.; Koźmiński, G. Przestrzenny rozkład częstości posuch atmosferycznych na terenie Polski. *Czas. Geogr.* **1967**, *38*, 321–325. (In Polish)
25. Łabędzki, L. Problematyka susz w Polsce. *Woda-Sr. -Obsz. Wiej.* **2004**, *4*, 47–66. (In Polish)
26. McKee, T.B.; Doesken, N.J.; Kleist, J. The relationship of drought frequency and duration to time scales. In Proceedings of the 8th Conference of Applied Climatology, Boston, MA, USA, 17–22 January 1993; Volume 17, pp. 179–183.
27. Łabędzki, L.; Bąk, B. Monitoring suszy za pomocą wskaźnika standaryzowanego opadu. *Woda-Sr. -Obsz. Wiej.* **2002**, *2*, 9–19. (In Polish)
28. Łabędzki, L. Estimation of lokal drought frequency in central Poland using the standardized precipitation index SPI. *Irrig. Drain.* **2007**, *65*, 67–72. [CrossRef]
29. Łabędzki, L.; Bąk, B. Meteorological and agricultural drought indices used in drought monitoring in Poland: A review. *Meteorol. Hydrol. Water Manag.* **2014**, *2*, 3–14. [CrossRef]
30. Łabędzki, L.; Bąk, B. Standaryzowany klimatyczny bilans wodny jako wskaźnik suszy. *Acta Agrophys.* **2004**, *3*, 117–124. (In Polish)
31. Evarte-Bundere, G.; Evarts-Bunders, P. Using of the Hydrothermal coefficient (HTC) for interpretation of distribution of non-native tree species in Latvia on example of cultivated species of genus Tilia. *Acta Biol. Univ. Daugavp.* **2012**, *12*, 135–148.
32. IMGW: Drought monitoring system "POSUCH@". Available online: http://posucha.imgw.pl/ (accessed on 14 July 2021).
33. ITP: System "Monitoring and Forecasting Water Deficit and Surplus in Agriculture". Available online: http://agrometeo.itp.edu.pl/ (accessed on 14 July 2021).
34. Ustawa z Dnia 7 Lipca 2005 r. o Ubezpieczeniach Upraw Rolnych i Zwierząt Gospodarskich. Dz. U. 2005. Nr 150 poz. 1249 z Późn. zm. 2005. Available online: https://isap.sejm.gov.pl/isap.nsf/DocDetails.xsp?id=WDU20051501249 (accessed on 14 July 2021). (In Polish)
35. Ustawa z Dnia 25 Lipca 2008 r. o Zmianie Ustawy o Ubezpieczeniu Upraw Rolnych i Zwierząt Gospodarskich oraz Ustawy o Krajowym Systemie Ewidencji Producentów, Ewidencji Gospodarstw Rolnych oraz Ewidencji Wniosków o Przyznanie Płatności. Dz. U. 2008. Nr 145 poz. 918. 2008. Available online: http://orka.sejm.gov.pl/proc6.nsf/ustawy/646_u.htm (accessed on 14 July 2021). (In Polish)
36. Schmuck, A. Susze atmosferyczne na Dolnym Śląsku w ostatnim 60-leciu. *Zesz. Nauk. WSR Wrocław* **1956**, *3*, 147–152. (In Polish)
37. Prawdzic, K.; Koźmiński, C. Susze atmosferyczne na terenie województwa szczecińskiego. *Szczec. TN, Wydz. Nauk Przyr.-Rol.* **1966**, *28*, 1–45. (In Polish)
38. Konopko, S. Częstotliwość występowania okresów posusznych w rejonie Bydgoszczy na podstawie wieloletnich obserwacji. *Wiad. IMUZ* **1988**, *15*, 103–113. (In Polish)
39. Łabędzki, L. Ocena zagrożenia suszą w regionie bydgosko-kujawskim przy użyciu wskaźnika standaryzowanego opadu (SPI). *Wiad. Melior.* **2000**, *43*, 102–103. (In Polish)
40. Mizak, K.; Pudełko, R.; Kozyra, J.; Nieróbca, A.; Doroszewski, A.; Świtaj, Ł.; Łopatka, A. Wyniki monitoringu suszy rolniczej w uprawach pszenicy ozimej w Polsce w latach 2008–2010. *Woda-Sr. -Obsz. Wiej.* **2011**, *11*, 95–107. (In Polish)
41. Kalbarczyk, R.; Kalbarczyk, E. Potrzeby i niedobory opadów atmosferycznych w uprawie ziemniaka średnio późnego i późnego w Polsce. *Infrastrukt. I Ekol. Teren. Wiej.* **2009**, *3*, 129–140. (In Polish)
42. Wójcik, I.; Doroszewski, A.; Wróblewska, E.; Koza, P. Susza rolnicza w uprawie zbóż jarych w Polsce w latach 2006–2017. *Woda-Sr. -Obsz. Wiej.* **2019**, *19*, 77–95. (In Polish)
43. Mitosek, H. Letnio-jesienna susza 1959 r. *Post. Nauk. Roln.* **1960**, *27*, 53–64. (In Polish)
44. Kupczyk, K. Charakterystyka suszy 1969 w Polsce. *Wiad. Służ. Hydrol.* **1970**, *6*, 53–55. (In Polish)
45. Bobiński, E.; Meyer, W. Susza w Polsce w latach 1982–1992. Ocena hydrologiczna. *Wiad. IMGW* **1992**, *15*, 3–23. (In Polish)
46. Czaplak, I. Posucha 1992 roku w Polsce a ogólne prawidłowości rozkładu dni bezopadowych. *Wiad. IMUZ* **1996**, *18*, 85–93. (In Polish)
47. Zawora, T.; Ziernicka-Wojtaszek, A. Susza 2005 roku na obszarze Polski. *Acta Agrophys.* **2007**, *9*, 817–826. (In Polish)
48. Hohendorf, E. Niedobory i nadmiary opadów w Polsce. *Gosp. Wodna* **1948**, *8*, 276–287. (In Polish)
49. Dzieżyc, J.; Nowak, L.; Panek, K. Dekadowe wskaźniki potrzeb opadowych roślin uprawnych w Polsce. *Zesz. Probl. Post. Nauk Roln.* **1987**, *314*, 11–33. (In Polish)
50. Farat, R.; Kępińska-Kasprzak, M.; Kowalczak, P.; Mager, P. Susze na obszarze Polski w latach 1951–1990. *Mater. Bad. IMGW Gosp. Wodna I Ochr. Wód* **1995**, *16*, 1–140. (In Polish)
51. Ziernicka-Wojtaszek, A.; Zawora, T. Niedobory i nadmiary opadów dla pszenicy ozimej w latach ekstremalnych na obszarze Polski (1971–2000). *Woda-Sr. -Obsz. Wiej.* **2005**, *5*, 375–382. (In Polish)
52. Przybylak, R.; Oliński, P.; Koprowski, M.; Filipiak, J.; Pospieszyńska, A.; Chorążyczewski, W.; Puchałka, R.; Dąbrowski, H.P. Droughts in the area of Poland in recent centuries in the light of multi-proxy data. *Clim. Past* **2020**, *16*, 627–661. [CrossRef]
53. Liszewska, M.; Konca-Kędzierska, K.; Jakubiak, B.; Śmiałecka, E. *Opracowanie Scenariuszy Zmian Klimatu dla Polski i Wybranych Regionów*; Report 2, KLIMADA Project; ICM: Warszawa, Poland, 2012; pp. 1–54. (In Polish)
54. Osuch, M.; Romanowicz, R.J.; Lawrence, D.; Wong, W.K. Trends in projections of standardized precipitation indices in a future climate in Poland. *Hydrol. Earth Syst. Sci.* **2016**, *20*, 1947–1969. [CrossRef]

55. Raza, A.; Razzaq, A.; Mehmood, S.S.; Zou, X.; Zhang, X.; Lv, Y.; Xu, J. Impact of climate change on crops adaptation and strategies to tackle its outcome: A review. *Plants* **2019**, *8*, 34. [CrossRef]
56. Hari, V.; Rakovec, O.; Markonis, Y.; Hanel, M.; Kumar, R. Increased future occurrences of the exceptional 2018–2019 Central European drought under global warming. *Sci. Rep.* **2020**, *10*, 12207. [CrossRef] [PubMed]
57. AghaKouchak, A.; Farahmand, A.; Melton, F.S.; Teixeira, J.; Anderson, M.C.; Wardlow, B.D.; Hain, C.R. Remotesensing of drought: Progress, challenges and opportunities. *Rev. Geophys.* **2015**, *53*, 452–480. [CrossRef]
58. Hu, X.; Ren, H.; Tansey, K.; Zheng, Y.; Ghent, D.; Liu, X.; Yan, L. Agricultural drought monitoring using European Space Agency Sentinel 3Aland surface temperature and normalized difference vegetation indeximageries. *Agric. For. Meteorol.* **2019**, *279*, 107707. [CrossRef]
59. Arléry, R.; Grisollet, H.; Guilmet, B. *Climatologie Méthodes et Practiques*, 2nd ed.; Gauthier-Villars: Paris, France, 1973; pp. 1–434.
60. Olechnowicz-Bobrowska, B. Częstość dni z opadem w Polsce. *Pr. Geogr. IG PAN* **1970**, *86*, 1–75. (In Polish)
61. Doroszewski, A.; Górski, T. Prosty wskaźnik ewapotranspiracji potencjalnej. *Rocz. AR Poznań* **1995**, *271*, 3–8. (In Polish)
62. Penmann, H.L. Natural evaporation from open water, bare soil and grass. *Proc. R. Soc. Lond.* **1948**, *193*, 120–145.
63. Kondracki, J. *Geografia Regionalna Polski*, 3rd ed.; PWN: Warszawa, Poland, 2011; pp. 1–468. (In Polish)
64. 2019 European Heat Wave. Available online: https://en.wikipedia.org/wiki/2019_European_heat_wave (accessed on 10 June 2021).
65. Casty, C.; Raible, C.C.; Stocker, T.F.; Wanner, H.; Luterbacher, J. A European pattern climatology 1766–2000. *Clim. Dyn.* **2007**, *29*, 791–805. [CrossRef]
66. Harris, I.; Jones, P.D.; Osborn, T.J.; Lister, D.H. Updated high-resolution grids of monthly climatic observations-the CRU TS3. 10 Dataset. *Int. J. Climatol.* **2014**, *34*, 623–642. [CrossRef]
67. Poljanšek, K.; Marin Ferrer, M.; De Groeve, T.; Clark, I. *Science for Disaster Risk Management 2017: Knowing Better and Losing Less*; Publications Office of the European Union: Luxembourg, 2017; EUR 28034 EN. [CrossRef]
68. Spinoni, J.; Naumann, G.; Vogt, J.V. Pan-European seasonal trends and recent changes of drought frequency and severity. *Global Planet. Change* **2017**, *148*, 113–130. [CrossRef]
69. Gudmundsson, L.; Seneviratne, S.I. A comprehensive drought climatology for Europe (1950–2013). In *Drought: Research and Science-Policy Interfacing*; Alvarez, J.A., Solera, A., Paredes-Arquiola, J., Haro-Monteagudo, D., van Lanen, H., Eds.; CRC Press: London, UK, 2015; pp. 31–37.
70. Spinoni, S.; Naumann, G.; Vogt, J.; Barbosa, P. European drought climatologies and trends based on a multi-indicator approach. *Glob. Planet. Chang.* **2015**, *127*, 50–57. [CrossRef]
71. Spinoni, J.; Naumann, G.; Vogt, J.; Barbosa, P. *Meteorological Droughts in Europe: Events and Impacts: Past Trends and Future Projections*; Publications Office of the European Union: Luxembourg, 2016; EUR 27748 EN. [CrossRef]
72. Stagge, J.H.; Kingston, D.G.; Tallaksen, L.M.; Hannah, D.M. Observed drought indices show increasing divergence across Europe. *Sci. Rep.* **2017**, *7*, 14045. [CrossRef] [PubMed]
73. Spinoni, J.; Vogt, J.V.; Naumann, G.; Barbosa, P.; Dosio, A. Will drought events become more frequent and severe in Europe? *Int. J. Climatol.* **2017**, *38*, 1718–1736. [CrossRef]

MDPI
St. Alban-Anlage 66
4052 Basel
Switzerland
Tel. +41 61 683 77 34
Fax +41 61 302 89 18
www.mdpi.com

Atmosphere Editorial Office
E-mail: atmosphere@mdpi.com
www.mdpi.com/journal/atmosphere

www.ingramcontent.com/pod-product-compliance
Lightning Source LLC
LaVergne TN
LVHW070135170726
843515LV00007B/1800
* 9 7 8 3 0 3 6 5 2 8 3 8 0 *